Sustainable Textiles: Production, Processing, Manufacturing & Chemistry

Series Editor

Subramanian Senthilkannan Muthu, Head of Sustainability, SgT and API, Kowloon, Hong Kong

This series aims to address all issues related to sustainability through the lifecycles of textiles from manufacturing to consumer behavior through sustainable disposal. Potential topics include but are not limited to: Environmental Footprints of Textile manufacturing; Environmental Life Cycle Assessment of Textile production; Environmental impact models of Textiles and Clothing Supply Chain; Clothing Supply Chain Sustainability; Carbon, energy and water footprints of textile products and in the clothing manufacturing chain; Functional life and reusability of textile products; Biodegradable textile products and the assessment of biodegradability; Waste management in textile industry; Pollution abatement in textile sector; Recycled textile materials and the evaluation of recycling; Consumer behavior in Sustainable Textiles; Eco-design in Clothing & Apparels; Sustainable polymers & fibers in Textiles, Sustainable waste water treatments in Textile manufacturing; Sustainable Textile Chemicals in Textile manufacturing. Innovative fibres, processes, methods and technologies for Sustainable textiles; Development of sustainable, eco-friendly textile products and processes; Environmental standards for textile industry; Modelling of environmental impacts of textile products; Green Chemistry, clean technology and their applications to textiles and clothing sector; Eco-production of Apparels, Energy and Water Efficient textiles. Sustainable Smart textiles & polymers, Sustainable Nano fibers and Textiles; Sustainable Innovations in Textile Chemistry & Manufacturing; Circular Economy, Advances in Sustainable Textiles Manufacturing; Sustainable Luxury & Craftsmanship; Zero Waste Textiles.

Miguel Ángel Gardetti ·
Rosa Patricia Larios-Francia
Editors

Sustainability Challenges in the Fashion Industry

Civilization Crisis, Decolonization, Cultural Legacy, and Transitions

Springer

Editors
Miguel Ángel Gardetti
Center for Sustainable Textile
Buenos Aires, Argentina

Rosa Patricia Larios-Francia
Centro de Estudios e Investigación para el Desarrollo, la Innovación y Sostenibilidad
Universidad Peruana de Ciencias Aplicada
Lima, Lima y Callao, Peru

ISSN 2662-7108 ISSN 2662-7116 (electronic)
Sustainable Textiles: Production, Processing, Manufacturing & Chemistry
ISBN 978-981-99-0348-1 ISBN 978-981-99-0349-8 (eBook)
https://doi.org/10.1007/978-981-99-0349-8

This Springer imprint is published by the registered company Springer Nature Singapore Pte Ltd.
The registered company address is: 152 Beach Road, #21-01/04 Gateway East, Singapore 189721, Singapore

Preface

The environmental, social and economic crisis evidently shows that sustainability-related advances are not enough, but rather changes that are too slow. An increased number of counter-hegemonic discourses arise from different sources that report failures in the dominant system, and inspire new horizons.

Reviewing the fashion system and, most specifically, progress in fashion sustainability, it is noticed that changes are actually slow and "superficial," intended to further fuel the prevailing production and consumption system. A comprehensive and systematic approach to sustainability is needed; otherwise, sustainability will remain superficial, with solutions focused on materiality and labor conditions within the same never-changing system, which only purpose is economic growth. Slow progress is believed to relate to a poor approach to the problem, entrenched in obsolete logics, where not only fashion but also the system and life are unsustainable. And from there, reality is narrow-sighted. To ponder about this situation, some questions should be raised. What would the new approach be like? Is it true that we who take the time to think about this are always the same? What about the unheard voices? What if they were heard to set up collaboration dynamics and include sustainable practices? The path along this closed, biased, Western line of thinking leads to no change and, specially, to continue with the same logics that brought us here. For this reason, varied perspectives will join if we open up to voices marginalized by the prevailing narrative. Each of them will contribute with new models, practices and a different approach to fashion and objects and, most importantly, to the human–nature "uniqueness."

This book starts with an "Introduction" by Miguel Angel Gardetti called "The Other Fashion." This is an introduction to vital concepts to change fashion: crisis of civilization, decolonization, interculturality, intercultural justice, plurinationalism and the relationship among biocentrism, design, fashion and textiles.

In the second chapter entitled "Retrofitting Etro: Robe Design Tradition Through Transition Spaces", Annette Condello explores through the case of ETRO, the adaptation, transition and unconventional cultural combinations when readapting luxury garments through the use of different techniques such as digital printing, with which the designer expresses his admiration between the culture of the West and the East.

Next chapter, “Communal Design: A Pedrana Approach to Sustainable Fashion” by Emily Oertling and Kim Hiller, introduces the communal design approach, as a collective approach to garment production, analyzed through the case of the women of San Pedro who enhance the Mayan culture of Guatemala.

Following, Prof. José Teunissen, in his chapter “Towards an Inclusive Fashion System”, explores the ability to redefine and modify the Western conceptualizations of the Eurocentric Fashion System into a more inclusive and alternative system that uses de-linking, precedence and aesthetics, pluri-versality as a new terminology to capture a new generation of fashion designers.

With the following chapter entitled, “Frankie Welch’s Cherokee Alphabet Design: Cultural Appreciation or Cultural Appropriation?”, Alyssa Opishinski, Linda Welters and Susan Jerome present a case study of Welch’s Cherokee alphabet-based textile design, which serves as a discussion tool for design, fashion history and business programs, as well as museum studies, in recognizing past designs as cultural appropriation or appreciation, including the cultural groups concerned, respecting their views and compensating them fairly.

In the next chapter “Cultural Fashion a Matter of Human Rights. Cultural Misappropriation as a Human Rights Violation. What is Wrong with the Fashion and Legal Systems and How Can We Make It Right? The Oma Ethnic Group of Nanam Village in Laos Have an Answer”; Monica Boţa-Moisin presents through a case study, an interdisciplinary solution designed as a response to the fashion company Max Mara using a textile design of the Oma ethnic group of Nanam Village in Laos without consent, credit or compensation. The model challenges the understanding of the notion of “public domain” and draws attention to the value of traditional textile knowledge in redesigning a fair, equitable and sustainable fashion system.

Then Daan van Dartel, in his chapter “Fashion and the Ethnographic Museum: A Case Study from The Netherlands”, emphasizes the importance of including different systems of knowledge about dress practices and the idea of fashion as a system from which the gears can be shifted toward a better relationship with our planet. Addressing it through museum practices and projects in which this responsibility is shared with the people whose cultural heritage is present in the museum’s collections.

Following this, Rosa Patricia Larios-Francia, Agustina Micaela Burgos and Joan Jefferson Sanchez Jimenez, in the chapter “Other Voices: Dynamic Tradition, Empowerment and Andean Fashion in Peru”, generate a reflection on the condition of dynamism or staticity of the textile tradition in Peru; they present the conceptualization of the evolution of Andean fashion and textile art throughout history and identify the contribution of designers as well as entrepreneurs to the empowerment of textile artisan communities, by revaluing Andean textile art, generating identity and sense of belonging in the artisans.

The next chapter is “Cultural and Cultural Appropriation Challenges of Indigenous People in the Global Fashion Industry”; Dr. LeeAnn Teal-Rutkovsk makes an analysis of those challenges specific to the fashion industry, regarding the current and potential impact of the population and culture of indigenous peoples, and the relationship of the industry on them, having the power to add value in indigenous communities, leading to sustainable innovation.

Subsequently, Clelia O. Rodriguez, in the chapter entitled "The Anatomy of One Size Fits All", presents an urgent call to politicize the notions of bodies, autonomy, voice, culture, colonialism, truth, land and, why not, history, everywhere. The author proposes to give an explanation or a form of justification for the immoderation of the "one-size-fits-all" situation. The anatomy of measurements is X-rayed to detect where adjustments need to be made in the body. The one-size-fits-all and its flesh, blood and bones, inspire the promise to always question numbers, all the time.

We continue with "Conversations on Decoloniality and Fashion: Hosting, Listening, (Un)Learning" by authors Erica de Greef and Angela Jansen, who through their chapter narrative acknowledge the teachings of a community, from a shared legacy, guided in mapping their own understandings (and misunderstandings), while practicing in relation to others a "politics of repair."

In the next chapter entitled "Fashion and Identity in Virtual Spaces", Kyoko Koma through a case study of a virtual game analyzes the kind of self-identity that is achieved in the discursive space of multiple actors, including social network users, through the use of virtual clothing and the denial of the body. It identifies how representations are constructed and examines the relationship between fashion and identity in the virtual space created by digital society in the era of liquid modernity.

Then, V. Nithyaprakash, Niveathitha and Thambidurai, in the chapter entitled "Interventions in Traditional Clothing Systems Through Anthropological Perspective", highlight the challenges fashion poses to the environment in the face of changing needs of the world; they report on the institutionalized meaning of fashion and its emergence vis-à-vis the origins of traditional clothing, questioning the Western perspective of traditional clothing systems.

In the present chapter, "Foregrounding the Value of Traditional Indian Crafts: Voices from the Fringe", authors Shalini Gupta and Ishi Srivastava through their study attempt to highlight the importance of those on the margins of the fashion system, why their unheard voices are important; it exposes how indigenous practices that support social enterprise and ecologically sensitive production can contribute to the growing need for a more sustainable and responsible industry.

Subsequently, Laura Camila Ramínez Bonilla, Edward Bermúdez Macías and Ileana Jalil Kentros, in the chapter, "Change, Imitation and Novelty. Notes on Cosmotechnics and Mass Communication in the Face of Fashion and Its Political Possibilities", have the purpose of reflecting on the phenomenon of fashion as a constant of imitation and change, made possible and accelerated by the reciprocal relationship with technological development and multiple cosmologies.

The next chapter, "Telling the Indigenous Ghanaian Fashion Cosmovision: The Case of Royal Ahenema Sandals"; Osuanyi Quaicoo Essel, presents the model of the indigenous Ghanaian fashion system, its evolution in terms of gender issues in production and commodification, and discusses how it functions using the case of the Royal Ahenema sandals, an indigenous Ghanaian yet international fashion accessory, to illustrate how it works.

In the chapter, "Artisans, Creativity, and Ethics: "Skill Regimes" in a Mumbai Fashion Export House", Matthew R. Webb presents the results of his ethnographic research conducted with fashion designers and producers in Mumbai, which explores the stakes of "artisanal fashion" as a juncture of ethical regimes, analyzing the production of a line of garments to traverse global markets. They suggest the decolonial potency of "craft fashion," seen as a genre of style and practice that allows collaborative participants to confront the constraints of the global marketplace by reflexively unpacking their own position and working to design projects that "remix" cultural elements and stereotypes.

Finally, Carla Costa, Carolina Casarin João, Dalla Rosa Júnior, Heloísa Helena do Olivera and Santos Mi Medrado, in their chapter entitled "The Fashion Crossroad Method: Political and Epistemological Practices", shed light on the theoretical engagements of fashion and decoloniality; the authors propose a graphic representation of the methodological steps to undertake the decolonial turn in fashion research; based on a multidisciplinary perspective, "the fashion crossroad method," whose political and epistemological practices show how to transgress the Western canon and the type of fashion studies taught in countries still facing colonialism.

Buenos Aires, Argentina — Miguel Ángel Gardetti
Lima, Peru — Rosa Patricia Larios-Francia

Contents

About the Editors

Miguel Ángel Gardetti (Ph.D.), founded the Centre for Sustainable Textile. He was guest co-editor for several publications in journals and books and renowned professor at different Latin American universities. Between 2012 and 2015, he was also a member of the Consulting Board of the "Future Fashion" Project of MISTRA (a Swedish government foundation that conducts research into environmental topics). He participated in the activity "Changing the World Through Fashion," organized by the United Nations for the 4th Conference on Sustainable Development (Rio+20). He is an active member of the Global Compact in Argentina—which is a United Nations initiative—and was a member of its governance body—the Board of The Global Compact, Argentine Chapter—for two terms. He was also part of the task force that developed the "Management Responsible Education Principles" of the United Nations Global Compact. This task force was made up of over 55 renowned academics worldwide pertaining to top Business Schools.

Rosa Patricia Larios-Francia, Ph.D., in Strategic Management with specialization in Business Management and Sustainability by the Consortium of Universities, Master in Industrial Engineering by the Ricardo Palma University and Industrial Engineer by the University of Lima. With a specialization in innovation by the International Program of High Specialization in Innovation Management. Researcher in the areas of innovation management, MSMEs management, cluster, biodiversity, sustainability, circular economy, humanitarian logistics, technology in the development of new materials and processes in the textile industry, fashion, and handicrafts. She has been Director of the Textile Innovation Center of the University of Lima; author of scientific articles and books on innovation, sustainability, fashion industry, handicrafts and small and medium enterprises; professor at the Faculty of Industrial Engineering of the University of Lima and at the Faculty of Design of the Universidad Peruana de Ciencias Aplicadas—with more than twenty years of experience in executive positions in the textile manufacturing sector—Member ISO/TC 279 Innovation management/WG1 Innovation management System; ISO/TC 279/STTF Spanish translation task force innovation management; Member ISO/TMBG/SAG_ESG Strategic Advisory Group on Environmental, Social, Governance (ESG) Ecosystem);

Member ISO/TC 133 Clothing sizing system—size designation, size measurement methods and digital fittings; Chairman of the Technical Committee of Standardization—INACAL for Research, Technological Development and Innovation Management. Secretary of the SC-CTN Apparel; Member of the CTNs of MSMEs Management, Textiles and Apparel, and Quality Management and Assurance; Director of the Peru Chapter of the International Association of Innovation Professionals—IAOIP; Member of the Latin American Association of Technology and Innovation Management—ALTEC; Member of the Peruvian Association of Textile Technicians—APTT; Member of the Sustainable Fashion Cluster; Member of the Advisory Council of the Sustainable Fashion Association; and Member of the International Alpaca Association.

Introduction

THE "OTHER" FASHION: Civilisation Crisis, Decolonisation, Cultural Legacy, and Transitions

Miguel Angel Gardetti

Abstract The so-called Western civilisational crisis is the expression used in recent years to characterise the extraordinary vicissitudes that the contemporary world is going through. This crisis is jeopardising its very existence as well as the continuity of life on planet earth. Moreover, it is now known that development, as a re-edition of the lifestyles of the core countries, is unrepeatable at the global level. This consumerist and predatory lifestyle is also endangering the global ecological balance and marginalising more and more masses of human beings from the supposed advantages of the "desired" development (or bad development?).The productivist vision of development is associated with an anthropocentric conception of nature, which conceives of humans as external and not integrated with it [1–3]. This seems to be the "crux" of the matter. That is why Ehrenfeld [12] suggests that we become aware of where we stand, what we want to achieve, and how we obtain it by continually improving, but not from the Western/capitalist perspective of progress. Fashion, and the growth of fashion, is presented as the manifestation of a process of civilisation, within a capitalist culture (capital understood as material possessions) that has become global, imperialist, and racist, of which—in an economic sense—the fashion "system" functions as one of its main instruments of exploitation. Moreover, to move from anthropocentrism to biocentrism and achieve a conjunction between nature and human beings, we must incorporate the peripheral stakeholders who are the voices we never hear and whose wisdom can make great contributions to the change we so desperately need. And this would mean "getting out" of the current Western system through these unheard voices. This chapter—which is really an introduction to the book—is intended to take the reader on a journey through the most important themes of this volume.

The "Other" Fashion was the original title of this book. The author wishes to thank Dr. Subramanian Senthilkannan Muthu—Series Director—for his time and kindness.

M. A. Gardetti (✉)
Centro Textil Sustentable [The Sustainable Textile Centre], Paroissien 2680, 5to B C1429CXP, Buenos Aires, Argentina
e-mail: mag@ctextilsustentable.org

M. Á. Gardetti and R. P. Larios-Francia (eds.), *Sustainability Challenges in the Fashion Industry*, Sustainable Textiles: Production, Processing, Manufacturing & Chemistry,
https://doi.org/10.1007/978-981-99-0349-8_1

Keywords Civilization crisis · Decolonization · Cultural legacy · Transition · Fashion · Design · Textiles

According to Quintero [42], over the past few years, *civilising crisis* (of the West [32]) is the phrase used to fully depict the remarkable ups and downs experienced by the contemporary world. This crisis is jeopardising its very existence as well as the continuity of life on planet Earth. Lander [28:27] explains it as follows:

> We are experiencing the terminal crisis of an anthropocentric,[1] monocultural and patriarchal civilizational pattern, based on endless growth and on systematic war efforts against the conditions that make life on planet Earth possible.

It should be noted that the global civilising model in question is the result of a very long process that can be traced back to the "conquest and colonisation" of the Americas, within a historical movement that would—eventually—globalise capitalism, racism, patriarchy, and Eurocentrism under the umbrella of coloniality [42].

According to Quijano [40], the coloniality of power is founded on a structural dependency between two main focuses. Firstly, a domination system based on an asymmetrical classification of global population, supported by the shaping and naturalisation of the ideas of "race," "class," and "gender/sexuality." Secondly, it articulates all known forms of exploitation of labour (and accumulation of capital) into a single structure to produce merchandise for the global market based on the current western system hegemony. Moreover, the crisis is deeply rooted in the current environmental decay [42]. In this connection, Gudynas [25:20–21] asserts that:

> The anthropocentric category refers to positions that are centred on human begins, regarding them as the starting point for any assessment… Therefore, only human beings can be valuable subjects; hence, all the other elements surrounding us, such as plants and animals, are valuable objects. (The underscored text belongs to M. A. Gardetti)

1 Decoloniality, Interculturality, Intercultural Justice, Plurinationality, and Racism

The "modernity/coloniality research programme" project should be understood as a different way of thinking, since it focuses its own approach on the very borders of the

[1] In diagnostic terms, the Anthropocene establishes the idea of a critical "threshold" with respect to already evident problems such as global warming and biodiversity loss—a concept that reveals the limits of nature and questions both the prevailing development strategies and the cultural paradigm of modernity [50:5].

ways of thinking and research into non-Eurocentric ways of thinking.[2,3] Therefore, this group suggests that the other way of thinking, the other knowledge—and the other world, in the spirit of Porto Alegre World Social Forum—are indeed possible [13]. According to this author (p. 61):

> The main conclusions are, firstly, that modernity/coloniality is the relevant analytical unit required to analyse modernity—in short, there is no modernity without coloniality, since the latter is a constituent of the former. Secondly, that the "colonial difference" provides for an epistemologically and politically privileged space. The vast majority of European theorists (...) have been blind to the colonial difference and the subalternation of knowledge and cultures. An emphasis on the colonial modern world system also helps identify, besides internal conflicts—conflicts between powers with the same world view—, those that unfold outside the borders of the modern/colonial world system (i.e., conflicts with other cultures and world views)."

Mignolo [34] argues that decolonial thinking emerged from the very foundations of modernity/coloniality as its counterparty, pointing out the decolonial twist as the openness and freedom of thought and of "other" ways of life; the restructuring of the coloniality of being and knowledge[4]; the separation of the rhetoric of modernity and its imperial imaginary in terms of the rhetoric of democracy. The essence—as well as the purpose—of decolonial thinking is the decoloniality of power, i.e. of the colonial matrix of power. Then, Quijano [41] argues that the alternative is clear: to destroy the coloniality of world power. To that end and first of all, epistemological decolonisation is necessary to clear the way for an original intercultural communication, a mutual exchange of experiences and meanings that lay the foundations of a different rationality—which may claim some universality—; and the claim that the

[2] The modernity/coloniality group was one of the most important critical thinking collectives in Latin America. It was a multidisciplinary and multigenerational network of intellectuals, including sociologists Aníbal Quijano, Edgardo Lander, Ramón Grosfoguel, and Agustín Lao-Montes, semiologists Walter Mignolo and Zulma Palermo, pedagogue Catherine Walsh, anthropologists Arturo Escobar and Fernando Coronil, literary critic Javier Sanjinés, and philosophers Enrique Dussel, Santiago Castro-Gómez, María Lugones, and Nelson Maldonado-Torres.

[3] Institutional support for the growing interest in the colonial question has been slow to manifest itself. It is especially important to cultivate a space in which the colonial question can be operationalised as a distinctive framework for research. Another working group, the colonial, postcolonial, and decolonial working group, has been set up to meet this need. The activities of this working group are not bounded by allegiance to any particular set of concepts (e.g. neo-colonialism), methodological tradition (e.g. postcolonial thought), or ethic (e.g. decoloniality), nor to any particular time period or substantive area of research. Instead, what unites this group as a research community is the posing of the colonial question as central to the research. Source: https://cpdbisa.wordpress.com/ Aceeso: August 10, 2022.

[4] The de/decoloniality of knowledge and being are two concepts introduced by the modernity/coloniality/decoloniality collective from 1998 onwards, which now meet the de/decoloniality of aesthetics to bring together different genealogies of re-existence in artistic practices around the world. This concept—that of de/decolonial aesthetics(s)—refers to current artistic projects that respond to and emerge from the dark side of imperial globalisation. The de/decolonial aesthetic(s) seeks to recognise and open up options for liberating the senses, while constituting itself as the terrain where artists from various parts of the world are interpellating the legacies of modernity and its reconfigurations within postmodern and altermodern aesthetics. Source: Transnational Decolonial Institute—https://transnationaldecolonialinstitute.wordpress.com/ Access: August 10, 2022.

world view of a particular ethnic group is imposed as the universal rationality, even if it is Western Europe, is to hope that a given provincialism is granted the title of universality.

The huge contribution of decolonisation (or independence)—according to Mignolo [34]—is that it put up for discussion the issue of decolonial pluriversality—a world where many worlds fit. This is reinforced by the slogan of the 2001 World Social Forum: "Another World is Possible."

Moreover, Dussel [10] rewrites universal history "from Latin America" as another take on the decolonial approach. This author suggests a historically and archaeologically reasonable "reconstruction" to correct the Eurocentric deviation. To that end, Dussel turned to the history of civilisations written by Western Europe in an attempt to find—both rationally and historically—the place in history that belongs to Latin America, since the American indigenous ethnic groups fail to be part of world history as context for the discovery of the Americas.

Moreover, González [23] explains that the troubling framework that gives rise and witnesses the development of interculturality can be described as: (a) a questioning of how the relationship between culturally diverse collectives is understood (i.e. what is different is inferior; however, it should be on an equal basis thanks to interculturality); (b) a commitment to equality between cultures, the ideal of symmetry, dialogue, and equal involvement (although this is seldom true); and (c) a special focus on cultural groups' knowledge, powers, and world views. Likewise, González argues that interculturality was addressed and implemented in different ways in terms of the relevant social, disciplinary, or institutional setting. Ideas such as hybridisation, transculturation, heterogeneity, miscegenation, melting pot, and syncretism may be regarded as specific classes or subgroups within the broader "interculturality" sphere.

Fornet-Betancourt [20] explains that any person or culture can achieve interculturality if their lifestyle is deliberately focused on nurturing the bond to one another in a comprehensive and all-encompassing way. This means to let oneself be "impacted" by the other person in our daily interactions. Along this line, Estermann [17] explains that intercultural philosophy, rather than a specific force with certain contents, is both conception and a commitment, an intellectual habit that penetrates all philosophical efforts. Above all, it is an "interculturality philosophy": a consideration about the conditions and limits of a dialogue between different cultures. True interculturality (and intercultural philosophy) rejects any claim of monoculturalism and ethnocentrism made by philosophical thinking. Moreover, it confirms minimum commensurability among cultures, as well as the highly ethical nature of intercultural dialogue.

Besides, Walsh [55] explains that interculturality is founded on the need to build relationships between groups, as well as between different practices and types of knowledge, with a desire to confront and transform the power relations that have naturalised social asymmetries.

In Latin America, there is an evident need of interculturality. Therefore, Fornet-Betancourt [20] explains that interculturality is related to the history of the conquest and colonisation of the sub-continent. It is the story of devastation. It is the destruction and oppression of cultural differences. Since the difference of the "other" is fully subdued, the other becomes a colonised object, i.e. socially, culturally, and politically neutralised. The other is subject to westernisation and doomed to be an outcast.

It should be understood that interculturality requires to break away from the traditional idea of the sources of law in order to include other legal systems [45]. Therefore, in [46], Astrain explained that "justice" would mean to secure the social, material, and spiritual conditions required for the existence of the emerging and recognised "other."

> ...an approach to "intercultural justice" calls for embracing the way humans are, make assessments and behave, which makes up the fabric of mixed-race cultures, in order to create spaces for debate to validate principles —this is the only way to shape a consistent proposal of intercultural justice to address both dated and emerging conflicts, old and new asymmetries. [45:47]

Sousa Santos [48] points out that one of the great innovations of western modernity is the liberal symmetry in which there is absolute rule of law and such rule of law belongs to the State. This is a complex symmetry. Not only does it disregard the vast diversity of legal systems independent of the State in different societies, but it also supports rule of law independent of politics.

The same author believes that plurinational constitutionalism breaks this paradigm since it asserts that the unity of the legal system does not necessarily assume consistency. Within the framework of plurinationalism, the constitutional recognition of indigenous law is a major aspect of both interculturality and self-government of native indigenous communities. Legal systems—Eurocentric, Indocentric and, in some countries or situations, Afrocentric systems—are autonomous, though not isolated: the relationships between them pose a demanding challenge. The partial separation of law from the State in order to bring it closer to the life and culture of the peoples will be no easy endeavour. Along the same line as [43, 48] explains that plurinationalism should change the—otherwise invariable—idea that a State should only have one nation.

According to Van Dijk [54], several studies show that racism against African descendants and indigenous peoples is a significant social issue. The above author continues (p. 21).

> In this connection, Latin America—regretfully—follows the lead of Europe and North America, among other regions where white Europeans are the dominant ethnic group. While we do not hold that "white" peoples are inherently racists—an essentialist position that goes against the fact that there are many "white" people fighting racism—the ubiquity of "Euro-racism" in the world is, rather, the historical consequence of centuries of European colonialism. In this system of domination, the non-European others were systematically perceived and treated as different and inferior—an ideology that served to legitimise slavery, exploitation and discrimination.

In words of Alma es Azul,[5] the first Argentine indigenous model [9]:

---

[5] Before her birth in an Argentine hospital, Alma's parents had decided to migrate from Peru in search of a better future—a journey that became a painful move. Good jobs seemed to be banned for them: her mother could only get odd housekeeping jobs, while her father worked double shifts to make ends meet [9].

> Racialised people grow up and evolve surrounded by dominant people that despise and look down on us, even to make us go away. Between words and prejudices which nobody wants to claim, but everybody replicates and accepts. Racism is often invisible to those who do not experience it.
>
> …
>
> Historically, our faces have been overlooked, and we have been denied any possibility of beauty: we are associated to the ugly, the bad and the marginal.

2 Crisis of Civilisation, Decolonisation, Design, Fashion, and Textiles

The current crisis of civilisation has reopened portals which, at first glance, were hidden to us: issues that lay the foundations of the space/time of the present that we are building, one of these issues being—undoubtedly—the *development* issue. Within the material and intersubjective modernity/coloniality universe, development is one of the practical notions that has driven the (re)production of the main colonial and imperial agenda of the civilising pattern [42]. Not only does development occur amid the deployment of well-known capitalist dynamics, but it also results in a consistent classification of global population through an image of the planet geographically divided in terms of ontological differences, based on the alleged "degrees of development" reached in each territory (first world, third/underdeveloped world.)

According to Acosta [1], the search for new alternatives to development began once problems destroy the myths surrounding it. Several "surnames" were added for differentiation purposes: economic development, social development, local development, global development, rural development, sustainable development, eco-development, ethno-development, human-scale development, endogenous development, gender equality development, co-development… ultimately, *development.*

Moreover, now we know that development—as to lifestyle in the main countries—cannot be replicated on a global scale.[6] Also, this consumerist and predatory lifestyle is jeopardising the global environmental balance, while increasingly depriving a huge number of human beings of the alleged advantages of the "coveted" development (or poor development?). On 8 December 2021, Nina Gbor—sustainable fashion educator from Sidney, New South Wales, Australia—posted the following text on her Eco Styles blog [22]:

> Black Friday (November 26th) marked the annual initiation into this season's global over-consumption ritual. It usually starts with Black Friday, goes into Christmas, gets hotter on Boxing Day, then New Year's and all throughout January. It's a lot! Our modern culture has set up this period as the festival of superfluous consumption. So it's primed for voracious

[6] Celso Furtado argued that one of the indirect conclusions of the Meadows Report was that the life style advocated by capitalism should be reserved to a minority (i.e. the industrialised countries and, among underdeveloped countries, to the dominant minorities), since any generalised attempt would collapse the system. And, along this line, back in 1975, Fundación Bariloche argued that the devastating use of natural resources and environmental degradation were not closely related to increased population, but to the high consumption of richer countries [49:67].

> use of material things far beyond any other time of the year. This is why environmentalists say Christmas is the greatest environmental disaster.
>
> Marked by the biggest displays of excessive advertising, manipulative sales tactics, oversupply and mindless buying with almost no consideration for the environmental cost to the earth, the wellbeing of humans involved in supply chain and raw materials extracted to manufacture the products.
>
> 80% of clothes purchased on Black Friday are reportedly thrown away after one wear. Some environmentalists say Christmas is the world's greatest environmental disaster.
>
> To be fair, for the underprivileged & marginalized groups, the season is an opportunity to afford little luxuries and needed items beyond regular price range.
>
> However, there's more than enough resources to cater for everyone's basic needs if we commit to taking care of people and planet. Especially if we embrace systemic global equity, equality and let go of capitalism's principles of placing profiteering well over the wellbeing of humans and the earth.

This is aligned with decolonial thinking and doing, which are not meant to be "put into practice," but they imply the very act of thinking while shaping ourselves, with dialogue and community. This is not a method, but a way, a path to reshape ourselves in the search for new ways of life and of governing (ourselves) in which we do not live to work, produce, and mainly, consume [32].

As explained above, the productivist view of development is grounded on an anthropocentric idea of nature, which regards the human being as an outsider who is not integrated with it [1–3]. This seems to be the "heart" of the question. Therefore, Ehrenfeld [12] suggests that we should become aware of where we stand, what we want to achieve, and how we will achieve it with continuous improvement. But not from the western/capitalist perspective of progress. Bear in mind that Fletcher [18:18] argues that "*fashion is consumption, materialism, commercialisation, and marketing*." And this has clearly an impact not only on the environment, but also on people involved in the production process of textile and clothing, either fashionable items or not.

Fashion—and its growth—is presented as the expression of a civilising process within a capitalist culture (capital understood as material processions) which has become global, imperialist, and racist, out of which—at an economic level—the fashion industry (or system) has been an important instrument of exploitation [57].

Active commodification results in a loss of consumer's—and even producer's—sensitivity, since the logic of the current market tends to homogenise the most basic aspects of symbolic content [29]. In the overview of his book about Lash and Urry's book [29], Ulises Ponce Mendoza [39] argues that:

> Everything in circulation (objects and subjects), and even the environment where they circulate (time and space), are depleted of meaning. This emptying out of the local and substantial meaning of time and space was necessary so that markets could have a global stretch, thus becoming abstract entities.
>
> Subjects are emptied out of their meaning of class, union, religion, etc.

The money value attributed to a cultural object causes part of the social fabric that has made up different social collectives for years, or even centuries, to be subject to indiscriminate exploitation of their identity heritage. Therefore, González Tostado

[24] defines cultural appropriation as the phenomenon of taking elements from a minority culture and using them without their original meaning in a different context, more often than not, for commercial purposes.

3 Towards Design, Fashion, and Textiles Biocentrism: Wisdom from the Periphery

Our current economic system is only focused on the main stakeholders.[7] To move from anthropocentrism to biocentrism in order to create a bond between nature and the human being since, according to Gudynas [25:60], in the latter:

> …emphasis is more comprehensive and focused on bestowing value on life, either on individuals, species or ecosystems.

This path should engage unheard voices, which wisdom can make great contributions to the much needed change. This means to "get out" of the current system.

Since early 2019, the Centro Textil Sustentable [Centre for Sustainable Textile][8] has been devoted to learning and understanding the world view and the spiritual realm of the aboriginal peoples from the Valles Calchaquíes [Calchaquí Valleys][9] in Catamarca and Tucumán.

Why aboriginal communities? Not only because they are unheard voices, but also because, design-related programmes (mainly textile, clothing, and fashion design programmes) usually come to simplistic conclusions alienated from indigenous nations' diversity that often lack context. Besides, as explained by Hrenyk and Salmon [26]:

> It is imperative that business students learn how to interact with Indigenous Peoples and people, within both ethical and legal frameworks…recognizing the historical (and ongoing) systemic injustices that these communities suffer.
>
> …without cultural safety in mind, these discussions can do more harm than good.

In this sense, indigenous cultures are civilising reservations of humanity [21]. For indigenous cultures, nature is both a respected source of production and the centre

[7] For example, employees, competitors, the media, the academia, the government, international regulations, unions, the financial sector, insurance companies, etc.

[8] www.ctextilsustentable.org.

[9] The Calchaquí Valleys form a system of valleys and mountains in the northwest of Argentina, stretching over 520 km in the north from La Poma (Salta) to Punta de Balasto (Catamarca) in the south, and in the west from the Sierras de Quilmes (or Sierras del Cajón), up to San Francisco mountain range and Sierras del Aconquija in the east.

They are regarded as one of the most remarkable and beautiful sites in Argentina. Natives from the large *diaguita-cacano* ethnic group of the Calchaquí people settled here, which resulted in the emergence of a variety of cultures, such as the agro-pottery culture Santa María, Candelaria, Cóndor Huasi, and Famabalasto.

of their universe, the heart of their culture and the origin of their ethnic identity. And the idea that everything—living and non-living things—is inherently bound to the human nature lies at the core of this deep bond. As put by Sarasola ([47]:23):

> Thanks to this particular relationship with Mother Earth, aboriginal peoples are one of the main stakeholders that warn us about our crisis. And since they have a better understanding of Mother Earth, they have a special ability to take care of it. Today, many voices are raised to say that if we value indigenous groups we are also protecting million hectares of biodiversity.

In this connection, in their 2019 paper, Earth Logic (pp. 35–36), Fletcher, and Tham explain [19]

> The notion of care can radically transform our engagement with fashion. Care is an Earth Logic paradigmatic shift away from the binary construct of production versus consumption staged by market thinking. Care is intrinsically relational, implying unfolding practices that nurture, grow, maintain, heal as opposed to the abrupt constructs of 'selling, buying, binning.' Care is also contextual and situated, that is rooted in the local environment and community. Care strengthens, expands and invigorates relations between humans, materialities, natural world.

As already discussed, the view of aboriginal peoples invites us to take an environmental approach to understand the world—a world where human beings and nature coexist in harmony. In February 2021, in Santa María, Valles Calchaquíes (Catamarca, Argentina), Luis Enrique Maturano explained:

> I surrender to the supreme beings. There is a mother, Mother Earth. A superior state which I am part of... We are all children of Mother Earth. ... Nothing is small, and nothing is supreme in the context of nature.

This approach, different from that embraced by our society, can also be found at the beginning of the Universal Declaration of Rights of Mother Earth's Preamble [52][10]:

> We, the peoples and nations of Earth: considering that we are all part of Mother Earth, an indivisible, living community of interrelated and interdependent beings with a common destiny; gratefully acknowledging that Mother Earth is the source of life, nourishment and learning and provides everything we need to live well; ...

Moreover, seen from the community world view of the Indigenous Peoples from Latin America and the Caribbean, the United Nations Sustainable Development Goals (SDGs) [27] offer an opportunity, a bridge of dialogue between intercultural and favourable development models to revitalise the entire humanity. They encourage us to understand a world where human beings and nature live in harmony, where the development proposal should be understood as a new way of social and environmental coexistence.

Therefore, all the above means to have a "decolonial" view that—according to Meschini and Porta [31]—could be understood simply as disassembling, dismantling,

[10] The Universal Declaration of Rights of Mother Earth was first introduced in 2010 World People's Conference on Climate Change and the Rights of Mother Earth. According to Svampa and Viale [51], the above summit was one of the milestones for decolonising thinking based both on a holistic view of indigenous peoples and criticism of the ecological debt of countries from North to South.

or reversing coloniality. In this connection, Mignolo [33:10] offers a more radical stance:

> Decolonisation is no longer understood in terms of "taking over the state," but as a more extreme task: to dismantle the entire system of knowledge that supports and legitimises the modern/colonial and modern State associated with economic coloniality and subjectivity control.

Against this framework, we should note the **Research Collective for Decoloniality & Fashion (RCDF).**[11] This initiative is

> …is an experimental platform beyond institutional, disciplinary and geographical boundaries established in 2012. It aims to critique the denial and erasure of a diversity of fashioning systems due to unequal global power relations based on the modern-colonial order, the Euro-American canon of normativity and the exploitation and abuse of cultural heritages, human beings and natural resources.

And as to design? Escobar [15:39] challenges us:

> Can design be extricated from its embeddedness in modernist unsustainable and defuturing practices and redirected toward other ontological commitments, practices, narratives, and performances? Moreover, could design become part of the tool kit for transitions toward the pluriverse?

And the author goes on Escobar [16:xvi]:

> The collective determination toward transitions, broadly understood, may be seen as a response to the urge for innovation and the creation of new, non-exploitative forms of life, out of the dreams, desires, and struggles of so many groups and peoples worldwide. Could it be that another design imagination, this time more radical and constructive, is emerging? Might a new breed of designers come to be thought of as transition activists? If this were to be the case, they would have to walk hand in hand with those who are protecting and redefining well-being, life projects, territories, local economies, and communities worldwide.
>
> In this context, designs for the pluriverse becomes a tool for reimagining and reconstructing local worlds. Today, the difference is embodied for me most powerfully in the concept of the pluriverse, a world where many worlds fit.[12]

Moreover, Whitty[13] [56:355–356] explains that, despite fashion's claim to be situated in the "now" or an idealised future, much of what we think and know about clothing from production, design, aesthetics, use and disposal speaks to, and is from, another time and place—the past—when our relationship with and understanding of our place in nature and the earth was markedly different. And she says (355).

---

[11] Source: https://rcdfashion.wordpress.com/. Accessed: 12 January 2022.

[12] All of us—and everything—live in the pluriverse [14:60].

[13] Professor of Fashion Systems and Materiality at The New School Parsons. Originally from Ireland, she has worked as a designer and educator in Paris, London, Wellington, Copenhagen, Dublin, Limerick, Tallinn, Xi'an, Shanghai, and Beijing. Her work has been exhibited and published globally in both academic and non-academic contexts. She is focused on developing new mindsets, models, and systems for alternative ecologies of fashion practice that work within planetary boundaries.

> Western exceptionalism has led to a superiority of thought and action, as we have deemed ourselves to be above and apart from nature. We behave like the planet's resources and capacity are infinite and limitless, ours for the taking. An anthropocentric, reductionist, modernist, colonial, capitalist, materialistic growth logic has ruled our thinking, actions and conception of time and place. It is clear that the construction of this 'place' or system does not consider planetary boundaries or a multiplicity of voices, particularly Indigenous voices.

Note that Escobar and Whitty are not the only authors who study/research into these fields. Some other examples are Monteiro [35], Coombs et al. [5], and Chock [4].

We should also mention the **Decolonising Design Collective**, which Editorial Statement dated 27 June 2016 states [7]

> …design theory, practice, and pedagogy as a whole are not geared towards delivering the kinds of knowledge and understanding that are adequate to addressing longstanding systemic issues of power.
>
> This planet, shared and co-inhabited by a plurality of peoples, each inhabiting different worlds, each orienting themselves within and towards their environments in different ways, and with different civilizational histories, is being undermined by a globalized system of power that threatens to flatten and eradicate ontological and epistemological difference, rewriting histories and advance visions of a future for a privileged few at the expense of their human and nonhuman others.
>
> To date, mainstream design discourse has been dominated by a focus on Anglocentric/Eurocentric ways of seeing, knowing, and acting in the world, with little attention being paid to alternative and marginalized discourses from the non Anglo-European sphere, or the nature and consequences of design-as-politics today.

From a more institutional perspective, we can mention **"The Other Design," an initiative of Universidad Iberoamericana Ciudad de México** [Ibero-American University, Mexico City]. According to the university [53]:

> In view of the civilising crisis, Otro Diseño es Posible [Other Design is Possible], is an interdisciplinary research group of Universidad Iberoamericana to explore and implement other ways of life. The purpose of the group is to create and interact with other fields of knowledge from a horizontal political and social stance based on a decolonial, depatriarchal and ontologically questioning approach.

This initiative states that Universidad Iberoamericana Ciudad de México [37:1–2]:

> Our most immediate reaction is to think about the vast number of actions that we witness in university education, as well as in professional practice, that replicate social values and interpretations of reality that we recognise as the causes of the crises converging in the world.
>
> ….
>
> We have created a "black box" around design and its moral ability—which is still seldom questioned as it only captures concerns to turn them into products, while turning people into players limited to their consumer role.

4 To Conclude: Should Design Programmes Be Decolonised and Indigenised in Terms of Clothing, Fashion, and Textiles?

According to Martínez Sarasola [30], the western mind-set usually tends to exclude or deny what cannot be understood. Moreover, "standardised" thinking is also quite common: things are one way or another, but they cannot be two different ways at the same time. This dogmatic and "narrow-minded" perspective tends to exclude, deny, or—in the best-case scenario—discredit anything that cannot be easily fitted in these frameworks.

For the indigenous world, the search for harmony and complementarity is natural: things, people, and all living beings are regarded as constantly interrelated and part of totalities ruled by integration.

Therefore, as explained by Saavedra [44], there is a need to devise and redeploy educational processes that make possible, easy, and viable true and effective experiences of dialogue among various fields of knowledge in order to rebuild "other" ways of thinking, different from the reductionist and dichotomic perspective inherited from western tradition. It is about creating and consolidating intercultural dialogue and learning processes to underpin and invigorate every denied or silenced knowledge and, thus, to design theoretical, methodological, and didactic proposals that are truly participatory, interactive, reciprocal, and democratised. According to de Santos [8], the university should be at the service of social forces that fight for a post-capitalist, post-colonialist, and post-patriarchal society.

References

1. Acosta A (2013) El Buen Vivir. Sumak Kawsay, una oportunidad para imaginar otros mundos. Icaria Editorial, Barcelona
2. Acosta A (2014a) El Buen Vivir, más allá del desarrollo. In: Delgado Ramos GC (ed) Buena Vida, Buen Vivir – Imaginarios alternativos para el bien común de la humanidad. Mexico, Centro de Investigaciones Interdisciplinarias en Ciencias y Humanidades, Universidad Nacional Autónoma de México
3. Acosta A (2014b) El Buen Vivir: una alternativa al desarrollo. In: Quintero P (ed) Crisis Civilizatoria, Desarrollo y Buen Vivir. Buenos Aires, Ediciones del Signo, pp 127–133
4. Chock SC (2020) Design justice—community-led practices to build the worlds we need. The MIT Press, Cambridge
5. Coombs G, McNamara A, Sade G (2019) Undesign—critical practices at the intersection of art and design. Routledge, New York
6. Daly HE (1996) Beyond growth: the economics of sustainable development. Beacon, Boston
7. Decolonizing Design Collective (2016) Editorial statement. Source: https://www.decolonisingdesign.com/statements/2016/editorial/. Accessed 8 Aug 2022
8. de Santos B (2021) Descolonizar la universidad – El desafío de la justicia cognitiva global. Buenos Aires, CLACSO (Consejo Latinoamericano de Ciencias Sociales
9. Durn C (2022) De chica la humillaban e intentó blanquear su piel, hoy es modelo indígena argentina para grandes marcas: "Nuestros rostros han sido invisibilizados". Buenos Aires, La

Nación. Source: https://www.lanacion.com.ar/lifestyle/de-chica-la-humillaban-e-intento-blanquear-su-piel-hoy-es-modelo-indigena-argentina-para-grandes-nid09072022/. Accessed 11 July 2022
10. Dussel E (2012) 1492. El encubrimiento del otro. Editorial Docencia, Buenos Aires
12. Ehrenfeld JR (2008) Sustainability by design. Yale University Press, New Haven
13. Escobar A (2003) Mundos y conocimiento de otro modo. Tabula Rasa 1:51–86, January–December 2003
14. Escobar A (2014) Sentipensar con la tierra. Medellín, Ediciones UNAULA
15. Escobar A (2016) Autonomía y diseño – La realización de lo comunal. Editorial Universidad del Cauca, Popayán
16. Escobar A (2018) Designs for the Pluriverse—radical interdependence, autonomy, and the making worlds. Duke University Press, Durham
17. Estermann J (2015) Filosofía Andina. Sabiduría indígena para un mundo nuevo. Ediciones Abya-Yala, Quito
18. Fletcher K (2015) Other fashion systems. In: Fletcher K, Tham M (eds) Routledge handbook of sustainability and fashion. Roudledge, New York, pp 15–24
19. Fletcher K, Tham M (2019) The earth logic, fashion action research plan. The JJ Charitable Trust, Londres
20. Fornet-Betancourt R (2017) La filosofía intercultural. In: Dussel E, Mendieta E, Bohórquez C (eds) El pensamiento filosófico latinoamericano, del Caribe y "latino", 1300-2000. Historia, corrientes, temas, filósofos. Siglo Veintiuno, Mexico, pp 639–646
21. Gardetti MA, Gabriel M (2021) The unheard voices in fashion accelerating sustainability. Paper and lecture given at the 23 international on-line conference, 15–21 March 2021, Accelerating Sustainability in Fashion, Clothing, Sportswear & Accessories.
22. Gbor N (2021) Christmas is the greatest annual environmental disaster. Australia, Eco Styles. Source: https://www.ecostyles.com.au/blogs/christmas-is-the-greatest-annual-environmental-disaster. Accessed 23 December 2021
23. González M (2021) Clase 3. Interculturalidad, Primera Guía. Within the framework of the seminar on miscegenation, interculturality and decoloniality: Possibilities and issues around three approaches to the Latin American cultural differentiators (Master's on Latin American Studies-CEL-), Universidad Nacional de San Martin, Faculty of Humanities
24. González Tostado FJ (2020) Sobre el dilema de la apropiación cultural: arte, diseño y sociedad. Estudios sobre Arte Actual Nro. 8(2020):311–320
25. Gudynas E (2015) Derechos de la naturaleza. Ética biocéntrica y políticas ambientales. Buenos Aires, Tinta Limón
26. Hrenyk J, Salmon E (2021) Complicating the stories we tell in business schools about indigenous peoples. PRiMEtime
27. Iniciativa Multipaís Derechos a la Tierra y el Territorio para Pueblos Indígenas y Afrodescendientes de la Coalición Internacional para el Acceso a la Tierra de ALC et al (2019) Los Objetivos de Desarrollo Sostenible: una mirada desde Latinoamérica en la cosmovisión de los Pueblos Indígenas. Iniciativa Multipaís Derechos a la Tierra y el Territorio para Pueblos Indígenas y Afrodescendientes de la Coalición Internacional, para el Acceso a la Tierra de ALC, Grupo Mayor de Pueblos Indígenas y los ODS- IPMG, Fondo para el Desarrollo de los Pueblos Indígenas de América Latina y El Caribe – FILAC y Centro para la Autonomía y Desarrollo de los Pueblos Indígenas – CADPI
28. Lander E (2014) Crisis civilizatoria, límites del planeta, asaltos a la democracia y pueblos en resistencia. In: Quintero P (ed) Crisis civilizatoria, desarrollo y buen vivir. Buenos Aires, Ediciones del Signo, pp 27–57
29. Lash S, Urry J (2001) Economías de signos y espacio – Sobre el capitalismo de la posorganización. Provincia de Buenos Aires, Amorrortu Editores
30. Martínez Sarasola C (2010) De Manera Sagrada y en Celebración – Identidad, cosmovisión y espiritualidad en los pueblos indígenas. Buenos Aires, Editorial Biblos
31. Meschini P, Porta L (2017) Acerca de la distinción decolonial/descolonial. In: Hermida ME, Meschini P (eds) Trabajo social y descolonialidad. Epistemologías insurgentes para la intervención en lo social. Mar del Plata, EUDEM, pp 20–24

32. Mignolo W (2014) Prefacio. In: Quintero P (ed) Crisis civilizatoria, desarrollo y buen vivir. Buenos Aires, Ediciones del Signo, pp 7–9
33. Mignolo W (2017a) Prefacio. In: Palermo Z (Comp) Des/decolonizar la universidad, Buenos Aires, Ediciones del Signo, pp 7–12
34. Mignolo W (2017b). El pensamiento decolonial, desprendimiento y apertura. In: Dussel E, Mendieta E, Bohórquez C (eds) El pensamiento filosófico latinoamericano, del Caribe y "latino", 1300-2000. Historia, corrientes, temas, filósofos. Siglo Veintiuno, México, pp 659–672
35. Monteiro M (2019) Ruined by design—How designers destroyed the world, and what we can do to fix it. Middletown, Ani King & Erika Hall Editors
36. Oviedo Freire A (2013) Buen Vivir vs Sumak Kawsay -Reforma capitalista y revolución alternativa – 3era. Edición. Buenos Aires, Ediciones CICCUS
37. Otro Diseño es Posible (2021) Proyecto político cultural del diseño. Grupo de investigación interdisciplinar, Universidad Iberoamericana Ciudad de México, México
39. Ponce Mendoza U (2001) Economías de signos y espacios, sobre el capitalismo de la posorganización – Reseña. Región y sociedad vol 13, no 22, July/December 2001, pp 185–191
40. Quijano A (1998) La economía popular y sus caminos en América Latina. Lima, Mosca Azul
41. Quijano A (2014a) Textos de Fundación. In: Palermo Z, Quintero P (eds) Anibal Quijano – Textos de Fundación. Ediciones del Signo, Buenos Aires
42. Quintero P (2014b) Introducción. In: Quintero P (ed) Crisis civilizatoria, desarrollo y buen vivir. Buenos Aires, Ediciones del Signo, pp 11–25
43. Ramírez S (2015) Matriz constitucional, estado intercultural y derechos de los pueblos originarios. Ad-Hoc, Buenos Aires
44. Saavedra JL (2008) Descolonizar e indianizar la universidad. Temas Sociales (Revista de la Carrera de Sociología, Universidad Mayor de San Andrés, nro 28 (edición virtual)
45. Salas Astrain R (2015) Hacia una justicia intercultural para el reconocimiento social y en un medioambiente sustentable. Justiça do Direito 29(1):33–48, Jan/Apr 2015
46. Salas Astrain R (2019) Justicia, pluralismo y democracia. In: López EA, Niño Castro A, González LT (eds) Claves de la Justicia desde América Latina, vol 6. Tirant Lo Blanch, Valencia, pp 163–182
47. Sarasola CM (2010) De manera Sagrada y en celebración, Identidad cosmovisión y es ritualidad en los pueblos indígenas. Buenos Aires, Editorial Biblos
48. De Sousa Santos B (2012) Refundación del estado en América Latina – Perspectivas desde una epistemología del sur. Grupo Editorial Siglo XXI, Mexico
49. Svampa M (2019a) América del Sur. In A. Kothari et al (Eds), Pluriverso – Un diccionario para el posdesarrollo. Barcelona, Icaria Editorial SA
50. Svampa M (2019b) Antropoceno. Córdoba, Editorial Cartonera de la Facultad de Filosofía y Humanidades de la Universidad Nacional de Córdoba
51. Svampa M, Viale E (2020) El colapso ecológico ya llegó – Una brújula para salir del (mal)desarrollo. Buenos Aires, Siglo Veintiuno Editores
52. Universal Declaration of Rights of Mother Earth's. The world people's conference on climate change and the rights of mother Earth, Cochabamba, Bolivia, 22 April–Earth Day 2010
53. Universidad Iberoamericana de la Ciudad de México (2020) Otro Diseño es Posible. Source: https://ibero.mx/prensa/otro-diseno-es-posible-una-apuesta-de-la-ibero-para-transformar-al-mundo-. Accessed 9 February 2022
54. Van Dijk TA (2007). Racismo y discurso en América Latina. In: Van Dijk TA (ed) Racismo y discurso en América Latina. Barcelona, Editorial Gedisa, pp 21–34
55. Walsh C (2002) Interculturalidad, reformas constitucionales y pluralismo jurídico. Publicación mensual del Instituto Científico de Culturas Indígenas 4(36), March 2002. Source: Interculturalidad, reformas constitucionales y pluralismo jurídico (nativeweb.org). Accessed 22 January 2022

56. Whitty J (2021) Fashion systems of earth logic and transition for this time and place. Fashion Style Popular Culture 8(4):355–375
57. Wilson E (2007) Explaining it away. In: Barnard M (ed) Fashion theory a reader. Abingdon, Roudledge, pp 15–24

Retrofitting Etro: Robe Design Tradition Through Transition Spaces

Annette Condello

Abstract Unravelling their own cultural legacy into global garments, Italian luxury label Etro accommodates unique design and rich patterned textiles by extending the robe's design tradition through transition spaces. Traditionally, luxurious robes and the non-fashion places where people were permitted to be slothful, that is by reclining on sofas covered with carpets for rest and relaxation evolved spaces for the transition to occur. The idea of the robe's fluctuating function as an item and transition space frames this chapter. Armoires or wardrobes and other upholstered pieces of furniture are always made to last but the vacation of these items within a particular transition space, such as a roof terrace or desert landscape, has altered global fashion through resort collections. Through the transition, it argues how the robe as an indispensable object and artefact cultivated a space for architecture through upholstery history, which popularised the invention of resort wear. Garment incorporation of unique textiles and embroideries, worn inside and outside in private became public, are indebted to Armenian traditions. As a global garment, the robe allows the body freedom of movement and as a piece of furniture not only stores it but also creates the voyage impetus to dress informally, trading outmoded fashion with environmental design's transient nature of sustainable spaces. Appealing to Gerlolamo Etro in the 1960s, "house coats" provide cues as to why a generation of Italian fashion designers were attracted to the East by figuratively housing the robe. By looking at Etro's idea of "new tradition," the chapter explores the West–East admiration of the robe, countercultural adaption of the paisley (or "boteh") motif and the invention of resort collection as the transitional space for architecture and fashion. Interpreting the use and role of Eastern influences on robe design in both the architectural and fashion realms, this chapter will explore how Italian fashion culturally marks the dressing gown as a sustainable transitional garment. Rampant in Western Europe in the nineteenth century, robe design, on a socio-cultural level, popularised the kimono because of the effects of sybaritism and Japanism. Expressing fluctuating transitions,

A. Condello (✉)
Faculty of Humanities, School of Design and the Built Environment, Curtin University, GPO Box U1987, Perth, WA 6845, Australia
e-mail: A.Condello@curtin.edu.au

M. Á. Gardetti and R. P. Larios-Francia (eds.), *Sustainability Challenges in the Fashion Industry*, Sustainable Textiles: Production, Processing, Manufacturing & Chemistry,
https://doi.org/10.1007/978-981-99-0349-8_2

Etro's unconventional cultural combinations retrofit luxury garments through digital upholstering.

Unravelling their own cultural legacy into global garments, Italian luxury label Etro accommodates unique design and rich patterned textiles by extending the robe's design tradition through transition spaces. Traditionally, luxurious robes and the non-fashion places where people were permitted to be slothful, that is by reclining on sofas covered with carpets for rest and relaxation evolved spaces for the transition to occur. The idea of the robe's fluctuating function as an item and transition space frames this chapter. Armoires or wardrobes and other upholstered pieces of furniture are always made to last but the vacation of these items within a particular transition space, such as a roof terrace or desert landscape, has altered global fashion through resort collections. Through the transition, it argues how the robe as an indispensable object and artefact cultivated a space for architecture through upholstery history, which popularised the invention of resort wear. Garment incorporation of unique textiles and embroideries, worn inside and outside in private became public, are indebted to Armenian traditions. As a global garment, the robe allows the body freedom of movement and as a piece of furniture not only stores it but also creates the voyage impetus to dress informally, trading outmoded fashion with environmental design's transient nature of sustainable spaces.

Appealing to Gerlolamo Etro in the 1960s, "house coats" provide cues as to why a generation of Italian fashion designers were attracted to the East by figuratively housing the robe. By looking at Etro's idea of "new tradition," the chapter explores the West–East admiration of the robe, countercultural adaption of the paisley (or "boteh") motif and the invention of resort collection as the transitional space for architecture and fashion. Interpreting the use and role of Eastern influences on robe design in both the architectural and fashion realms, this chapter will explore how Italian fashion culturally marks the dressing gown as a sustainable transitional garment. Rampant in Western Europe in the nineteenth century, robe design, on a socio-cultural level, popularised the kimono because of the effects of sybaritism and Japanism. Expressing fluctuating transitions, Etro's unconventional cultural combinations retrofit luxury garments through digital upholstering.

1 Introduction

The fashion world constantly experiences transition, heedless of the climate, but what precisely is this and how has it changed the built environment in terms of enabling viable-sustainable operations? In modifying the growing needs of people and the planet, the spaces within and outside architecture require us to requalify our reuse of existing artefacts for places and to consider variegating retro practices. One might follow a natural-vintage artefact trend and reuse existing buildings, but the wardrobe is one item that requires laser-focus attention. The wardrobe (or in the Italian language: guardaroba, literally "watching the robe") can follow you like a scent does, such as a bottle of Serge Lutens' Borneo 1834 with mesmerising patchouli notes, but not wrapped in silk as it once did for Parisians to eradicate moths and other insects from their scarves in trunks. Or you may follow it via visual platforms. Why does a holiday season, mid-season, or trans-seasonal matter other than a ploy to entice people to consume more of what we already have? In our age of uncertainty, retrofitting matters.

Recently, at the E/Motion: Fashion in Transition exhibition (2021–2022) held at the Fashion Museum in Antwerp, Belgium, curator Elisa De Wyngaert observed the way fashion has served as a visual form of uncertainties. The third exhibition space, entitled "Home," provided a labyrinthine space world, uncovering issues of cultural appropriation, origin, and nostalgia, among other themes. De Wyngaert states that "subconsciously or otherwise, these garments functioned as a kind of architecture designed to protect the wearer, a form of self-preservation" [2, p. 63]. Historically, she notes that "designers have found inspiration in other cultures, often appropriating other religions, identities and civilizations, without contextualizing or crediting their sources" [2, p. 63]. In addition, De Wyngaert suggests that "a buoyant transition has been instigated by young designers who, each in their own unique way, are contributing to the decolonization of the fashion industry. These designers use their voices and collections as important platforms for activism, to tell different stories and represent different communities" [2, p. 63]. Though this may be the case today many of the high-end Italian luxury fashion labels have been enduring the most of criticism for cultural appropriation, but this has always taken place for centuries across the globe, not only in Italy but elsewhere.

What is more at stake is how the fashion transition creates the climate adaptability challenge for architecture as well. De Wyngaert's "Home" exhibition also brings the architecture and fashion realms together, something that Christian Lacroix established and that Etro and Dries Van Noten take on board effectively. For Dries Van Noten, "people are starting to see that the whole system of seasonality maybe a bit outdated… Every time, you add things to your wardrobe that do not go out of fashion. The whole concept of fashion is maybe a bit outdated" (interview in Arnold et al., p. 262). The seasonal wardrobe itself, figuratively, requires reconsideration, including its outdated spaces but not the role of the imaginary in the creative process as this foundation supports the act of retrofitting and we wear it.

Commenting on the wearer of fashion practice Ella Sampson articulates the role of fantasy and the imaginary in the creative process. She writes: "unlike a material artefact, the imaginary artefact is never truly graspable and thus never truly complete. It is ephemeral and transitory; its essence eludes capture" [21, 351]. Suggesting the transitory nature, time factor involved in manufacturing as well as the transition of garments produced from one place to another, when these textiles are displayed in space, the present alternative materialised spaces are usually the ones we cannot see. Mobile applications have already replaced the walk-in-robe to out-date fashion, and to make do with what we have (including the tapestried footstool), to another world where one is able to switch from one platform anytime you like, to any place wherever on the globe you find culture(s) (Fig. 1).

When we rethink of the meaning of "transition" what comes to the fore is the fluctuating circumstances of where one is on the globe at any given moment and what to wear and question the garment's appropriateness for the environment to switch to someplace else with ease. Since the post-pandemic, life is unpredictable, and fashion stuffs have become obsolete to a certain extent. Yet the space where garments are

Fig. 1 Go through your wardrobe: make-do and mend, ca 1940s advertisement. *Source* https://commons.wikimedia.org/wiki/File:Go_Through_Your_Wardrobe_-_Make-do_and_Mend_Art.IWMPST4773.jpg

exhibited or the place where they could be worn in the environment continues to present opportunities to resort to as a form of escapism. Considering the dressing gown as a retro garment for the luxury sector, the garment has been "upholstered," that is, through the act of covering furniture with their patterned textiles, which have impacted ideas about resort wear for the contemporary home and hotel. Merging the incompatible global garment with what is acceptable to the masses might insult the minority or those who do not want to understand fashion, but the robe accommodates a different world of its luxury.

2 Etro's "New Tradition" and Vintage Garments

Etro implies retro. Globally recognised for its Pegasus logo, borrowed from ancient Greek culture, and the paisley motif, derived from the Persia word "boteh," Italian luxury label Etro was established in 1968. The name Etro, for some, infers a "river" in the Persian language. With this cultural background in mind within Etro's photo-shoots of the 2000s, for instance, there are shared motifs inspired with Armenian, Bornean, Moghul Indian, Persian, Navajo Indian, Peruvian, Patagonian, Japanese, and Mexican devices. More specifically the house garment appears slothful and therefore luxurious, relaxed. The firm's global design tradition straddles between the architectural and fashion realms. In particular, the Eastern encounters within Etro's collections have received no attention in relation to the retro to real spaces where they are paraded in markets or for their advertising campaigns on roof terraces and out in the desert.

Referring to Italian writer Molho's [17] story of Etro's world and their "culture," in looking at the retro origins of the company, their influences and clothing will demonstrate Etro's penchant for slothful living or relaxed lifestyle. Intriguingly, one does not know whether an Etro collection is new or not, only timeless and "un-second-hand," even if neo-vintage (like Christian Lacroix's and Dries Van Noten's eclectic and elegant garments). As far as traditional design is concerned, Etro has profound respect for the cultural origins of clothing by injecting a level of vivaciousness. By undoing what has been designed before they encapsulate a transition to what has not been explored, merged, or created before.

Renata Molho has studied Etro's methods and techniques and their "new tradition" clothing. Etro's family company started in 1968 by the Gimmo Etro and his children (Jacopo, Kean, Ippolito and Veronica) are avid collectors of art with a penchant for "satoriality and a taste for craftsmanship" [17, 10]. The design of the garments "instil the sentiments and the need for beauty and ethics that are proper to Etro, transforming the adventure of the everyday into something that is both unique and universally useful. To translate a personal utopia that everyone can share" [17, 11]. Etro categorises their collections into cultural traditions of other cultures, but without appropriation for they archive everything. The nature of their methods shares the importance and presence of sustainable luxury. The Etro company takes risks, and they accomplish this technique by adapting the design of the robe and in doing so have established a new tradition. Etro always adds something new that lends itself to ignite the transition.

Since designers and fashion critics have reacted to the oversupply of clothing the intelligent inquisitiveness Etro have maintained is the family company tradition in reusing similar motifs in many of their collections in a spontaneous but in an informed manner. With a sense of immediate recognition, Etro incorporates into their clothes a kaleidoscope of paisley in action. Etro's core concept is the "new tradition" by "respecting the strict canons of dressmaking" [17, 281]. Commenting on Etro's artisanship, Molho states that their diligence "are canons that cannot be overlooked, for it is thanks to them that each piece of clothing contains the experience of a generation of dressmakers who have made history" [17, 281]. The difference Etro bring to fashion is their avant-garde approach, and the concept of the "new tradition" is that they transform the boteh motif. Some of Etro garments had stitched the labels with the following messages: "This garment can be recycled endlessly—your (great-) grandchildren will be proud of you" or "For Your Hands Only" [17, 293]. A sense of recognition of the first-hand clothing in the family or family members or through inheritance to the next generation remains as this "new tradition." The second-handedness elements of the garment are not tainted with hobo or boho aesthetics as such, but despite this have reacted to embrace an heirloom quality prolonging the transitory garment for future generations to appreciate.

Yet second hand, that is, a passed-on garment and other clothing goods from one person the next (other than that have been worn or discarded) are treated in the same way as furniture from a previous owner have been attested since the sixteenth century. Etro's design tradition is "distinguished from the idea of 'clothing' by the high dose of risk and innovation that they contain and express" [17, 81]. The company reworks existing motifs, including the harlequin lozenge-pattern, into something that go beyond all known traditions. This is where their uniqueness resides. More importantly, in their creative attempt, "in shaping the fabric and inventing the representation of a fashion object, we can hear the echo of the knowledge of thousands of things distant from each other" [17, 81], as with the 2022 Liquid Paisley collection.

Garments that make one think is what is important to Etro as they thrive on observing and finding innovative ideas, and they show a deep appreciation for the depth of the origins of patterns by abstracting them digitally into something new. Their range of intense colours for example are derived from the inspiration of Italian surrealist Giorgio de Chirico "for the background of their clothing campaigns and advertisements as well as in the clothing—'dressing up art'" [17, 133], the start of the architectural embedded into their designs. For Molho, " Etro, as always, works with the materials to achieve unique solutions filled with meaning: precious scarves, points of encounter between the most diverse ethnic origins and techniques, are knotted in a very personal hand game, while embroidery isn't sewn, but rather poured directly onto the clothes" [17, 100].

Culturally speaking, the theme of Etro's nomadic culture in the 2000s collections leads one to concentrate on their linkage with Eastern traditions and the silk re-routing into their clothes and wardrobes. Their cultural legacy embraces what we can learn from the negative aspects of cultural appropriation. Instead Etro appreciates

Armenian traces not to be forgotten. Surfaces of their garments are reminiscent of medieval Armenian costumes as we locate in Sergei Parajanov's films. This is especially the case when observing the juxtapositions of simple designs, bold fabrics, and exotic prints that take us to Mongolia or Uzbekistan, Etro connects silk with nature and fashion as well as the colours of the earth's surface. Being earthbound, their "techniques are elaborate, including tweed patchwork, leather stitching, topstitching, and the application of ribbons or inserts. The colours are those of the fall, spicy ones, and of the earth, of the ever-changing paths to be taken, just like the places to be reached and the cultures to be encountered" [17, 173]. From this perspective, from their trip from the late 1970s, Etro introduced flowing kimonos and to date their luxury homewares characteristically highlight paisley in a unique way accompanied with fine embroideries through upholstery. Additionally, Etro concentrates on the idea of nomadism, housing, and celebrating their collections in a yurt and with their approach to encompass falconry in their 2020 resort collection, merging the cultural with the novel in their designs, specifically when looking at their simple robes embroidered with hope and expressing relation as a slow retro "take-away" (Fig. 2).

In considering luxury artefacts, Gerlolamo (or Gimmo) Etro offers clues as to how the vintage robe transformed fashion through textiles and the collection of antique pieces from the 1980s onward. He recalls visiting his grandmother and being attracted to her dressing gown and would "look in the dresser drawers: among other things, they contained Paisley shawls and scarves, some of them vintage" [17, 329]. His wife, Roberta "was an antiques dealer, started collecting Paisley shawls. Not to sell them, but for own personal pleasure" [17, 329]. Intriguingly, the relation between the dressing gown, dresser drawer and paisley as artefacts of pleasure are luxury items that have pinned Etro's transition from anti-fashion to encompass the architectural realm. With profound respect for the vintage fabrics and the designs on the scarves, his involvement in future collections came directly from the East. Gimmo continues: "in many cultures, from the most primitive to the most evolved, social affirmation was often manifested through one's clothes. I continue to be fascinated by the history of costume, from ancient canonicals to Eastern traditions—places with a history to tell and an incredible artistic and artisanal production" [17, 331]. The Etro couple has collected these textiles mostly from the nineteenth century. As a result, the robe and dresser drawers offer an insightful connection as to extension of the vintage fashion-architectural transition to incorporate homewares.

3 Accommodating the Robe as a Transitional Garment

The term "robe" implies many material things, but in the Italian language "roba" does not necessarily mean dress. Since the sixteenth century "roba" has been associated with good stuff, luxurious goods handed down, or taken away, from one generation to the next. A fine gown or expensive robe made from luxurious textiles infers something exotic and obtains erotic associations which are derived from the East.

Fig. 2 Photomontage demonstrating the transition space between a medieval Persian miniature, a robe with falcon, with Etro's 2020 resort collection with a similar inspiring theme. *Source* Image created by the Author (8 September 2022)

Curiously, there is an Italian phrase that suggest that roba means material or property and "bona-roba" signified a courtesan [18, p. 93], especially ones dyed with saffron. Today, it is associated with a well-dressed woman who is not a prostitute. Yet in nineteenth-century Venice the port centred on importing luxury goods from the East where saffron robes made from silk became popular. Evidence of this was found more than a decade earlier. Deriving from luxurious materials, it appears the "bona-roba" expression was linked with the ancient Sybarites whose lifestyles were associated with moral excess. As I recount in my book The Architecture of Luxury, in terms of pleasure and the control of excess, in the sense of eradicating immoral behaviour, I conceptually discuss what is admissible in a luxury sense or when luxury is forbidden in architecture and its dispersal of sybaritism to run amok. In addition, I argued in my chapter on venues for sybaritic parades in Italy and beyond, one of the places where luxury fashion began was in ancient Sybaris (now in the Calabrian province—the toe of Italy) where the Sybarites clothed themselves with saffron garments. The ancient Sybarites also invented the clothing patent. These so-called "sybaritic luxury" customs, informed by Italian writer Romualdo Cannonero in 1876, reappeared as a myth in Renaissance Venice [4, p. 156]. Variations of the sybaritic robe affected theatrics, the shopping experience and fashion consumption from Lyon and later in Saint-Germain, France.

In contrast to the Western perception of the robe in Italy, in the East there is within Buddhist attire an expression of "taking the robe" which was identified as a full commitment to one's religion. One English writer, for instance, observed in 1860 how a priest "who keeps the superior ordinance may not lean on any place, or make his robe into a seat, or take hold of a piece of cloth fastened to a tree" [22, p. 107], habits that appear distant or obscure but reappeared in a strange form today. (Here, I am thinking of the ghastly crocheted remnants dotted around some cities choking trees for no real reason.) Other robes were stolen from people's graves. Moreover, robes were made out from cast-off remnant pieces of fabric. As an early upcycling practice, these salvageable pieces were used to make robes for they were a display of ascetic modesty.

Robes evolved from undergarments into house coats worn inside evoking Eastern traditions. Many of Etro's designs used paisley as the definitive universal characteristic. Initially, their travels to India inspired them to incorporate it into their collections. Other collections were inspired by Persian and Armenian motifs. Regarding Etro's clothing line, "paisley landed in Europe in the mid-seventeenth century, after the East India Company's trade expeditions. It was a triumph and conquered all classes of the population. There were those who saw it as a talisman to ward off bad luck, those who instead turned it into a symbol of rebellion, and those who were enthralled by its beauty, mystery, and the myriad of meanings it held within" [17, 334]. Etro made the paisley more accessible to fashion in the 1970s and transformed paisley shawls into long caftans as well as retro housecoats that can be worn outside too. Today anything is acceptable within the realm of anti-fashion within (or without a) reason.

4 The Robe's Fluctuating Function and Transitional Spaces

In terms of the transition space within the architectural realm, "fashion poses a doubled view" in that "those in power want to present that culture; and a view of how, at least within a range of choices, a culture wishes to depict itself" [16, 142]. The mechanism of fashion "renovates these options into more socially prolific phenomena, capable of permeating every stratum of a society without regard to economic or ideological diversities" [16, 142]. In undressing architecture, according to Mary McLeod: "fashion may again offer a model for architecture, though one quite different from that which prevailed during the first part of the century. The fluid boundaries in contemporary women's fashion underscore the continued rigidity in architecture discourse, an avant-garde moralism that is all too evident in the peculiarly puerile debates… that continue to plague the profession" [16, 90]. With its fluid cultural frontiers, Italian luxury label Etro is constantly transitional and, in effect, seasonal.

Whereas in the fashion realm "retro" might seem to be something outdated or closer to the historical types of garments, this phenomenon has opened the debate and mindsets for the change, environmentally, as a return to nature. That is, by transiting or reusing what remains. Some fashion companies embody the capacity for clothing to have resilience, for them to last fever. This is the mindset of the Etro company, a mindset which is rare these days. Other designers "choose to create fashion that has resilience. This means not only designing pieces able to withstand the test of time and stress, but also building social resilience, connecting communities, and encouraging conversations where divisions are rife" [8, 156–157]. Here, I am thinking of the English label By Walid, encouraging the resilience of textiles and their reuse of tapestry remnants into new garments or sneakers. Sustainable fashion practices and social resilience will only work if companies change their mindset of the customer and the industry to seek retro in a new light.

The robe's resilience has taken generations to develop into something worth keeping and this item has enabled the transition of clothes and spaces within architecture to evolve as "retro." People's habits have also shaped how transitions of place and space to occur, that is, through lounging. Yet how did loungewear begin to be popular in the West? It is peculiar to consider slothful habits. For "sloth makes all things difficult, but industry all easy…." [11, 477]. The fashion industry is filled with advertisement scenes of human sloths in robes—where life has become too easy for some while not for others. It is as if human sloths surround us, languishing around at home, lounging around plush sofas in shopping centres in public spaces. Lying on the floor sleeping at airport custom points with nothing to declare, without care and without any concern about how hygienic the interiors are or how they are retrofitted.

At a time where the undergarment has become an outer garment, the robe in the form of a dressing gown was introduced into Italy through France. In Fabric: the Hidden History of the Material World, Victoria Finlay writes that when Armenians migrated to Marseilles in France and to Venice at the end of the sixteenth century, because of the battle between the Ottomans and the Persians, many refugees ran cafes, fabric print-shops and dye-works. She states that "their expertise in textiles came in part from a particular element of Armenian Christianity that involves the ritual concealing and revealing of the altar with screens and curtains. By the tenth century, Armenian monasteries had developed ways of printing cotton with stories from the Bible—a portable equivalent of stained glass, but in cloth" (p. 91).

In the mid-seventeenth century, the printed cotton dressing gown influenced European fashion. Finlay writes.

> The French called it dressing a' l'Armenian after the people who both sold them and wore them, and the gowns themselves were called indiennes. In Britain, they were called bayans, after a Hindi word for merchant. Their origins can be traced to 1609 when the Dutch east India Company was granted trading rights with Japan and began to bring an adapted form of silk kimono back to Europe. These comfortable, stylish gowns, worn by both men and women, reflected what must have felt like a new era: less fussy, led riffled, less lacy clothes marking a new, plainer century... But the silk ones were expensive, and very quickly versions made with Indian chintz became popular too, not only in Europe but also in Japan. (p. 91)

At the time dressing gowns a' l'Armenian, undergarments, and outer garments, became a form in transit. However, "the gown's name had little to do with actual Armenian fashions and was more likely inflected by the fact that Armenians merchants negotiated and controlled a large part of the painted and printed chintzes intended for European…." [14, p. 146]. One such case was during the Japanism movement in France and Italy where it was worn for relaxation inside the house. There, as in France and elsewhere, the dressing gown or robe was worn for covering up or for camouflage, but not to be worn outside but for protection against harsh climates (Fig. 3).

So when rethinking about the transition issue from East–West and vice versa in the luxury sector and the current environment and climate adaptability, the role of fantasy in creating a garment within the architectural realm go hand in hand. This is the case of the Etro company's global garments, resembling the so-called a' l 'Armenian dressing, the contents of the Fortuny Palace's ethos as well as its Venetian aesthetics.

In tracing the origins of Italian fashion, Sofia Gnoli outlines how artist-couturier Mariano Fortuny's Venetian garments evoke the East. Fortuny fascinated Italian writer Gabriele D'Annunzio as with his collection of Japanese artefacts displayed at his Lake Garda Estate in Lombardy. She mentions it piqued French writer Marcel Proust's interest where Fortuny's richly decorated gowns were particularly reminiscent of the East. Citing Proust, Gnoli writes:

> These Fortuny gowns faithfully antique but markedly original, brought before the eye like a stage setting, with an even greater suggestiveness than a setting, since the setting was left to the imagination, that Venice loaded with the gorgeous East from which they had been taken, of which they were, even more than a relic in the shrine of Saint Mark suggesting the sun...

> the fragmentary, mysterious, and complementary colour. Everything of those days had perished, but everything was born again, evoked to fill the space between them with the splendour of the scene and the hum of life, by the reappearance, detailed and surviving of the fabrics worn by the Doges' ladies [12, 24].

Fig. 3 Cesare Vecellio's engraving of an Armenian robe worn as an undergarment. *Source* https://commons.wikimedia.org/wiki/File:Cesare_Vecellio._Armenian_woman_regional_costume._16th_cen.jpg

"To fill the space between" suggests the literal, an architectural sense to the transition into something or its reappearance as retro. In other words camouflage into another culture as a good: new or as a transitory artefact to impact the future of fashion. The Fortuny gown is to be worn well and well-worn, a motto Etro incorporate into their collections. Curiously, Mariano Fortuny studied Japanese methods of hand-printing, designed a scarf for the American dancer Isora Duncan as well as tea gowns and Arabic caftans at a time when the less formal attire became popular. The caftans, no doubt, were a derivative of the Armenian dressing gown and all these aspirations are rebound in Etro's outmoded robe designs. The Armenian robe ought to gain more attention because of its impact and dispersal (Fig. 4).

Whereas transitional clothing within any given wardrobe extends across the seasons from four to two or two, all depending on where it is located globally, in terms of wear and tear the seasonality of clothing is impacted by the interior or exterior architectural spaces. So the clothing and its seasonal changes depend on many factors of which include the seasons' interchangeability and unpredictability of specific climatic conditions of a particular space. And this is affecting the global fashion industry, particularly the luxury brand sector. When we closely observe the transitional wardrobe, it is a little absurd or ambiguous because the object itself cannot move, only the artefacts on the hangers. At any given moment, only the person can move and select through the wardrobe—or the dressing room as an architectural feature—wherever they are destined to be and how they feel due to the weather, the weather, or the type of day it is. Such fluctuating circumstances of

Fig. 4 Photomontage showing the Etro's housing of robes resembling Venice's Fortuny Museum's exhibition. *Source* image created by the Author (8 September 2022)

the weather affecting the wardrobe governs how sustainable luxury practices have become popular today through a transition of garments as a real space.

Transition space as a real space also transforms a mindset that lures the customer directly towards luxury garments to extend the excess of a textile source through slothful appearance. That is, without any relation or association with moral sensibilities, because garments draped and worn on the body in private or tried on in the dressing room exist in two ways. One, as if nobody is watching, without any of the wearer's inhibition, and the other through a mirror, but not for the purpose of exhibition but for the testing of time for a garment to last in transitory spaces.

5 Armoires as Transitory Spaces

Historically, armoires are the transitory spaces where clothing and architecture merge into space, housing the wardrobe. "Freestanding, upright, and decorated," the armoire, according to Henry Urbach "evoked the clothed human body" [23, 64]. Armoires are like "volumetric objects with unambiguous spatial presence" [23, 64] For Urbach, in America they became outmoded cabinets in the mid-nineteenth century because they were overrun by the invention of the built-in closet, which concealed its contents for storing linens, clothes, and other excesses, which changed the domestic planning of house and apartments. Whereas in the hotel, the lobby and roof terrace are the transitory spaces for storing and outlaying the resort wear to the public.

Intriguingly, there is an Armenian presence in Naples that alludes to architectural transitory spaces that link with the armoire. The cultural district of San Gregorio Armeno established in the nineteenth century showcases a little street between Spaccanapoli a little street between Spaccanapoli and Via dei Tribunal renowned for its artisanal shops selling Presepe (or nativity scenes), which once inspired Christian Lacroix, nicknamed "crib street" where one can find waxed figurines of arms. Moreover there are reliquaries displaying scared relics such as altar curtains, and it is possible that armoires transformed into linen closets that housed reliquary arms and tapestried church textiles.

And then there are the gold and silver arms housing fractured bones, such as the philosopher-and-Saint Thomas' reliquary Arm glorified in an awkward pose at the Basilica of San Domenico Maggiore. In any case, off the coast of Naples on the Isle of Capri, the place became a fashion haven for many tourists from the 1910s onward. And today, for example, Etro's 2022 resort wears at the Punta Tragara Hotel (a building where Le Corbusier produced a series of revamp sketches for Emilio Enrico Vismara in the 1930s) on the Isle of Capri. Etro's resort collection displays a spectacular awe-inspiring range of homeware within the Pegasus hotel suite and upholstered furniture covered with paisley motifs on its roof terrace embracing views out to the famous Faraglioli rock formations in the Mediterranean Sea.

Curiously, bedroom furniture comprises the guardaroba—to look after one's clothes. It appeared in Europe in the 1820s as chests with stored clothes in the form of an armoire or wardrobe to store clothes, as in the use of armoire trunks. These pieces of furniture were fitted against a wall and were, and still are, considered as valuable property. Armoires or chifforobes were places for the recycling of textiles. There is an ancient link with the collection of precious silks from the East and other leftover lightweight fabrics to be stored in such wardrobe trunks stuffed with layers of patchouli scent to eradicate moths, something we would relish again when thinking of Serge Lutens' powerful and alluring Borneo 1834 fragrance. Soon these products were made for luxurious voyages. Louis Vuitton, for instance, created his first wardrobe-trunk in 1875, which would prove itself a modern luxury classic [19, p. 88].

Turning from the Mediterranean to Armenian crafts in the Ottoman Empire to see how the robe and its cultural exchange between Italy and Armenia created new cultural dimensions, it is through the sultan's interior where armoires transformed. Here these armoires became the transition space.

Nora Khatcherian suggests that material culture from the Ottoman period, especially the robes made from luxurious silks were in fact Armenian [15, 501]. Additionally in the case of Uzbekistan "robes made from luxury fabrics such as silk and brocade also comprised a longstanding feature of political ceremonial throughout the Persianate world, from Qajar and Safavid Iran to Mughal India, and Central Asia was no exception" [20, 4]. Moreover, "Armenian silks were introduced to the West at the Ottoman Empire booths at international trade expositions beginning with the London Exhibition of 1851" [6, 35]. It is necessary to observe how this material culture developed in the context of architecture and style in France.

Evidence of the Armenian material culture of the robe appears in France again and this time at a fairground which became fashionable shopping grounds. Curiously in her chapter on a new kind of shopping, DeJean [7] discusses how the attitude towards collecting within an interior in France that made today's passion for antiques possible started to take shape in the last decades of the seventeenth century. Both shopping experiences are quests for just the right style: antiques bring the standards of couture into the home. Collecting began on a larger scale and the antique emerged as a category when people began to shop for antiques as they shopped for designer clothes. According to DeJean, late seventeenth century Saint-Germaine fairgrounds established "posh shopping in elegant surroundings" where evidence of the Armenian robe is apparent. "Within these novel settings, improvised farces in the Italian commedia Dell'arte tradition, luxury shopping became a for the first time ever a truly public event. Here people dressed to the nines. Merchants transformed into antique dealers" [7, 235–236]. It was in the comedies where merchants started to deal in exotic wares, and this is evident in Jean-Francois Regnard and Florian Dancourt's descriptions "distributed in France after its East Indies trading company was established" (Dejean, 238). For instance, "Oriental imports such as silks and coffee was served "at the stand of an Armenian named Harouthioun." Here, the owners of the café were dressed an Armenian—in the elegantly exotic flowing garments worn by those who served refreshments at the real life fair and in the original posh cafes that became a fixture at the Saint-Germain neighbourhood. [7, 238–239]. Regnard's comedy frontispiece shows a Harlequin figure" trying on an 'Armenian' robe for size" (239). In addition, the Armenian robe was renowned as a strange outfit, [7, 139] and linked to Italy.

Turning from Armenian silks to the fashion a la Turque or Turkomania, for example, which was widely circulated in artworks and literature in eighteenth-century France, the undergarment, caftan and purple fabrics, publics in pattern books formed new "middle ground" trends across Western Europe. But this was not really the mode in Italy as such since in Venice the fashion for a Turkish fountain and marble seating featured in luxury palaces for the satraps, such as the Ca-Dario three centuries earlier [5, 56]. Trade between the Ottoman Empire with Venice affected the fashion transition. Inal Onur explores the interchange between Eastern and Western worlds in the past in the context of the globalisation process. For her part, cities were the touch base "between Ottoman and European societies and came to provide an alternative domain for cultural and economic intervention to the rest of the empire, not only because of their privileged economic and political positions, but also because of their multicultural and tolerant environment." And "they functioned as 'middle grounds' between the Ottoman and Christian European worlds" [13, 248]. Additionally, "Ottoman elite women and townswomen became willing to consume foreign goods in this period. European costumes, together with comestibles, furniture, and other household items, were among the goods that Ottoman women consumed, since such consumption defined social identity and rank" [13, 268]. As the interchangeable process occurred between the Eastern and Western perceptions of the globe became evident in architecture and fashion, with the creation of new devices, such as mobile furniture, became popular.

Turquerie suddenly appeared influencing, painting architecture and artefacts as well. In terms of luxury and the transition of Ottoman fabrics and textures in future fashion, the "appeal of Turqueries lay in their making luxury permissible by recasting it as a form of culture…Turquerie helped to circumvent this common critique by dressing up luxury in foreign garb" [3, 102]. Theatrical play and stage sets were the venues where one could experience "Ottoman indulgence in luxury, as they regularly took advantage of the visual excitement produced by Ottoman fabrics, jewels and furniture" [3, 102]. Around the 1750s, the sofa, a new sybaritic commodity whose arrival was met with the criticism of moralists, is one such example." [3, 103]. They associated the Ottomans with these "new trappings of luxurious leisure" and "manufacturers gave Turkish names to enhance the appeal of their new products." Moreover, "luxurious clothing likewise drew inspiration from Ottoman models…as it was difficult to recline comfortably on a padded sofa in a corset" [3, 103–104]. Furniture and fashion immediately became the cultural forebears for the rest of the world. "The study of Turquerie proves that histories of transmission are not just amusing curiosities in the background of an object or idea's 'adult' social life, they are all-important genealogies of culture" [3, 118]. The Ottomans introduced the sybaritic sofa and upholstery into Italy.

Originally recognised as the "up-holder," that is, the art of needlework, the art of upholstery originated in the East. "It is known that from the earliest times the Egyptians used upholstered couches and chairs and the Assyrians, their neighbours, who were masters in the art of weaving, introduced the rich woven draperies used both for covering the seats of furniture and for the draping of walls" [1, p. 6]. Meanwhile in England, "when the gentlewomen of those days acquired the art of needlework the 'up-holder' came into his own. The needlework was used at first only to drape the walls of aristocratic mansions, and thus the artisan who was proficient in this work came to be known as the 'up-holder' of tapestries and, years later, the upholsterer" [1, p. 6].

The armoire was originally the item that was generated from the veneration of arm reliquaries in Venice and in Naples since there were many Armenian migrants who lived there and established a new system of keeping the tradition of the dressing gown alive. Etro appears to have accommodated all these clothing and housing garment practices, especially when viewing all the collections through time and their "new tradition" of designing inspiring robes to be worn out in the desert. By accommodating many cultures, the countercultural adaptation of the paisley or boteh motif and their fresh digitised forms throughout their garment designs and other merchandise fitted out at a hotel, when travelling or out in the desert have provided outlets for resorting permitting freedom to dream in lighter garb. Armenian robes transformed outmoded fashion, permitting the freedom of movement of the body through transitional spaces (Fig. 5).

6 Retrofitting Etro

From a "new tradition" to a renewed sustainable engagement, Etro unremittingly embraces Armenian tradition of the silk and cotton garment from a middle ground. The novel aspects of the structure of Etro's garments for inside or outside are no doubt influenced by foreign filmmakers, especially the imaginative stage settings depicted in the Armenian director Sergei Parajanov's epic film The Colour of Pomegranates [25]. The film is an exquisite series of tableaux vivants of the priest and poet called Sayat Nova, and it is visual extravaganza of medieval Armenian costumes, lace makers, and a surreal expose of robes without wardrobe. The wardrobe are the people, changing garments from one's hand to the next in the form of an invisible armarium (a closet for storing clothes in antiquity). The richly decorated gowns passed on from hands in unusual ways featured in Parajanov's film were reused from a previous film he made in Armenia. One scene shows a white robe magically appearing to cover a black robe and another with gorgeous rugs, furniture, and gilded picture frames from churches to living quarters. At this point in time, the Armenian robe design influence has come to the fore within Etro's work as this new transition space (rather than scenes we find in Lady Gaga's 911 video clipping Parajanov's scenes together with Alejandro Jodorowsky's 1970 Mexican acid art-film El Topo).

Having established this historical West–East perspective, the chapter observed some of the socio-cultural aspects that Etro's robe design covers, from the dressing gown to the Fortuny gown, or houses that are in continual transition as non-spaces. Retrofitting Etro is a laser focal-point that casts architecture through the fashion realm. I also argued how fashion architecture expanded how Etro's collections have

Fig. 5 Photomontage of Etro's 2022 resort collection in the desert, a "non-place" a retrofitting transition equated with a scene of the black robe from Sergei Parajanov's the Colour of Pomegranates. *Source* Image created by the Author (8 September 2022)

expanded to embrace outmoded fashion. Etro's new tradition' was fitted out to embrace another way of conserving fashion and sustainability in the context of new ways of coding, categorising luxury without a robe or abode.

New and second-hand information about the history of garments therefore uncover some of the architectural, Eastern impact, and sources relating to the robe or dressing gown. Remnant fabrics that have survived are now venerated so much more than the past and cast as more valuable because of their ethnicity. Perishable textiles and their patterning remain important for recognising the true essence of the wardrobe and its seasonal changes that it encases or ruins.

Etro transports our thoughts elsewhere. Inspired by the armoires of history, Etro has fashioned the other spaces that integrate this transition. Their collections uncover a retrofitting future. Etro's expression of culture, however, neglects its own Indigenous traditions—of Calabria—the place where Armenian culture had some traction and not only in Puglia and Campania. In retrospect, the architectural realm unveils a rich history relating Etro to non-fashion. To date, Etro's new creative director Marco de Vincenzo should further connect the company's cultural origins to encompass Italy by embracing not only Sicilian but also Calabria's origins. Cutuli Cult has commenced a transition by focusing their attention to upholstering laser garments that use experimental spatial techniques. And Calabria's silk production has been revitalised which hopefully will uncover forgotten techniques embracing past and future visions.

The theme of retrofitting Etro entices other luxury brands to perceiving the other in fashion through these new transitional spaces. Their designs embody retroism with second-hand nature—but then inject a spirit of sartorial elegance. Although their archive reminisces elements for the East, ambiguous details arise the design of their robes. This is specifically the case when observing the robes of rich silk from Armenia, recognised as one of the countries to officially adopt Christianity and which influenced Italian culture in the Veneto and Calabrian regions. In other words, clothing needs seasoning with that extra sartorial sapience, acclimatised by both the architecture and the landscape. With climate change, when it comes to outdated fashion what will be the core of clothing humanity when its fluctuating transitions have adapted the climate? Retrofitting as an idea of improvising the robe into future designed outfits will inevitably fit-out fashion. The fine-grained needlework Etro produces digitally exemplifies architectural fashion in transition. In a way, this can be defined as "digital upholstery" that fits-out the outfit, retroactively.

References

1. Reporter unknown (1925) The history of comfort. The Manchester Guardian, February 26, 6p
2. Arnold R et al (2021) E/Motion: fashion in transition, MOMU, fashion museum Antwerp. Lannoo Publishers, Belgium
3. Bevilacqua A, Pfeifer H (2013) Turquerie: culture in motion, 1650–1750, Past and Present, no 221, November, Oxford, pp 55–118
4. Condello A (2012) In fashion: Venues for sybaritic parades in Italy and beyond. In: Brownie B, Petican L, Reponen J (eds) Fashion exploring critical issues (Part 4). Inter-Disciplinary Press, Oxford, United Kingdom, pp 155–164
5. Condello A (2014) The architecture of luxury. Routledge, Farnham
6. Condra J (ed) (2013) Encyclopedia of national dress: traditional clothing around the world, vol 1. ABC-CLIO: Santa Barbara, California
7. DeJean J (2005) The essence of style: how the French invented high fashion, fine food, chic cafes, style, sophistication and glamour. Simon and Schuster, New York
8. Ehrmann E (2018) Fashioned from nature. V & A Publishing, London
9. Fausch D, and Singley P et al (1994) Architecture, in fashion. Princeton Architectural Press
10. Finlay V (2021) Fabric: the hidden history of the material World. Profile Books, London, Great Britain
11. Franklin B (1834) Memoirs of Benjamin Franklin, vol 2. McCarty & Davis, Philadelphia
12. Gnoli S (2014) The origins of Italian fashion. V & A Publishing, London

13. Inal O (2011) Women's fashions in transition: Ottoman borderlands and the Anglo-Ottoman exchange of costumes. J World Hist 22(2):243–272
14. Lasic B (2017). Ethnicity. In: McNeil P (ed) A cultural history of dress and fashion in the age of enlightenment. Bloomsbury, London, New York, pp 139–160
15. Khatcherian N (2019) Armenian crafts in the Ottoman empire: Armenian identity and cultural exchange. Nat Identities 21(5):485–505
16. McLeod M (1994) Undressing architecture: fashion, gender, and modernity. In: Fausch D, Singley P et al (eds) Architecture, in fashion. Princeton Architectural Press
17. Molho R (2014) Etro. Rizzoli, Milan, New York, Paris, London
18. Nares R, Wright T (1888) A Glossary: Or collection of words, phrases, names and allusions to customs, proverbs, etc. Reeves and Turner, London
19. Pasols PG (2005) Louis Vuitton: the birth of modern luxury. Abrams, New York
20. Roosien C (2022) I dress in silk and velvet: women, textiles, and the textile-text in 1930s Uzbekistan. Central Asian Surv 41(1):1–21
21. Sampson E (2018) The cleaved garment: the maker, the wearer and the "Me and Not Me" of fashion practice. Fashion Theory 22(3):341–360
22. Spence R (1860) Eastern monachism. An account of the origin, laws, discipline, sacred writings, mysterious rites, religious ceremonies, and present circumstances of the order of mendicants founded by Gotama Budha. Williams and Norgate, London
23. Urbach H (1996) Closets, clothes, disclosure. Assemblage, no 30, August, 62–73
24. Warke V (1994) Architecture: observing the mechanisms of Fashion. In: Fausch D, Singley P et al (eds) Architecture, in fashion. Princeton Architectural Press
25. Parajanov S (Director) (1969) The colour of pomegranates. Armenia. Restored at L'Immagine Ritrovata, Bologna'

Communal Design: A Pedrana Approach to Sustainable Fashion

Emily J. Oertling and Kim Hiller Connell

Abstract Communal design is a collective approach to garment production, where community knowledge and collaborative work efforts jointly influence the product. Observations on communal design practices arose from a qualitative study conducted with the Tz'utujil—Maya residents of San Pedro La Laguna, Guatemala, in the summer of 2020. Participants' accounts regarding the production and use of the blusa Pedrana are the foundation of this sustainable strategy. Pedranas' (the women of San Pedro) engagement with the fashion system is not intended to be sustainable. Their design framework is rooted in an Indigenous Knowledge System that yields a product and process with great social significance [5]. From ideation to use, the social and cultural customs of the municipality inform the production of local fashion. Ultimately, this chapter argues for the inclusion of culture within sustainable fashion methodologies, specifically the centering of community and collective knowledge, while recognizing that the work of Pedranas cannot be replicated nor exploited in the name of progress.

1 Introduction

The global fashion industry is complex. The simplest of garments are often the product of manufacturers in multiple countries, with roots in the agricultural and petrochemical industries. The production of fibers and yarn, followed by textiles, then clothing is a twisted chain that remains opaque to the end consumer; and the environmental and social impacts of this supply chain, although difficult to measure, abound.

Upwards of 8,000 chemicals, many of which are toxic, are used to produce clothing. Insecticides (agriculture), solvents and lubricants (textile manufacturing),

---

E. J. Oertling (✉)
California State University—Sacramento, Sacramento, CA, USA
e-mail: oertling@csus.edu

K. Hiller Connell
Kansas State University, Manhattan, KS, USA

M. Á. Gardetti and R. P. Larios-Francia (eds.), *Sustainability Challenges in the Fashion Industry*, Sustainable Textiles: Production, Processing, Manufacturing & Chemistry,
https://doi.org/10.1007/978-981-99-0349-8_3

dyes (fiber, textile, and clothing production), repellents (finishing), and other chemical applications are under-regulated, degrading human health and the environment [22]. Water, a key natural resource, is required to produce clothing and is highly polluted at all stages of manufacturing. The number of goods produced per year (approximately 100 billion pieces) and manufacturing speeds (an estimated 50 cycles each year) are inherently unsustainable (Schumacher & Forster). Sixty-five million people and the planet support the fashion supply chain; and far too often, the product of their work is discarded within a year after it hits the market (Schumacher & Forster).

The variety and mass of environmental, social, economic, and cultural issues resulting from the production of textile and clothing goods are met with ample and differing sustainable design methodologies. Practitioners have sought to combat problems along the supply chain (i.e., water pollution or sweatshop labor) and in the product lifecycle stages (i.e., laundering or recycling) [6]. This chapter aims to illustrate one approach to fostering a socially and culturally informed sustainable fashion design system, as modeled by the residents of San Pedro La Laguna (San Pedro), Guatemala.

In San Pedro, the approach used to design and construct a local garment, referred to as the blusa Pedrana or the blouse of San Pedro, is referential to many sustainable design frameworks. The process exemplifies aspects of co-design (collaboration between designer and wearer), emotionally durable design (design focused on product relationships), slow fashion (design and consumption of quality goods), participatory design (knowledge and involvement in production), customization (personalized design), and localization (locally connected production) [6, 10, 11]. Pedranas' (the women of San Pedro) engagement with the fashion system is not intended to be sustainable, as prescribed by these other methods. Instead, it is their design framework, rooted in an Indigenous Knowledge System[1], that yields a product great social significance [5].

Observations on design practices discussed in this chapter arose from a qualitative study conducted with the residents of San Pedro La Laguna in the summer of 2020. The primary data source is 21 interviews with women ages 19–46. Participants' accounts regarding the production and use of the blusa Pedrana are the foundation of the communal design approach.

This chapter continues with a discussion on the role of culture in practicing sustainable design. Next, the chapter introduces San Pedro, its Tz'utujil Maya residents, and municipal dress. Then, the authors use the production and acquisition of the blusa Pedrana to explain the role of culture and community in the local fashion system. Following, the cultural significance of the garment is discussed in terms of sustainability. The chapter concludes with a summary of communal design and argues for the application of culture in the design processes.

---

[1] Indigenous Knowledge Systems are empirical, normative ways of knowing and ways of being in the world that guide everyday relational life between living and nonliving things and protect cultural continuance in many Indigenous communities [20, p. 80].

2 Culture and Sustainability

Achieving sustainability, defined as "meeting the needs of the present without compromising the ability of future generations to meet their own needs" [3, p. 16], is dependent on cultural practices. Culture is a shared "...set of beliefs, sentiments, technology, language, social organization, and worldview created, learned, and transmitted by a group" [19, p. 74]. Worldviews shape the relationships people have with things—people, groups of people, physical objects, norms, and social structures [2, 15].

The design and the significance of objects, like clothing, change as society changes [7]. They are a product of people, influenced by culture. As society progresses, cultural beliefs are reinforced, subsequently meriting the reproduction of things—garments, aspects of garments, and manufacturing processes—that best reflect the community [1, 13]. Chandler [5] suggests that the maintenance of Indigenous societies is dependent upon these processes, described as ways of knowing and being, rather than the material outcomes.

Western fashion design practices, when informed by sustainability, are often an action influenced by cultural patterns and obligations. It is a response to the fact that the current fashion industry is negatively impacting the planet. Sustainable design methods yield outcomes created with positive intentions for the future. However, they do not always reflect the needs of the present. When scaled, sustainably manufactured garments follow the dangerous path of the current system, yielding mass quantities designed for the masses—again, negating the social and cultural practices of the present population. The disconnect between clothing and culture can throttle sustainable progress.

3 San Pedro La Laguna

Over 5000 years, the Maya people have developed their culture [19]. The practices discussed in this chapter is one of many reproductions. Once the largest civilization in the Western hemisphere, the Maya have continuously adapted to environmental degradation, regional conflict, colonialization, religious conversion, and genocide [8, 23]. Dress and dress practices, a dominant visual signifier of Maya heritage, also have changed regarding these macro- and micro-level forces. Nevertheless, human agency and creativity have prevailed, leading to today's contemporary attire [26].

Maya culture cannot be defined in a manner that accurately represents the over 18 million people living in Guatemala [25]. Nor, can the meaning and purpose of Maya dress practices be generalized to encompass the 340 municipalities in the country [18]. Within each municipality's borders are complex cultural practices unique to the residents [24]. The collective actions of the municipalities' residents determine the appearance and popularity of locally manufactured styles [17, 23].

This chapter attends to the dress practices of women living in the municipality of San Pedro La Laguna (San Pedro). This town, of approximately 12,000–18,000 residents is in the department of Solalá, on the southern shore of Lake Atitlán, in the southwestern part of Guatemala. It is one of five Tz'utujil municipalities in the Guatemalan highlands. Tz'utujil is a Maya ethnolinguistic group. The term ethnolinguistic brings attention to the importance of language and ethnicity as the basis for a group's unity [12]. There are 22 recognized Maya ethnolinguistic communities in Guatemala [18]. The 2018 census indicated 72,000 people spoke Tz'utujil in Guatemala, with 106,012 identifying with the ethnolinguistic group [21]. In 2020, individuals with Maya heritage comprised 41.7% of the country's population. Residents of San Pedro or Tzunun Ya' in Tz'utujil, are most likely to identify with their municipality and refer to themselves as Pedrano/as, or person of San Pedro [23, 24].

There is a significant international influence in San Pedro. Foreign residents constitute an estimated 10% of the population; and their presence is apparent in the dress, cuisine, and entertainment (the culture) of the town. Unlike the neighboring Tz'utjuil municipalities of Santiago Atitlán and San Juan La Laguna, San Pedro's economy is not based in textile production. Rather, cultural and ecological tourism are the primary attractions for temporary visitors who stay in the many low-cost housing accommodations and participate in the town's active nightlife. San Pedro's reputation as a Spanish language learning hub is also a central part of the community [21]. Many San Pedro residents are employed as teachers in one of the thirteen Spanish language schools, some of which also teach Tz'utujil and English.

4 Municipal Dress

Two categories of dress used in San Pedro denote Maya heritage.[2] These are traje and típica. Traje is the term used to describe a Maya dress ensemble worn by men and women associated with a specific municipality [13, 16, 27]. It is a set of garments designed to represent the community's heritage and geographic space. Traje, translated as "suit", is more indicative of heritage when worn as a unit (see Fig. 1). Design elements, precisely the garment's colors, are explained using culturally-based justifications. Rosa, a participant in the study, recalled themes repeated by nearly all participants when asked about traje:

> The traje of San Pedro and when we use the traje of San Pedro is because *we want, we have, we are* very proud to be from here in San Pedro La Laguna. The colors, they are specified for this suit. For example, green means the mountains, the volcano of San Pedro, the trees of San Pedro. White is purity, purity, and red it means blood. Because of the culture before, the ancestors shed blood for us, and I believe that every color has something special (Rosa, Tz'utujil – Maya, lives in San Pedro La Laguna, personal communication, July 2020).

[2] Most women younger than 40 years old will wear Western dress in public. Women between the ages of 40 and 60 years commonly wear Western dress in private.

Fig. 1 María Juana Chavajay in the traje of San Pedro La Laguna. Woman pictured wearing the güipile (blouse), corte (skirt), faja (belt), and tzute (shawl) of San Pedro La Laguna. Photograph by Abner Mariano González Tay, 2020

Design elements and their justifications are influenced by recent historical memory and periodically shift.

Típica (typical clothing) is a phrase popular in San Pedro that indicates that the Maya style garments adorning the body have different geographic origins, sourced from multiple communities or factories. These ensembles denote Maya heritage but are not necessarily affiliated with a specific municipality [9]. For example, the weave of a garment may originate in one town and be worn in the style of another (Regina, personal communication, September 2020). Likewise, factory-made garments can represent a broad, pan-Maya identity [8]. These garments mimic the traje's form but have weaving patterns and colors that follow the latest trends.

Traje and típica are made in San Pedro.[3] Textiles for güipiles (a garment fashioned from two rectangles of cotton that extends from the shoulder to waist), fajas (a closureless wrapped belt), and the tzute (a rectangle of fabric suspended from the shoulders) are hand-woven in the community. Weaving is a skill practiced and stewarded by the elders [4, 14]. Residents of all ages and genders understand the value of the güipile. The garment, worn daily by the oldest women in the community, has the greatest historical and cultural value of all garments worn today in the municipality. The güipile signifies respect for the elders and the ancestors' traditional dress practices.

Blusas, or blouses, are also made in San Pedro. The term "blusa" is common throughout Central America and indicates a generic-style torso covering without direct municipal affiliation [14]. The garment can be categorized as típica and is not considered traditional. The blusas made and worn in San Pedro are representative of the municipality, to the extent that they are referred to as the "blusa Pedrana" or blouse of San Pedro. Unlike the güipile, it is constructed with factory-woven textiles. Therefore, the garment is disconnected from the weaving process and the community's elders. Although its form has a close likeness to the güipile, this disconnect means that the garment is not subject to the same historic or geographically based signifiers (see Figs. 2 and 3).

5 Change and Choice

The work of Rosado-May et al. [20] highlights the positioning of innovation in Indigenous practices. They recognize that, "Individuals' momentary decisions to follow customary approaches are themselves part of what maintains a community's cultural practices…" [20, p. 82]. However, they also recognize that individual decisions, multiplied by the collective, change the process. The knowledge and outcomes of that knowledge—such as dress—transform [20]. The dress of San Pedro, including discussion about dress, production, use, and styling, is continuously transforming.

Residents assume that others (outsiders and insiders) are interested in the most traditional methods and garments. In preparation to discuss their daily dress, women spoke with their mothers and grandmothers about weaving and the history of each

[3] Traje and típica are also purchased in regional marketplaces.

Fig. 2 Petrona Rodríguez Chavajay wearing a modern güipile. Photograph by Abner Mariano González Tay (2020)

garment. Many anticipated that they would be questioned about symbolism, ceremony, or nature. Some residents had difficulty understanding why their dress could be significant; despite it adorning a Tz'utujil woman, it was not Mayan enough. These events demonstrate the lingering nature of ethnocentric essentialist perspectives, which reduce Maya people and culture to a static interpretation of tradition [9].

Residents favored the most traditional garments, but they were not the everyday choice. Instead, the garment centered in discussions was the blusa Pedrana. Pilar compared the meaning of the popular contemporary garment to the dress practices of her elders.

> I say now that it's modern, it's more about taste…not because it represents such and such God or such and such meaning or because it is Sunday. It is no longer for that, but for taste, for fashion. Now it is more of fashion - more fashion. For them [ancestors], they were going

Fig. 3 María Teresa González Rodríguez wearing a blusa Pedrana. Photograph by Abner Mariano González Tay (2020)

> that way for the connection to Gods of the Earth, of the Sun. Because it had meaning for their ceremonies, but now this corte or a white güipile of San Juan is normal. It doesn't mean that I'm married or that I'm going to get married... it's whatever you like.

Her explanation or the use of the garment should not be considered an act of defiance directed toward Pedrana culture [2]. Instead, using contemporary típica is a choice that allows women to engage with their municipality through fashion.

Fig. 4 Multiple generations of Pedranas wearing traje and típica. From left to right: María Juana Chavajay (traditional güipile), Petrona Rodríguez Chavajay (modern güipile), María Teresa González Rodríguez (very complex blusa), Mirna Josefina González Rodríguez (complex blusa), Catarina Rodríguez Chavajay (simple blusa). Photograph by Abner Mariano González Tay (2020)

6 Communal Design Practices

The design of the blusa is rooted in community knowledge and social expectations. In San Pedro, it is fashionable and expected for women to match colors within a Maya dress ensemble. The colors incorporated into a blouse (güipiles and blusas) are selected to match the wearer's corte (a wrapped garment covering a women's body from the natural waist to ankle) (see Fig. 4). Cortes are also the first garment selected when getting dressed and are often chosen each week. A corte will be used consecutively for 5–7 days and paired with different coordinating blouses.[4] The corte's appearance, therefore, is a key factor in the design process. Cultural pressure to wear a corte, the weekly dress patterns for the garment, and local ideals about appearance determine to the design of the blusa. It is common to bring the garment to stores, markets, and makers in the pursuit of matching accompaniments.

[4] A Maya dress ensemble includes the use of a corte. The garment symbolizes gender and represents ethnolinguistic affiliations and beauty ideals for Maya women. Cortes are factory-made. Therefore, women do not have input on the design. Women can select different colors from marketplaces in and outside of the community. Multi-colored cortes, capable of matching many blouses, are more valuable because of their utility within a wardrobe.

Designers and patrons are also keenly aware of trends, which spread quickly in the small town. For example, in March 2020, shop windows, fabric counters, and women in the street displayed blondas (large band of machine-made lace extending from the neckline opening of blusas) with a sunflower motif. Matching accessories, such as sunflower earrings and bracelets, could also be found in local shops. Five months later, the focus shifted to the neckline finish. Crochet replaced the overcast embroidery stitches that secure the blonda to the body of the garment. This trend, due to the pandemic, was less present on the street. Instead women advertised the favored technique on the town's buy-sell-trade Facebook group and on personal Instagram pages.

Independent designers[5] in San Pedro are interested in observing trends and creatively applying their findings.[6] Carolina, who aspires to own her own boutique, explains her design process as:

> It's like invented, we invented them ourselves because not all people use blusas – just that we made it up…I see other designs and either try to change a little bit of what is there, or I can make it similar. And sometimes I try to change it, an ornament, put a different one on it - so it doesn't look the same. But I try to imagine them and make them.

In this statement, Carolina credits the blusa to the community ("We invented them…") and the design development to herself ("I try to imagine…"). Her description highlights the positon of the blusa as a product of San Pedro and the work of the designer.

Early involvement in design and making, observation of techniques and others, and experimentation with the process are characteristics of an Indigenous Knowledge Systems [20, p. 80]. These ways of learning, compounded by the extensive history of garment design and production in Maya culture, shape relationships with apparel in the municipality. As a result, women seeking customized garments, regardless of practical making skills, are knowledgeable about design and manufacturing. María, who does not sew or embellish garments, describes the process of working with her mother and sister-in-law. Stating,

> I need to tell the woman who makes the blouse how I want my blouse. I give my idea, my colors and usually we give a measurement. For example, I have this one – when I want another one, I tell my mother-in-law, "Ok. This is my measurement. I want another one just like it, with different colors". When she makes another one, she also gives me ideas and she also has a connection with me. When we talk about the new design of a garment, for example, we both give new ideas and whether it is right or not? Then, we have a conversation, a good connection.

Her description illustrates her ability to assert her creative choices and take on the role of co-designer. María explained that she was disinterested in the production of her other clothes—traje from outside municipalities and second-hand Western dress—because she could not "give her ideas" on the design.

---

[5] A participant who self-identified as a fashion designer and earned income for their work.

[6] It is estimated that there are approximately three trends cycles in the blusa's design each year in the municipality.

7 Collaborative Construction

In San Pedro, blusas are often constructed through a small network of makers, connected through social and kin-based relations.[7] The process involves a sewing machine operator, who assembles the body of the garment and one or more women capable of hand-finishing the neckline opening.[8] The more complicated the design (additional embroidery, beading, and adornments), the more likely additional makers will be needed in the supply chain. Personal making skills, social connections, and the patron's available time influence the production process.

There are multiple approaches to making the blusa Pedrana. One method of production begins with the client, who brings the corte or pictures as a design reference to a sewer. The sewer may complete the garment herself, but it is more likely that she will outsource parts of the production to others. Each woman in the process acquires the supplies (fabric, thread) they need to complete their task from local fabric stores.[9] The final cost of the garment is divided among the group. Another, more cost-effective but lengthier strategy, requires the end-user to purchase all the needed materials, then move the garment through the supply chain herself. Under this approach, the final user brings the fabric and notions to a sewer, collects the stitched item, then transports it to a finisher, who attaches the blonda. The wearer then takes the garment to another woman who completes the remaining embroidery or other embellishments.[10] A final woman may be involved in the fitting of the garment. If the wearer is also a maker, the production process reflects their skills as well. A woman may outsource the cutting and sewing of the garment and finish the neckline herself or, conversely, stitch the garment and seek out a finisher or assistance with fitting. The more involvement the wearer has in the process, the lower the cost.

This style of manufacturing is dependent on the end-user knowledge of the community. Knowing the residents' skills, as well as the time and costs to complete each task, is essential. The garment can be made in a single day, but may be drawn out over a week. Therefore, the wearer must also plan ahead. Blusas that can be acquired readily from stores can be three-four times the cost. Rosa explains how to approach the task of acquiring a blusa:

> For example, you come to San Pedro, you want a blouse, and you tell someone I need a blusa. Ah, look at such a woman, such a woman in this house. In this shop, they sell it, or they make it. Everyone knows who works on it, and it's easier for you. And as it is a very small town here, we all know each other, and we all know who, who embroiders, who sews. Going like this is not that complicated.

---

[7] Blusas are available in local stores and sold by independent designers at a higher price.

[8] Women learn to machine sew through social connections. Although machine sewing has existed in the community for 70 years, it is not a traditional skill. Therefore, it is not passed from one generation to the next.

[9] The exact source of the factory-made materials is unknown. Signage in stores indicates the last known location of the product. Fabrics are labeled with the names of Guatemalan cities, like Xela. These materials may have been manufactured in one of Guatemala's cities or imported into the city from abroad, then transported to San Pedro.

[10] Local stores follow the same process, but stock the finished item at a higher price.

Women prefer to work with and purchase from known members of the community. Therefore, navigating the making of the blusa is not overtly clear to outsiders nor easy for those within the municipality who are disliked. Social connections, in addition to skill, are pivotal to the success of personal businesses and good design.

On the streets of San Pedro, clothing is a common point of discussion. "Hello, how are you? I like your blouse, who made it?" is the standard greeting between women. As a result, good design regularly receives praise in the community. The expectation of commentary (positive and negative), in part, drives women's attentiveness to their appearance. Designers and their involved clients are aware of this exchange and use the interaction to gain notoriety for their work. Women feel a sense of pride when they can take credit for the design of their blouse or when they can direct the onlooker to their family or friends, who can potentially craft them a garment. However, the ample number of designers means that a favorable garment is just as likely to be copied as it is purchased.

The limited interaction on the streets of San Pedro during the pandemic contributed to the popularity of online networks for retailing. Curfews led to an increase in making within the home, and closures of public spaces pushed vending online. More women began selling finished garments on social networking platforms like Facebook and Instagram to people in San Pedro and the neighboring Tz'utujil municipalities. Some women, or groups of women, created anonymous profiles to circumvent the tendency for the residents to purchase from known or liked people. Goods and currency are exchanged in person. Therefore, in due time, vendors lose their anonymity—restoring the importance of social connections and quality garments. It is too soon to determine how the ongoing pandemic will alter the production and acquisition of the blusa Pedrana.

8 Social Use and Cultural Significance of the Blusa

Pride in San Pedro is communicated with municipal-affiliated garments. There is an expectation that women wear traje and típica representative of the town during community events and in community spaces. Although uniformity among groups of women, for example, singing in a church choir, is socially expected, using the traje of San Pedro is not a requirement. It is inappropriate to suggest someone wear the attire because of cost.[11] Therefore, the blusa is an economically favorable option. Esperanza described this phenomenon:

> I feel that in San Pedro, the majority of women like to dress in uniform. If they wear the same color, the same color. If it's a party, the same güipile, the same güipile, the same color blouse or the same color blouse, that's what I think is the custom of the Pedranas…An example—in churches when there is an anniversary party or a wedding or a birthday, the group that is leading the activity…makes uniforms. All the women agree and say "we don't want to wear

[11] The güipile of San Pedro cost between 40 and 80 USD. Güipiles from outside of San Pedro are an upwards of 150 USD. A blusa costs between 8 and 22 USD.

> green or red"... And that happens in every church and in every school and in every social group in San Pedro...I don't know if you noticed it at Ban Rural, at the Bank of San Pedro, the women bank workers always have the same uniform—blusa. And that does not happen in other municipalities. They wear a uniform of the bank, but, in San Pedro—the blusa, but the same color.

The desire for uniformity extends beyond structured social groups. Women also coordinate their appearance with friends and family before entering unstructured settings, such as dining in a restaurant or going to the beach.

Uniformity is not absolute conformity but rather shared identification. Wearing the blusa Pedrana is a visual signifier and outward support of Maya culture. It is also a symbol of economic sustainability. To wear the garment means that you have participated in the production and acquisition process, actions that support the financial livelihood of women in San Pedro. Adriana, who predominately wears Western dress, explained her decision to wear a highly decorative blusa to an event[12]:

I was invited to a birthday party and my friend is from San Pedro and I know that traditional clothes are important for them. So, in order to fit in with all the local people, I decided to wear the traditional clothes and I like to wear them too. I'm just not used to using it every day...I like to live with other people and I realize that when I wear the típica clothes, they feel better talking to me with my típica clothes than with my normal clothes... it's more like a sign of acceptance toward the culture of the place, and more than anything, it's out of respect for the people.

The act of dressing in a blusa can lead to greater social and cultural accessibility for those unaccustomed to traje or financially unable to purchase the garments. Adriana's account also alludes to the potential social sanctions for not participating in local dress practices. Admiration for the work of the community, and the expectation of using locally made garments, perpetuates engagement with a local system—void of excess transportation, energy use, and chemical applications pervasive throughout the decentralized fashion supply chain.

The blusa Pedrana, and its many variations, exist within the boundaries of cultural expectations. It is representative of Maya culture and the municipality but is not restrained by traditional practices. Women felt that it is a clothing option that signifies their respect for the community and showcases their personality. Ester explains that ample design options mean that there is no need for Pedranas to dress alike. "People here—one invents styles of blouse and have different options to dress. Here, we should not all dress the same", she said. Others echoed her viewpoint, adding that Pedranas were not less Maya for exercising their creativity through their appearance. Valentina asserted, "I am Mayan because I am, because I was born Mayan, and not because of clothes. I'm going to change [my clothes]—Of course". The garment assists with the navigation of personal and ethnic identity in the municipality, subsequently sustaining cultural practices in a manner that meets the needs of many Pedranas.

Creative exploration in manufacturing and design is upheld by local knowledge of production, a cultural characteristic rooted in the history of the Maya and maintained

[12] Adriana identifies as Tz'utujil but was not raised in a Tz'utujil municipality.

by Indigenous Knowledge Systems. Although many women in the community are not strictly involved in the production of clothing, it is a familiar subject. Residents have observed and engaged with the making of clothing since childhood [20]. Processes related to weaving, dyeing, cutting, sewing, and embellishment are apparent in San Pedro and the neighboring municipalities. The transparency of the local fashion system is evident in everyday discussion on dress. Conversations reflect an advanced understanding of the topic and include specifics on textile quality (weave structure, fiber properties), remarks on finishes (mercerization), and colorfastness. Informed consumers, aware of the time and cost for each task along the supply chain, value the work of their collaborators. As a result, garments are cared for and not readily discarded.

The social nature of design and manufacturing in San Pedro—based on the collective knowledge of the community—has led to a high regard for locally made clothing.[13] The blusa Pedrana, and other municipal garments, are not only products of the culture. They are intrinsically connected to another person, someone who is also a mother, daughter, sister, aunt, niece, or friend. The reverence held for the garment is strengthened by these relationships. Consequently, it is the production of garments throughout the municipality that sustains social systems among the residents. Dissimilar to the plight of mass manufactured goods, culture has not been erased from the process or the products of Pedranas. As such, all iterations of the blusa Pedrana communicate the ethnicity, residency, uniqueness, skill, and the social prowess of its wearer.

9 Communal Design

The production and acquisition of the blusa Pedrana is a model for sustainable fashion. The garment can be considered the result of a communal design process. Communal design is a collective approach to garment production where community knowledge and collaborative work jointly influences the product. "Ingenuity and innovation feed new knowledge in Indigenous communities" [20, p. 82]. In this instance, community knowledge includes the history and presence of clothing production, as well as the continuous maintenance and construction of knowledge employed in the design and manufacturing processes [5]. Illustrated by its supply chain, the blusa is the outcome of collaborative work among a network of designers, makers, and users—and the collective expectations for appearance in the municipality. This work, reflecting the municipality's unique and current cultural practices, is intended for local consumption. The crux of communal design is its dependency on the community. From ideation to use, the social and cultural customs of the municipality inform the design process, subsequently fostering sustainable practices across the local fashion system.

[13] Residents hold the greatest reverence for traje, then típica. Women wear Western dress but do regard it as highly as locally made garments.

10 Culturally Informed Sustainability

The argument for the inclusion of culture within sustainable fashion methodologies is in opposition to the global yet homogenous, mass approach to garment design and manufacturing. The prevailing fashion system is void of cultural nuances, and therefore it does not attend to the environmental, social, or economic practices that benefit a place or its people. The negative impacts of consumerism, a culture itself, cannot be mitigated with a mass solution. Cultural issues must be met with culturally informed solutions. However, localization—most often discussed in terms of regional manufacturing—is only one part of the strategy. Sustainable methods must seek to understand the social patterns, behaviors, and worldviews that influence dress practices. In addition, practitioners must recognize that solutions based on culture are not generalizable. The work of Pedranas is sustainable because it is informed by history, culture, and society. Their worldview cannot be replicated nor exploited in the name of sustainable progress. However, it is a lesson that demonstrates the role of apparel knowledge in community connection. The communal design process in San Pedro is not a perfect model of sustainability—there are gaps in the supply chain—however, it is an informed approach that meets the needs of the current population and is mindful of the municipality's future.

References

1. Baudrillard J (1993) Symbolic exchange and death. Sage, Thousand Oaks
2. Blumer H (1969) Symbolic interactionism: perspective and method. University of California Press, Berkeley
3. Brundtland G (1987) Report of the world commission on environment and development: our common future. United Nations General Assembly document A/42/427. Oxford University Press, Oxford
4. Carey D (2001) Our elders teach us. Maya—Kaqchikel historical perspectives. University of Texas Press, Austin
5. Chandler MJ (2013) On being Indigenous: an essay on the hermeneutics of 'cultural identity.' Hum Dev 56:83–87
6. Chapman J (2015) Emotionally durable design: objects, experiences and empathy. Routledge, London
7. Dant T (1999) Material culture in the social world. Open University Press, Philadelphia
8. Fischer EF (2001) Cultural logics and global economies: Maya identity in thought and practice. University of Texas Press, Austin
9. Fischer EF, Brown RM (1996) Maya cultural activism in Guatemala. University of Texas Press, Austin
10. Fletcher K (2008) Sustainable fashion and textiles: design journeys, 3rd edn. Routledge, New York
11. Fletcher K (2012) Fashion and sustainability. Design for change. Routledge, New York
12. French BM (2010) Maya ethnolinguistic identity: violence, cultural rights, and modernity in highland Guatemala. University of Arizona Press, Tucson
13. Hendrickson C (1995) Weaving identities. Construction of dress and self in a highland Guatemala town. University of Texas Press, Austin

14. Hendrickson C (1996) Women, weaving, and education in Maya revitalization. In: Fischer EF, Brown RM (eds) Maya cultural activism in Guatemala. University of Texas Press, Austin, pp 156–164
15. Johnson K, Lennon SJ, Rudd N (2014) Dress, body and self: research in the social psychology of dress. Fashion Text 1:1–24
16. Little WE (2004) Mayas in the marketplace. Tourism, globalization, and cultural identity. University of Texas Press, Austin
17. Otzoy I (1996) Maya clothing and identity. In: Fischer EF, Brown RM (eds) Maya cultural activism in Guatemala. University of Texas Press, Austin, pp 141–155
18. Pan American Health Organization (2021) Guatemala. https://www.paho.org/salud-en-las-americas-2017/?p=3338#:~:text=The%20Republic%20of%20Guatemala%20extends,22%20departments%20and%20340%20municipalities. Accessed 18 Jul 2022
19. Rodríquz Guaján D (1996) Maya culture and the politics of development. In: Fischer EF, Brown RM (eds) Maya cultural activism in Guatemala. University of Texas Press, Austin, pp 74–88
20. Rosado-May F, Urrieta L, Dayton A, Rogoff B (2020) In: Suad Nasir N, Lee CD, Pea R, McKinney de Royston M (eds) Handbook of the cultural foundations of learning. Routledge, New York, pp 79–96
21. San Pedro La Laguna (2019) https://www.frommers.com/destinations/san-pedro-la-laguna
22. Schumacher K, Forster AL (2022) Facilitating a circular economy for textiles workshop report. NIST Special Publications 1500-207. https://nvlpubs.nist.gov/nistpubs/SpecialPublications/NIST.SP.1500-207.pdf. Accessed 18 Jul 2022
23. Smith C (1990) Race-class-gender ideology in Guatemala: modern and anti-modern forms. Comp Stud Soc Hist 37:723–749
24. Tax S (1937) The municipios of the Midwestern highlands of Guatemala. Am Anthropol 39:423–444
25. United Nations Population Fund (2022) World population dashboard: Guatemala. https://www.unfpa.org/data/world-population/GT. Accessed 21 Jul 2022
26. Virtual Tour (2020) Museo Ixchel del traje indígena. https://museoixchel.com/virtual-tour. Accessed 18 Jul 2022
27. Watanabe JM (1992) Maya saints and souls in a changing world. University of Texas Press, Austin

Towards an Inclusive Fashion System

Jose Teunissen

Abstract More and more non-Western brands and designers entering the fashion world introducing new values, narratives and aesthetics into a dominant Western Fashion discourse deeply intertwined with notions of conceptualism, modernism and postmodernism. In a moment of history where we are facing a failing fashion system, these new voices are introducing new values and new imaginations that try to transform the system into a more inclusive, fair and decentred discipline. Also, on an academic level transnational and decolonial theories are proposing new and fresh lenses on how to critically re-think the Eurocentric Fashion system using new terminologies as de-linking, precedence and aesthesis and the pluri-versal. Do these new perspectives give us the tools to redefine and modify the Western definitions of fashion into a more inclusive and alternative system?

Keywords De-Westernisation · Transnational · Decolonialism · Altermodernism · Fashion vs dress

1 Introduction

With the increasing globalisation and the raise of non-Western brands and designers entering the market, the conventional Eurocentric fashion discourse has started to include new voices with different traditions and more diverse cultural backgrounds. Therefore, new values, narratives and aesthetics are entering the fashion discourse, which are not originated from the conventional values and notions of the dominant Western Fashion system deeply intertwined with notions of conceptualism, modernism and post-modernism.

In a moment of history where we are facing a failing fashion system in all areas of production, environmental, overproduction and social justice, these new (designer) voices are increasingly important in re-thinking and re-defining the fashion system. These fresh voices from over the whole world are widening the scope by introducing

J. Teunissen (✉)
Johannes van der Waalsstraat 95-1, 1098 PH Amsterdam, The Netherlands
e-mail: j.g.l.teunissen@hva.nl

M. Á. Gardetti and R. P. Larios-Francia (eds.), *Sustainability Challenges in the Fashion Industry*, Sustainable Textiles: Production, Processing, Manufacturing & Chemistry,
https://doi.org/10.1007/978-981-99-0349-8_4

new values and new imaginations that might be able to transform the system into a more inclusive and more informed and more decentred discipline.

At the same time, transnational and decolonial theories are proposing new and fresh lenses on how to critically re-think the Eurocentric Fashion system using *de-linking, precedence and aesthesis, pluri-versal* as a terminology to replace the modernist terminology as *the contemporary, the radical new* and *the universal.* The question I would like to explore is: are we able to re-define and modify the Western definitions of a system, which is so strongly rooted in Western enlightenment, industrialisation, capitalism and modernism into a more inclusive and alternative system? I am taking the last twenty years of my work as curator and scholar as a journey to de-ravel the developments in thinking on globalisation, de-Westernisation, identity and inclusivity in fashion.

2 Fashion, Modernity and Focus on the New

Fashion is one of the faces of Western modern life. 'Fashion was the first manifestation of a passion characteristic of the West, the passion for what is "modern," ' describes Gilles Lipovetsky in his renowned book *The Empire of Fashion: Dressing Modern Democracy* [8]. 'Novel became a source of worldly value, a mark of social excellence. People had to keep up with whatever was new; they had to adopt the latest changes' ([8]: 24). This permanent drive for 'change' and for the 'new' is connected to modernity's permanent need to maintain the semblance of *economic* and *narratological* change, always both subverting and improving upon history ([5]: 231). Fashion as expression of modernity has often be perceived as an improvement of society. It is defined as a cultural expression of the effort of human beings to make themselves masters of the conditions of their own existence, a mark of being able to improve and modernise the world. A sign of being able to step out of their birth right and origins and being able to climb the social ladder. As a result, fashion as expression of modernity is rooted in a strong belief that it has made life better, smarter and accessible for everyone.

Exactly, these assumptions are questioned and under review now by a new global generation of designers who are facing and addressing societal changes such as sustainability, social justice and inclusivity. They are questioning and tempting the boundaries of the system, replacing them with new values, frameworks and terminology.

3 The Origin of Modern Fashion

The origin of modern fashion is often situated in the fourteenth century with rise of city states Burgundy, Florence, Venice and the rise of Merchant capitalism ([8]: 21–26) ([17]: 17–18). Suddenly change was no longer an accidental, rare, fortuitous phenomenon, it became a fixed law of the pleasure of only high society. Early nineteenth century modern fashion transformed in a more 'democratic' system supporting the needs of the modern masses of bourgeois citizens that arose as a result of the Enlightenment, urbanisation and modernisation in the nineteenth century. People acquired the freedom to indicate to which group they wished to belong by means of dress, whilst they retained the freedom to position themselves as individuals within that group ([8]: 82–83) [15].

What happened to the system and meaning of clothing when modern fashion came into life? In the era of the pre-modern, clothing was based on fixed codes, traditions and rules imposed by sumptuous laws where colours, ribbons, fabrics or buttons had fixed meanings communicating the social status of people, the ritual, what religion they embraced, or their civil status [8]. This pre-modern Western dress system offered society various guidelines, certainties and a sense of belonging and was primarily referring *to place.*

In contrast, modern fashion introduced an accessible, open system using a *self-reflective* system which is constantly changing in form and meaning, primarily referring to the (fashion) system itself and referring to *time* (skipping every reference to heritage, local culture and *place*). The consequence of this fundamental change is, that to become familiar with these ever-changing fashion codes, one must move in the right circles and be in-the-know. It is one of the reasons why (fashion) communication became essential. Initially information about new trends were distributed via fashion dolls and little manuals in the eighteenth century, followed by the newly invented (fashion) magazines during the nineteenth century [18]. In the same period a new ephemeral city life with department stores and public venues started for people to stroll and to show off. What defined modern fashion here is that it broke with conventional social roles of custom and tradition and automatically became a *signifier of the new.* Since the system of fashion (pretends) allowing everyone to access the system (provided one has enough money to enter the system) it was perceived as democratic, whereas pre-modern dress used to be a matter a of birth right ([22]: 44).

Lipovetsky argues that the modern fashion system over time has become even more inclusive and democratic foregrounding the 1960s as an example of a second democratisation wave where the youth culture overruled the hegemony of the French pre-scribed fashion trends, allowing different social groups to choose their own clothing style expressing their political ideas, their identity and sense of belonging. Since, the modern fashion system is accommodating different styles at the same time for different social groups and therefore the system (seems) more inclusive. In *Le Sacre de L'Authenticité* [9], Lipovetsky explores a third era of democratisation, defined as 'egomania,' that took place early twenty-first century with the rise of Internet, resulting in a global society obsessed with social media where individuality and self-realisation seem to be completely democratised and commercialised. In this new era, people are able to fully develop themselves without any social restrictions on background. In their search to be authentic, they only are responsible for themselves and the primacy of ego rules. Individuality and authenticity have become part of the consumer culture consisting of clicks and likes, being out there in social media and the urge to make a brand of yourself.

4 Modern Fashion: A System of Exclusion

Although Lipovetsky makes an argument that the modern fashion has become more global, individualised, affordable and accessible for a wider audience over time, the current modern fashion system is still *self-reflective* focusing on *change* and the *new,* and therefore, a system where people have to be in the known of the fashion rules and as a consequence can be classified either 'in' or 'out' of fashion which is fundamentally opposite to inclusivity. How modern fashion as a system of exclusion operates is convincingly explained by Otto von Bush in *The Pschyopolitics of Fashion* [3] where he uses the metaphor of a nation to describe the modern fashion system: a regime obsessed by it borders and who is 'in' and who is 'out.' Where Lipovetsky defines fashion and modernity to be inclusive systems open for everyone to join because global media provides worldwide access and no longer birth right (local culture) is allowing people access to the system, Von Bush argues convincingly modern fashion to be a system of exclusion. With new terms such as *Aesthetic Domination* governed by *supremacy of style* and the concept of institutional *aesthetic apartheid,* Von Bush is unravelling how this is covered under a soft regime of 'democratic' fast fashion, 'always affirming and positive, as it continuously broadcasts the message that "everyone" has equal access to the free expressions of self on offer' ([3]: 71).

In order to develop a truly inclusive fashion system, Von Bush proposes the concept of Deep Fashion, a system that increases agency to the wearer, is pluralistic, more honest and more experimental, through putting depth in the *relations,* in shared attention and in the praxis of togetherness. In order to achieve this, designers need to focus on the *process and passions* of themselves and wearers rather than on the material and the products. Designers are encouraged to use their passion to energise the emotional qualities of life, in a much richer way than consuming ready-to wear ([3]: 138–139). Von Bush proposes to focus on *relations, processes and passion—* as principles to make the system more inclusive. How effective this new approach is to understand a new generation of designers and how much this is in line with decolonial theories I will address later in this article.

5 Global Fashion, Local Tradition: Focus on Regional Heritage, Including Traditions

The first time that I was aware that conventional definitions of modern fashion with a focus on the new (*time*) instead of heritage and tradition (*place*) were not applying to a new generation of designers was when I developed the exhibition *Global Fashion, Local Tradition* (2005) and the aligned publication [19]. With the rise of Internet early twenty-first century, the scope of fashion had widened. No longer Paris, Milan, New York and London where the leading centres where the fashion business took place and where the new trends were launched. Suddenly Fashion Weeks cropped up all over the world promoting their new local talent with the purpose to put their own country on the cultural map cherishing to a certain extent their local (craft) heritage ([19]: 9) Under these new conditions, fashion designers took on national significance engaging with their local heritage craft and dress tradition, in such a way that it attracted a global audience. Falling back on tradition, using signs of heritage and cultural regional identity were at that time novel and outstanding, something that until than used to be perceived as belonging to the world of dress and not of modern fashion, which excluded any signs of geographical and local cultural background. An interesting example at the exhibition was the Coopa Rooca collective that started in 1981 with women from the Nordestinas using their unique skills in quilting, drawstring applique, macrame and crocheting, providing these handiwork techniques to well-known Brazilian designers such as Carlos Miele and Alexander Herchovitch, who used these handicrafts as signifiers for their Brazilian identity ([19]: 78).

The work from Coopa Rooca used by Carlos Miele and Alexander Herchovitch could no longer be framed and perceived as exotism ([14]: 3–4) or cultural appropriation into the domain of fashion that only deals with temporality, but it was a serious attempt of non-Western originated designers to bring non-Western cultures into the realm of the global fashion system using their cultural heritage and sense of belonging as a distinctive feature.

6 A Divers Eurocentric Fashion System

Another insight that made me question the adequacy of the modern fashion system was the publication of a dedicated Fashion Theory issue (2011, Vol 15, Issue 2) on identity and heritage of European Fashion called: ‘New European Fashion Centres: Dreams of small nations in a Polycentric Fashion World’ (ed: [16]) which focused on Fashion in small European countries and highlighted their diversity. How did they relate to domination of the modern fashion system, primarily dominated and invented in Paris? How were these countries included or excluded in the system? The contributions in the journal surfaced how Eurocentrism in Fashion needed to be understood as a more nuanced picture taking in account that Europe consisted of big power nations such as France and the UK versus small nations such as Sweden, Denmark, Ireland, Belgian and Holland who found themselves in a position that they were following fashion that was ‘dictated’ from Paris. From the nineteenth century with the rise of Haute Couture multiple voices and cultures within Europe had to adapt and modify the pre-scribed Paris fashion to a dress style that was acceptable in their own nation ([13]: 133–135) Riegel. In ‘Deconstructing Belgian and Dutch Fashion Dreams: From Global Trends to Local Crafts’ to the journal I highlighted the remarkable difference between the Dutch and Belgian attitude to modern fashion. Due to their proximity to France, the Belgian manufacturers were closely connected to the couture practice of Paris and the country, therefore Belgium easily adapted the modern fashion system ([20]: 165). The Dutch relation to fashion (or dress), however, was strongly enrooted in its egalitarian ideology, expressed in an iconic style of sobriety during the seventeenth century Dutch Golden Age, foregrounding the Calvinistic attitude. The same ideology deterred the Dutch in the nineteenth from participating in upcoming French haute couture being afraid to lose their own identity. Instead, the Dutch cultivated the local and regional dress in the nineteenth century with strict coding of belonging to escape from the influence of the frivolous French haute couture, modernisation, urbanisation and consumerism ([22]: 42, [20]: 69).

These examples illustrated that the Eurocentric Western Fashion system, normally perceived as one system, on a micro level included different cultures and approaches to the dominant 'French' Fashion system. Identical to the introduction of European fashion in many non-Western countries ([6]: 10) also in Europe the introduction of modern fashion did not threaten the continuity of local fashions, but rather increased a process of selection, appropriation, hybridisation, reinvention and redefinition. Therefore, we need to take into account that the dominant project of modernity has also suppressed diversity and narratives within Europe and therefore it is also important to include and to listen to these 'others of Europe' ([25]: 188). It made me aware that we needed to learn from the different cultures and heritages that were normally not included into the modern fashion system and history. For me, it raised a further question about a new generation of fashion design practices that took a different and more critical approach to modernity especially around *the new*, emphasising their local heritage and background.

7 The Future of Fashion is Now: Re-thinking Identity, Values and Heritage

In 2014, I developed a new exhibition and publication *The Future of Fashion is Now* (2014, 2015 Rotterdam, Shanghai, Shenzhen) addressing how sustainable thinking, digitalisation, a new approach to materials and craft had become new drivers for change in fashion. No longer the *new* (in the sense of a new silhouette or aesthetic) was a driving force, but a design for a better future reflecting new values, explored new narratives and new stories embedding more inclusive identities and political activism [21]. Again, designers from all over the world were invited to take part, and a few of them were given a commission via an open call.

Different from *the Global Fashion Local Tradition* exhibition where designers tried to become part of the modern fashion system, this generation started to re-think and reshape the classical notions of identities explored by the conventional fashion and art system addressing issues such as globalisation, de-colonisation, political, social and economic systems, whilst using traces of heritage, craft and values of their own cultural and social origins in a space where those elements come together in non-hierarchical ways. Tania Candiani's project *La Constancia Dormida* (2006) (Mexico) for example used a bankrupt textile factory for a period of thirty days where she set herself the goal to make one dress a day, where each dress functioned as a page from a diary in which anecdotes were embroidered from the time that the factory was still active, mixed with stories from visitors who dropped in to see Candiani. Thus, by means of the dresses Candiani not only made the history of this factory tangible, with its harsh working conditions, but she also exposed the nostalgia of the former workers and those who lived nearby. Her fashion designs referred to a social network on the one hand, but it also envisioned the daily practice of life in the factory ([21]: 160–161).

To understand these new approaches to fashion, I drew on Nicolas Bourriaud's notions of *altermodernism* and *radicant* (Bourriaud: 2009. *Altermodernism* is defined as a further iteration of modernity that does not concern itself with the past, origins and 'authentic' and 'national' identities, but with the future: premised on the destination rather than the source' ([2]: 40). Especially, the term '*radicant*' describing the contemporary subject caught between the need for a connection with its (cultural and local) environment and the forces of uprooting, between globalisation and singularity, between identity and opening to the other seemed an adequate term to describe these designers using their cultural background and values to weave them in the global fashion narrative; *Radicant* as an identity that was globally connected in culturally globalised world, whilst integrating local cultural backgrounds into the future using problem-solving thinking and approaches. As such, initiating a process that Jennifer Craik and Angela Jansen described as a process of selection, appropriation, hybridisation, reinvention and redefinition ([6]: 10; [2]: 22).

Whilst Bourriaud undertook a serious attempt to incorporate different cultures and backgrounds of a new generation into the *Altermodern* contemporary canon, it left one unsolved problem: within his (alter) modernistic ideology there was still a focus on 'novelty' and 'progress' (Bourriaud uses the term future) and therefore the system was not inclusive; still in- and excluding people and cultures.

8 State of Fashion: Searching for the New Luxury: Beyond Progress and Novelty

The exhibition *Searching for the New Luxury* (Arnhem, 2018), developed for State of Fashion (Arnhem Mode biennale), started as a further exploration of this disruptive transition to the fashion system. Via an open call supported by the Prince Claus Fund, we invited designers from all over the world to participate and to share ideas that offer new perspectives on this fashion theme. The selection of the panel as well as further research informed the selection of 50 contributors to the exhibition and deepened the themes.

The exhibition interrogated what constitutes fashion luxury in this era of time where we face big societal challenges and scarcity of our resources? And with what kind of luxury, we want to surround ourselves in the twenty-first century? In more theoretical terms, I explored what 'fashion luxury' meant in a context where we are re-defining the fashion discourse and fashion system taking in account globalisation, de-colonisation, de-Westernisation and the urgent need for more responsible consumption. It highlighted how the modern fashion system and role of the designer is fundamentally changing. Fashion designers are taking up different roles, not putting themselves in the centre as the creative genius, instead taking a supporting and collaborative role bringing together skills, ideas, narratives, heritage, artisans and engineers to create a sustainable, inclusive, ethical future ([24]: 15–26; [4]: 33). Living in a globally connected and culturally globalised world, the designers created new paths and practices whilst integrating their local cultural backgrounds (Fig. 1).

At the exhibition, Mia Morikawa (1983) and Shani Himanshu (1980) presented *11.11/eleven eleven* the Khadi Way project, a journey of kala cotton and khadi denim, from seed to stitch. Each product within this collection was handmade. The eleven garments had a product code that traces the human imprint on the product and helped to connect the maker and wearer. The project contributed to environmental sustainability by using organic materials and recycled waste materials. The project also contributed to social sustainability by cherishing values such as traceability, transparency and local craftsmanship. In their approach, the product becomes the materialisation of a novel relationship between consumer and the artisan, which is shifting the focus from the 'star designer' to the value of the artisanship, cultural origins and traces behind it whilst putting them in a new global context of ethical sustainability. The project surfaced how (due to the digitalisation) more information can shape new,

Fig. 1 11.11/eleven eleven

horizontal relationships between us and the producers of our garments. With a better understanding of the skills and craftsmanship that go into the creation of a garment (a knowledge that many of us have lost), the artisans involved in making the clothes gain not only recognition, but also better financial rewards.

The example showed how a new generation of designers are moving beyond the traditional concept of a successful famous star designer and profitable brand, to foreground social, inclusive and fair practices. Trying to strive to a world beyond an ephemeral and glamourous phantasy, replaced by an imagination that foregrounds social and embodied practices of fashion culture hand highlights concepts of neighbourhood, nationhood and moral economies ([1]: 7).

It is exactly here where Bourriaud's altermodern terminology no longer seemed to be adequate. Bringing in traces of culture and background values in the global fashion system that is still focusing on *novelty* and *progress* doesn't correlate with the aim of these generation of designers who are focusing on the product, the makers/artisans and the process and values behind it. Improvement for society is not necessary based on 'novelty' which is pushing consumption and technological progression, whilst wiping out cultural values other than only being a materialist consumer.

To be able to include more traditional, cultural embedded practices as sustainable and inclusive solutions there is a need to review the modern fashion system's framework fundamentally. Building on Von Bush concept of Deep Fashion ([3]: 121–137), a truly inclusive fashion system requires deeper listening to culture, heritage and background, bringing back values whilst questioning the overall modern fashion system embedded within an (alter)modernistic system. It is at this point where the decolonial lens is helping us to shape new notions for an inclusive future fashion system.

9 A Decolonial Lens on Fashion Studies

Decolonial theory originated as a critical theory addressing the logic, metaphysics, ontology and matrix of power created by the processes of colonisation and cannot be separated from a history of enslavement, discredited by the forces of modernity and capitalism. As a result, modernity and coloniality are fundamentally intertwined ([12]: 22-32). Therefore, the history of progress and civilisation cannot be separated from the history of enslavement. Decolonial theory offers a way to re-learn the knowledge, values and cultures that has been pushed aside, forgotten, buried or discredited. From a decolonial observation, modernity starts at the beginning of the colonial enterprise in 1492, where Europe as a 'unified ego' poses itself against and above the other, fundamentally excluding non-Western culture ([25]: 186). From the decolonial lens, it is therefore essential to move away from modernity, postmodernism, altermodernism and neo-liberal civilisation completely and to explore alternative way of thinking with different perspectives and other consciousnesses.

Decolonial theory is working towards a vision of human life that is not dependent upon or structured by the forced imposition of one ideal of society over another and those that differ, which is what modernity in essence does ([10]: 459). These visions of human life and values can be found in local histories that have been marginalised by the temporal and spatial colonial differences ([10]: 492). This process, coined as *de-linking,* does not only change the *content* of the discourse but also the *terminology* ([10]: 459). Crucial terms—where I would like to focus on—to endorse the decolonial thinking are *precedence, aesthesis* and *pluri-versal.*

Precedence is a useful term, because it introduces a different, more holistic concept of time referring to what precedes us, not immanent, not fully contained in the now and present but both ahead of us and before us ([11]: 193). This notion moves away from the conventional modern fashion system time concept focusing on the *contemporary,* the *new* and *the future* wiping out any history, heritage and alternative narratives. Instead, *precedence* deepens the sense and experience of temporalities by bringing already existing genealogies and paths into the to-come ([25]: 185).

Secondly, *aesthesis* as a term is helpful to overcome the restrictive modernistic term of aesthetics which frames aesthetics in a framework of the modern and seeks regulation through a canon, which means for fashion that it is focusing on a combination silhouette, colour and fabric. On the contrary, aesthesis is open and allows for the recognition of the plurality of ways to relate to the world via stories, the sensible, the tangible, the visual and other areas that have been silenced ([11]: pp. 9, 10). Where aesthesis is open for alternatives, aesthetics is restricted by a framework, a canon and norms.

The third and very valuable and useful term for a new inclusive fashion system is *pluri-versal,* which refers to an approach where each knot on the web of the genealogy is point of d*e-linking:* this is about re-introducing multiple voices, languages, memories, different economies, social organisations and double subjectivities and incorporates all viewpoints without hierarchy and without excluding anyone [10].

These new terms enabled me to define and capture the meaning and impact of the latest generations fashion designers that were presented at the latest State of Fashion Biennale: Ways of Caring, June–July 2022 in Arnhem.[1]

10 State of Fashion: Ways of Caring 2022: Multiple Voices

State of Fashion (Arnhem) is a critical platform, supported by the Dutch Government, with the aim to re-define the meaning of fashion. For the 2022, State of Fashion biennale two curator collectives were selected through an open call. The NOT_ENOUGH collective (Andrea Chehada Barroux, Mari Cortez, Marina Sasseron de Oliveira Cabral) and Fashion Open Studio (Tamsin Blanchard, Orsola de Castro, Filippo Ricci, Niamh Tuff) collectively decided to explore the theme Ways of Caring (Arnhem 2022) in different programmes. NOT_ENOUGH addressed caring by challenging the dominant view on being, by making space for unseen feelings and untold stories.

[1] Being a member of the board of State of Fashion I was closely involved in the strategy and the selection of topics and curatorial teams of Ways of Caring 2022.

Fig. 2 Bodies that make, bodies that consume 2022

Throughout the exhibition, they aimed to activate fashion along three themes. Via Transforming narratives they questioned the Eurocentric beauty ideals often exoticising other's identities and challenging notions of gender, identity and vulnerability. Through Exercising Compassion they encouraged collective practices instead of individual actions, using fashion as a tool to connect people via sharing and making. Finally, via Coexisting Knowledges they addressed formal ways of production and distribution of knowledge, creating space for diversity and enabling plural worldviews to coexist. Via an open call, designers from all over the world were invited to respond to those ideas (Fig. 2).

As a result, quite a few transnational collectives were granted to execute their ideas in a commission. Santiago Utima (Columbia), Siviwe James (South Africa) Wide Asari and Riyadhus Shalihin (Indonesia) co-created the installation *Bodies that make, bodies that consume (2022)* addressing the gap between 'the hands that make' and 'bodies that consume.' With a series of standard garments—polo-shirts, T-shirts—they foregrounded the various emotional and physical situations textile and clothing manufacturer workers from the global south endure using the label as a place where the manufacturer workers described in detail their effort and personal life.

Here, the three terms *aesthesis, precedence* and the *pluri-versal* helped to characterise and to unfold what has been silenced by the colonial fashion manufacturing practices. Not the aesthetics of the installation garments was crucial, but the aesthesis where the stories and traces of the workers was foregrounded in the stains on the clothes and the stories of the workers making the clothes. The installation was not focusing on the new or the canon of fashion but tried to unfold different temporalities highlighting the unfair mechanism of clothing production, combining different (pluri-versal) stories for each continent.

11 Conclusion

For decades, fashion has been promoting a dominant discourse with a focus on the *contemporary* and the *new* and a narrow-subscribed aesthetic focusing colour and silhouette. In order to do so, the system wiped out elements from heritage, craftmanship, cultural values and local (other) cultures and therefore excluded stories and cultures that where not part of the *self-reflexive* modern fashion system.

On the contrary, the decolonial terminology empowers and above affirms cultures by unfolding and helping to reveal untold and unseen practices in a non-hierarchical way. In a moment of history where the fashion system is failing in all areas of production, environmental, overproduction and social justice, the decolonial lens is very useful to highlight and endorse, new values and new imaginations around fashion, identity, inclusivity and sustainability who are crucial to unfold the failures in the systems and are helpful to define solutions.

Going back to my initial question: are we able to re-define and modify the modern fashion system, which is so strongly rooted in Western enlightenment, industrialisation, capitalism and modernism with its focus on *the contemporary, radical new* and *the universal*? The answer is no. We urgently need to replace the modern fashion terminology and theory by decolonial thinking using *de-linking, precedence and aesthesis, pluri-versal* as a new terminology to capture in more depth the meaning, values and imaginations of the new generation of ground-breaking, engaged fashion designers from all over the world. 'In order to develop a truly decolonial fashion discourse we need to replace fashion as a *noun* defined in terms of temporality and the new and therefore automatically inheriting a system (of power) and an industry (of capitalism) as part of modernity and coloniality,' states scholar Angela Jansen. Instead, we need to start to use the term fashion as a *verb* putting a focus on the act of fashioning the body, which includes all temporalities and geographies and operates beyond the colonial difference ([7]: 1).

Von Bush' concept of Deep Fashion with a focus on *relations, processes and passion* is based on comparable principles and therefore also a serious attempt to create a more relevant fashion system that is not only more inclusive but also centred around the act of fashioning the body to capture different layers of meaning, relations, belonging and deeper cultural values (around environment, inclusivity that fashion and dress traditions/cultures are able to offer us) in a non-hierarchic way ([3]: 138–139).

How exactly the fashion system will transform into a new inclusive fashion system is for future research and exploration, yet decolonial theory and the concept of Deep Fashion will lead us in the right direction.

'It is only through an act of listening to the other, of the understanding itself through the voice of 'others,' that the West can overcome the ignorance of Eurocentrism and recognise itself,' states Vasquez convincingly ([25]: 191).

References

1. Appadurai A (1996) Modernity at large. University of Minnesota Press, Minneapolis
2. Bourriaud N (2009) The Radicant. Sternberg Press, New York
3. Bush O (2020) The Pscychopolitics of fashion. Conflict and courage under the current state of fashion, London: Bloomsbury
4. Calefato P (2019) Fashionscapes. In: Geczy A, Karaminas V (eds) The End of Fashion. Bloomsbury, London, pp 31–47
5. Geczy A, Karaminas V (2016) Walter benjamin: fashion, modernity and the city street. In: Rocamora A, Smelik A (eds) Thinking through fashion, pp 201–231
6. Jansen A, Craik J (2016) Introduction. In: Jansen A, Craik J (ed) Modern fashion traditions. Negotiating tradition and modernity through fashion, London: Bloomsbury, pp1–22
7. Jansen A (2020) Fashion and the Phantasmagoria of modernity. Fashion Theory, vol 24(6):1–24. https://doi.org/10.1080/1362704X.2020.1802098
8. Lipovetsky G (1994) The empire of fashion: dressing modern democracy. Princeton University Press, Princeton, NJ
9. Lipovetsky G (2021) Le Sacre de L'authenticite. Gallimard
10. Mignolo W (2007) Delinking the rhetoric of modernity, the logic of coloniality and the grammar of de-coloniality. Cultural Studies 21(2–3):449–514. https://doi.org/10.1080/09502380601162647
11. Mignolo W, Vasquez R (2013) Decolonial AestheSis: colonial wounds/Decolonial Healing—social text, pp 16–18. https://socialtextjournal.org/periscope_article/decolonial-aesthesis-colonial-woundsdecolonial-healings
12. Quijana A (2010) Coloniality and modernity/rationality. In: Walter D (ed) Globalization and the Decolonial Option. Routledge, Mignolo and Arturo Escobar. New York, pp 22–32
13. Riegel M, Skov L (2011) Fashion theory, vol 15 Issue 2, Special Issue (guest ed. L. Skov and M. Riegels Melchior), 133–135
14. Said E (1977) Orientalism. Random House, New York
15. Simmel G (1904) Fashion, reprinted in the American J Sociol (1957), 62(6):541–558
16. Skov L (2011) Dreams of small nations in a polycentric fashion world. Fashion Theory Vol 15 Issue 2, Special Issue (guest ed. L. Skov and M. Riegels Melchior), 15:2, pp 137–152
17. Steele V (2008) Paris fashion, New York: Oxford 1998
18. Teunissen J (1990) Mode in Beweging. Amsterdam NFM
19. Teunissen J (2005) Global fashion/local tradition: on the globalisation of fashion. In: Brand J, Teunissen (eds), Global fashion, local tradition: on the globalisation of fashion, Arnhem: Terra, pp 8–23
20. Teunissen J (2011) Deconstructing belgian and dutch fashion dreams: from global trends to local crafts. Fashion Theory, 15:2, Special Issue (guest ed. L. Skov and M. Riegels Melchior), pp 157–77
21. Teunissen J (2014) The future of fashion is now: longing for a future society. In: Teunissen J, Nefkens H, Arts J, von der Voet H (eds), The future of fashion is now, Rotterdam: Boijmans, pp 12–72
22. Teunissen J (2017)'Clogs on high heels: dutch cultural heritage and fashion. In: Smelik A (ed) Delft Bleu, to Denim Blue. I.B Taurus, pp 30–50
23. Teunissen J (2018) State of Fashion: Searching for the New Luxury, virtual exhibition catalogue, https://issuu.com/stateoffashion/docs/stof_072018_catalogus_ nteractive. Accessed 3 June 2021
24. Teunissen J (2022) The transformative power of practice-based fashion research. In: Titton M, Gaugele E (eds) Fashion knowledge. Wenen, pp 15–26. Forthcoming 2024
25. Vasquez R (2018) The Museum, Decoloniality and the end of the contemporary. In: Lijster T (ed) The Future of the New, Artistic Innovation

José Teunissen is Professor in Fashion & Identity and Director of Amsterdam Fashion Institute (Amsterdam University of Applied Sciences. Next, she is working as a free-lance fashion curator. Her research is focusing on fashion as a culture phenomenon using curation as a tool. Her major exhibitions include *The Future of Fashion is Now* (2014 Boijmans Rotterdam), *Searching for the New Luxury* (2018 Arnhem Modebiennale, State of Fashion). She regularly writes about Transnational Fashion, the future of fashion (design) and the transformative power of Fashion-Tech for the future of the fashion discipline.

Frankie Welch's Cherokee Alphabet Design: Cultural Appreciation or Cultural Appropriation?

Alyssa C. Opishinski, Linda Welters, and Susan J. Jerome

Abstract Cherokee Alphabet is the best-known textile design of Frankie Welch (1924–2021), although she created other designs based on imagery and symbols from Indigenous American cultures. This study asks whether Cherokee Alphabet is a case of cultural appreciation or cultural appropriation. Welch self-identified as part Cherokee, but researchers have failed to confirm this through genealogical research, and inconsistencies in her business documents archived at the University of Rhode Island further complicate interpretation. The approach was to understand the case from a variety of viewpoints in order to prompt constructive discussion.

This chapter is a case study of Frankie Welch's best-known textile design titled Cherokee Alphabet. Her design is based on the Cherokee syllabary (i.e., a set of written characters representing syllables that serves the purpose of an alphabet), created in the 1820s by Sequoyah, who worked to unify the Cherokee nation and elevate his people. Welch, a home economics teacher turned designer and entrepreneur, printed the design on several kinds of textiles and in many colorways. She used it most often for scarves, but also on clothing, tote bags, and household textiles. Aside from the Cherokee Alphabet, Welch's body of design work consisted of commissions from corporations, professional organizations, political groups, and individuals. Some of her subsequent work was also based on Native American artifacts.

The Cherokee are a group of Indigenous Peoples native to the Southeastern Woodlands of the United States, including Welch's home state of Georgia. According to the Cherokee Nation website, the first record of Cherokee contact with Europeans

A. C. Opishinski · L. Welters (✉) · S. J. Jerome
University of Rhode Island, Kingston, RI, USA
e-mail: lwelters@uri.edu

A. C. Opishinski
e-mail: aopishinski@uri.edu

S. J. Jerome
e-mail: sjjerome@uri.edu

M. Á. Gardetti and R. P. Larios-Francia (eds.), *Sustainability Challenges in the Fashion Industry*, Sustainable Textiles: Production, Processing, Manufacturing & Chemistry,
https://doi.org/10.1007/978-981-99-0349-8_5

was 1540; however, their oral history extends for millennia [1]. After first contact, the Cherokee often chose collaboration and adaptation over conflict with European immigrants. They participated in trade, intermarried, fought for different European groups, and many also converted to Christianity. By 1828, the Cherokee Nation had a Constitutional government, developed their own syllabary, and published a newspaper in both Cherokee and English. Throughout the 1830s, however, the United States government worked to force the Cherokee off their land as a result of the discovery of gold there. Ultimately, many Cherokees perished in the process, while others ended up resettling in what would become Oklahoma. When Oklahoma's statehood was established in 1907, the sovereignty which the Cherokee had rebuilt for themselves was dissolved, as by default they became United States citizens. After a dark period of significant poverty, they found inspiration in the civil rights movement of the 1960s and eventually ratified a new Constitution in 1975. The Cherokee Nation has many longstanding traditions, yet their culture is ever evolving, and many Cherokee embrace a mix of modern and traditional aspects. They emphasize that beliefs and degree of cultural knowledge vary among individuals, and that "there is no universally agreed upon way to express Cherokee culture" [16]. Many spiritual and ceremonial facets of Cherokee culture, however, are not shared publicly.

Welch self-identified as part Cherokee. While other researchers have searched for evidence to support her Cherokee ancestry, currently that evidence is missing. Nevertheless, she has been featured in exhibitions and books as a Native American designer. Several objects with the Cherokee Alphabet design are housed in the collection of the National Museum of the American Indian. Most recently, a Frankie Welch dress utilizing the Cherokee Alphabet fabric was included in the Metropolitan Museum of Art's Costume Institute exhibition "In America: A Lexicon of Fashion" in 2022 [44]. Business documents from her archive in the University of Rhode Island's Historic Textile and Costume Collection (HTCC) complicate interpretation of her design today, as they provide inconsistent origin stories and attributions.

For fashion historians, curators, and educators the question of how to interpret the Cherokee Alphabet design arises. Is it an instance of cultural appreciation or cultural appropriation, as these terms are currently defined? Does the interpretation of a designed object shift over time, when considering the various perspectives and contexts involved? What did other members of the Cherokee nation think about this use of the syllabary, an important symbol of their independence and resistance to assimilation? The timing of the Cherokee Alphabet design – the late 1960s – coincides with the rise of civil rights organizations such as the American Indian Movement (AIM) as well as the popularity of ethnic-inspired fashions. Was there a connection to such organizations for Welch, or was she simply following the ethnic fashion trend? Did the design evoke empathy for Cherokee and Indigenous rights in consumers or was it a case of "redwashing"? By redwashing, we mean capitalizing on a Native American connection for a perceived advantage.

This case study should prove useful as a tool for discussion in both fashion history/design/business and museum studies programs. The chapter is illustrated with examples of Frankie Welch's designs that are housed in the HTCC at the University of Rhode Island. The HTCC also contains documents related to Frankie Welch's business including her shop in the Washington D.C. area. Titled the "Frankie Welch Archive," these records and artifacts inspired the study.

1 Biography

Frankie Welch (1924–2021), born Mary Frances Barnett, credited her semi-isolated, rural childhood on her family's farm in Rome, Georgia with fostering her creativity, as one had to make do with limited means [41]. She kept herself occupied with sewing, painting, and exploring the countryside [41]. At age eight, Welch's interest in Native American culture was sparked after finding four arrowheads, although in her book *Indian Jewelry* she reflects that: "Certain Indian ways and little traditions were simply a part of my childhood...." ([40], p. 8). She grew up hearing stories about her family's Cherokee ancestry, an attribution that she and her family never questioned [9].

Welch received formal training in design, earning a B.A. in clothing and design from Furman University [41]. At Furman, she established her first successful independent business venture, designing dresses for her wealthier peers [41]. Welch selected the pattern and materials, conducted fittings, and had them made up by local seamstresses; a 5% markup was paid for her own clothes [41]. Shortly after completing her junior year at Furman in 1944, Frankie married Pfc. William Welch, her childhood sweetheart [32].

In 1952, the Welches moved to Alexandria, Virginia, for William's job with the federal government [9]. It was here, near the nation's capitol, that Frankie found her calling as an entrepreneur, initially with fashion consultations and personal shopping services, and then as owner of a retail shop named Frankie Welch of Virginia, which opened in 1963 [41]. The shop was located on the first floor in an 18th-century historic house known as Duvall House; the family lived on the upper two floors [7]. She focused on American themes in her buying and merchandising and is often cited for her strong connection to "Americana" [9]. Welch preferred that her shop not be called a boutique (French for "shop"), because "we're all-American" [14]. In her marketing materials, she often included the story that Duvall House was where George Washington had alternately banked, dined, and was entertained, although in one interview she instead linked the building to Confederate General Robert E. Lee's family [6, 18, 35]. Both Frankie and her husband were interested in history; William's graduate studies were in history [9].

Women involved in the political scene of Washington D.C. (e.g., as elected officials, wives and daughters of elected officials, and activists) valued her store and its services. Welch created custom designs for many political candidates and special interest groups of both parties, and she is remembered for designing gowns for First Lady Betty Ford [18]. Welch avoided declaring a political affiliation, which, along with her vibrant personality, contributed to her success in Washington D.C. [9]. Additionally, Welch claimed to treat customers equally, recalling when she and her staff worked to simultaneously serve the wife of a British ambassador and an unidentified young wife and mother who had "been saving her nickels and dollars to buy a coat" [14]. Responding to her successes, in 1975 she changed the name of her business to Frankie Welch of America and added national and international corporate image consulting to her already numerous services [9, 36]. After nearly three decades, Welch officially closed her store in 1990, although she continued to design and teach [9].

As Welch's business grew, so did her sphere of influence. Welch donated archival material and objects from her business to several universities, an act which she considered "giving a talent back to the students" [41]. After befriending a University of Rhode Island faculty member who was on sabbatical leave in Washington D.C., she donated business records, garments, and representative scarves to the HTCC. The now-retired faculty member was interviewed for this case study, and the interview notes added to the Frankie Welch Archive. The donations arrived at the university from 1982 to 1985. Welch designed a scarf for URI in 1983, and that same year she gave a talk on campus, accompanied by a small exhibit of her work [18].

Welch received an honorary doctorate from the Miami International University of Art & Design [7]. From 1982 to 1985, she was Distinguished Visiting Professor of Merchandising and Design in the School of Consumer Science at Winthrop University [20]. Winthrop additionally honored her with their (now defunct) Frankie Welch Design and Merchandising Laboratory, while the Fashion Group International of Greater Washington named a scholarship after her in 1998 [21]. Welch was appointed by President George H. W. Bush in 1991 to the Textile and Clothing Board of the Department of Defense and was also slated to serve on the Export Retail and Wholesaling Board of the Commerce Department through 1996 [11]. Her legacy as a successful woman entrepreneur in an era when women were fighting for equal rights in the workplace is unquestionable.

2 Iconic Designs

The designs for which Welch is best remembered are the Cherokee Alphabet scarf, her "Frankie" hostess dress (Fig. 1), and her various presidential and political commissions. The versatile Frankie dress tied in eight different ways and was sold in a variety of fabrics including cotton, velvet, gray flannel, and, of course, the Cherokee Alphabet [9]. Anyone in the Washington D.C. social scene had a Frankie in the 1970s, and it was worn to barbecues, luncheons, and other casual gatherings [18].

The exact origins of the Cherokee Alphabet design are hazy, as each account differs. The most-repeated (and therefore most likely) statements are compiled here. Virginia Rusk, wife of the Secretary of State, is credited with commissioning Welch to create an "all-American" design for President Lyndon B. Johnson's administration to use as official gifts [2, 18, 36, 37]. This resulted in the Cherokee Alphabet scarf, which Welch later developed into a variety of textiles. The scarf is variously dated between 1966 and 1968 [6, 18, 19]. Welch herself credits a book about Floyd County, in which her hometown of Rome, Georgia is located, as her inspiration for the design

Fig. 1 Advertisement for the Frankie Dress, from Frankie Welch of Virginia. (ca. 1975) University of Rhode Island, URI Frankie Welch Archive, Donor: Frankie Welch

[40]. She states: "The book had a page picturing symbols of the alphabet invented by the great Cherokee Sequoyah, so the Indians could transcribe their own language. [...] From these symbols I drew the design for my Cherokee alphabet scarf ...." ([40], p. 12).

An article from 1969 has a different origin story, stating that the scarf was first designed to aid a Native American education fund, while other accounts imply her philanthropy (sometimes specifically to a Cherokee education fund) was secondary to its government commission [4, 8, 35]. *Women's Wear Daily* reported that a dollar from every scarf and twenty-five cents from every shopping bag printed with the design went to a higher education fund, with Welch stating, "Everyone should help the Indians, don't you think?" ([8], p. 23). On April 22, 1968, she presented a percentage of the proceeds to Congressman Roy A. Taylor of North Carolina for the "higher education fund for Cherokee Indians" ([12], p. 28). It is unclear exactly how the funds were distributed after this point.

Another significant part of the story is Welch's friendship with LaDonna Harris (Comanche), the wife of Senator Fred Harris of Oklahoma (1964–1973), who first visited Frankie's shop in the mid 1960s [9]. Harris modeled in Vera Maxwell's 1966 "Indian Summer" fashion show coordinated by Welch [9]. Harris later helped start the advocacy organization known as Americans for Indian Opportunity. According to Harris, Welch asked for her help in locating a "clear" copy of the Cherokee alphabet for the scarf ([9], p. 128, [24]).

Fig. 2 Cherokee Alphabet Scarf, 1967. Silk. 21.25″ wide × 21.375″ high (53.9 cm × 54.2 cm). University of Rhode Island, URI 1985.08.16, Donor: Frankie Welch. The Cherokee Alphabet Scarf was the first scarf designed by Frankie Welch

Welch worked with textile designer Julian Tomchin to prepare the design for production [9]. He cleaned up the lines and facilitated the manufacturing [9]. Originally, the scarves were screen printed on silk in a square format with a choice of three colors: Cherokee red, Sequoyah green, and Sabra brown (Fig. 2) [9]. Tomchin also helped her develop the first oblong version; a later oblong format with a striking use of stripes but without "Cherokee Alphabet" printed on it was known as "Running Cherokee" [9]. The Cherokee Alphabet scarf became a best seller, and it was reprinted 45 times on a variety of fabrics, including a new synthetic nylon called Qiana [9]. The design graced a wide range of products besides scarves including clothing (e.g., the Frankie, dresses, skirts, shirts, and linings), tote bags, plastic shopping bags, and home furnishing fabrics (Fig. 3) [9, 18]).

Scarves were a popular fashion accessory for professional women at the time. Vera Neumann, well-known for her silk scarves signed "Vera" alongside a ladybug, paved the way for signature scarves which she first offered in 1947 [9]. Couture and other luxury designers followed suit in the 1960s, offering scarves in various formats ranging from large squares to oblong shapes. Welch's initial success with

Fig. 3 Promotional bags printed with the Cherokee Alphabet. University of Rhode Island, URI Frankie Welch Archive, Donor: Frankie Welch. Welch used the Cherokee Alphabet widely for advertising

the Cherokee Alphabet scarf led to more design opportunities. In 1968, she was commissioned with creating a scarf for the First Lady's Discover America campaign and fashion show [13, 18]. The next year, Welch officially began offering her services to design scarves for government, businesses, and educational institutions (Fig. 4). Initially, she distributed 5000 brochures to potential clients and was very successful, continuing to design scarves into the early 1990s [9]. A formula that she used frequently consisted of 8″ by 8″ modules.

Welch's interest in Native American cultures did not stop with the Cherokee Scarf. Welch's extensive interest in Native American cultures began as a child. She described the day when she was eight and had found four arrowheads while playing

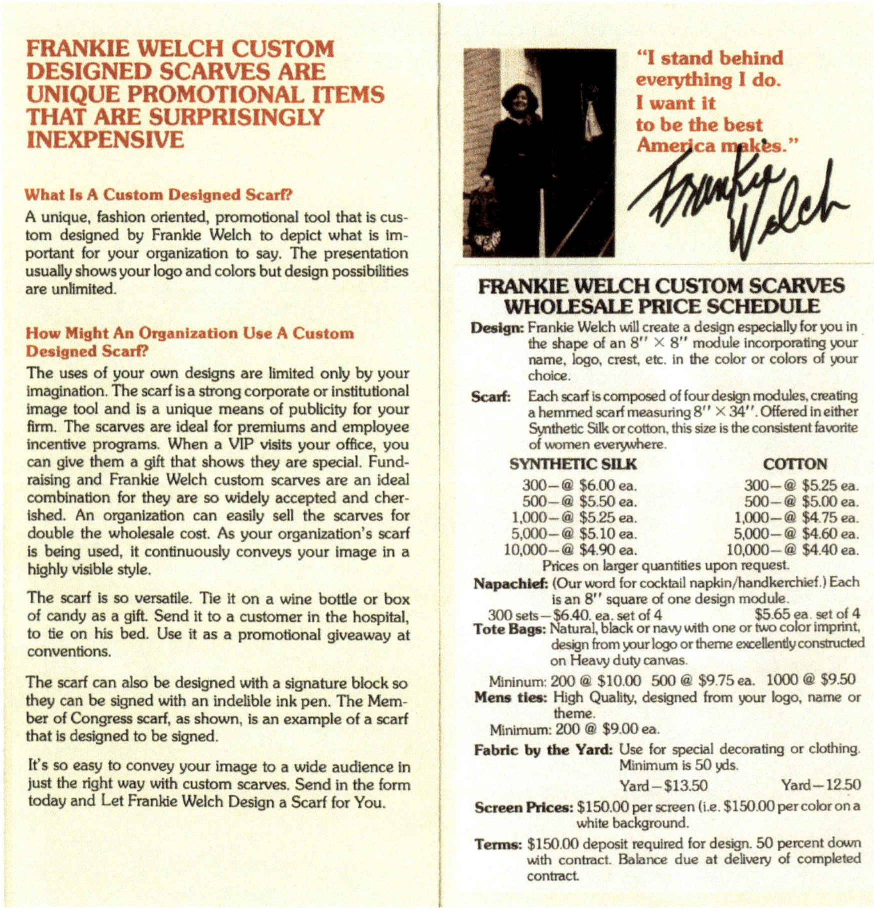

FRANKIE WELCH CUSTOM DESIGNED SCARVES ARE UNIQUE PROMOTIONAL ITEMS THAT ARE SURPRISINGLY INEXPENSIVE

What Is A Custom Designed Scarf?

A unique, fashion oriented, promotional tool that is custom designed by Frankie Welch to depict what is important for your organization to say. The presentation usually shows your logo and colors but design possibilities are unlimited.

How Might An Organization Use A Custom Designed Scarf?

The uses of your own designs are limited only by your imagination. The scarf is a strong corporate or institutional image tool and is a unique means of publicity for your firm. The scarves are ideal for premiums and employee incentive programs. When a VIP visits your office, you can give them a gift that shows they are special. Fundraising and Frankie Welch custom scarves are an ideal combination for they are so widely accepted and cherished. An organization can easily sell the scarves for double the wholesale cost. As your organization's scarf is being used, it continuously conveys your image in a highly visible style.

The scarf is so versatile. Tie it on a wine bottle or box of candy as a gift. Send it to a customer in the hospital, to tie on his bed. Use it as a promotional giveaway at conventions.

The scarf can also be designed with a signature block so they can be signed with an indelible ink pen. The Member of Congress scarf, as shown, is an example of a scarf that is designed to be signed.

It's so easy to convey your image to a wide audience in just the right way with custom scarves. Send in the form today and Let Frankie Welch Design a Scarf for You.

"I stand behind everything I do. I want it to be the best America makes."

Frankie Welch

FRANKIE WELCH CUSTOM SCARVES WHOLESALE PRICE SCHEDULE

Design: Frankie Welch will create a design especially for you in the shape of an 8″ × 8″ module incorporating your name, logo, crest, etc. in the color or colors of your choice.

Scarf: Each scarf is composed of four design modules, creating a hemmed scarf measuring 8″ × 34″. Offered in either Synthetic Silk or cotton, this size is the consistent favorite of women everywhere.

SYNTHETIC SILK	COTTON
300—@ $6.00 ea.	300—@ $5.25 ea.
500—@ $5.50 ea.	500—@ $5.00 ea.
1,000—@ $5.25 ea.	1,000—@ $4.75 ea.
5,000—@ $5.10 ea.	5,000—@ $4.60 ea.
10,000—@ $4.90 ea.	10,000—@ $4.40 ea.

Prices on larger quantities upon request.

Napachief: (Our word for cocktail napkin/handkerchief.) Each is an 8″ square of one design module.
300 sets—$6.40. ea. set of 4 $5.65 ea. set of 4

Tote Bags: Natural, black or navy with one or two color imprint, design from your logo or theme excellently constructed on Heavy duty canvas.
Mininum: 200 @ $10.00 500 @ $9.75 ea. 1000 @ $9.50

Mens ties: High Quality, designed from your logo, name or theme.
Minimum: 200 @ $9.00 ea.

Fabric by the Yard: Use for special decorating or clothing. Minimum is 50 yds.
Yard—$13.50 Yard—12.50

Screen Prices: $150.00 per screen (i.e. $150.00 per color on a white background.

Terms: $150.00 deposit required for design. 50 percent down with contract. Balance due at delivery of completed contract.

Fig. 4 Marketing brochure for "Let Frankie Welch Design A Scarf For You" campaign. University of Rhode Island, URI Frankie Welch Archive, Donor: Frankie Welch. The brochure declares, "Fundraising and Frankie Welch custom scarves are an ideal combination for they are so widely accepted and cherished"

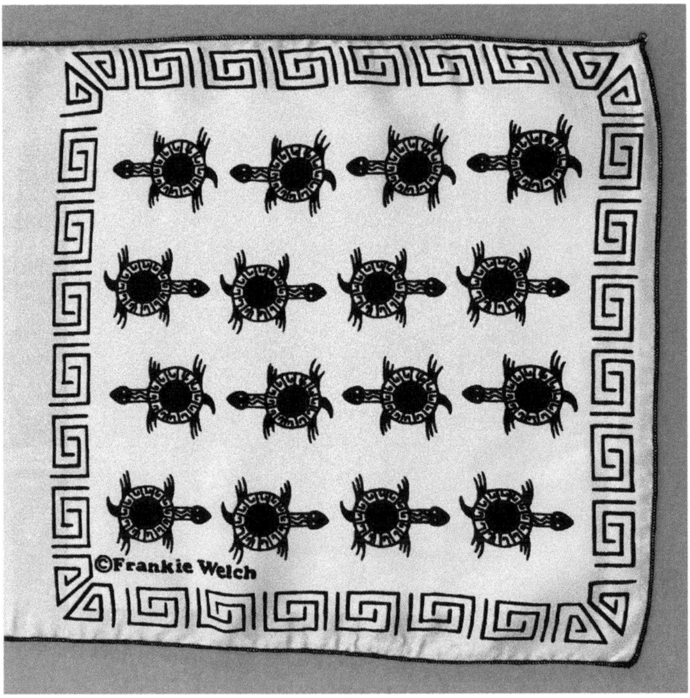

Fig. 5 Turtle design, 1969–70. University of Rhode Island, URI 1985.08.167, Donor: Frankie Welch. Welch used an individual turtle as a logo appearing next to her signature. This 8″ × 8″ square is from a silk four-module scarf measuring 8″ × 34″

in the woods, as "…the beginning of my life-long fascination with the Indians, their jewelry, baskets, pottery and history" ([40], p. 9). As an adult, she actively collected and researched the history of traditional Native American craftwork and created more designs using Native American motifs. The inspiration for Turtles (Fig. 5), created in 1969–70, came from a 16th-century piece of pottery from Arizona seen in a museum [18]. This popular design, produced as yardage as well as scarves, became her logo and appeared next to her name or initials, as Vera had done with her scarves [9]. Callahan [9] points out that Welch ignored the sacred significance of the turtle to Native Americans and ascribed it with her own symbolism: the traveler and risk-taker. Welch discusses the art of Central and South America in her book *Indian Jewelry,* revealing a curiosity in the cultures of Indigenous Peoples that extended beyond the United States. This interest is displayed with her El Tumi design, introduced in 1971 [18] and based on a sharp tool native to the Central Andes region (Fig. 6). Today, the tumi is the national symbol of Peru.

The Five Civilized Tribes Museum in Muskogee, Oklahoma, opened in 1966. Through Martha Griffin White, a friend of Welch's since 1967, the museum was given its own variation of the Cherokee Alphabet scarf to stock in its gift shop [4]. She also designed a scarf for the museum that reflected the culture of each "civilized tribe":

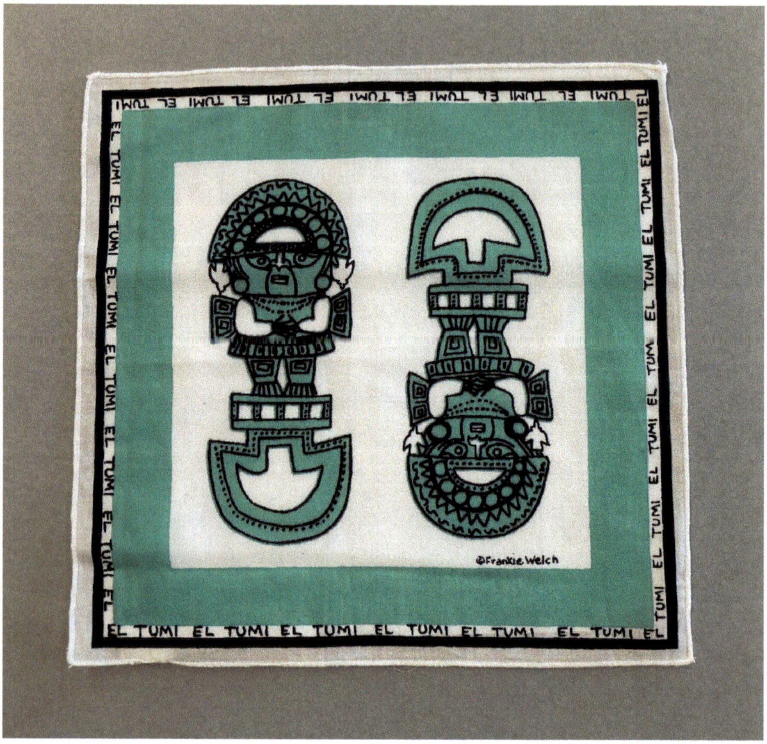

Fig. 6 El Tumi, 1971. University of Rhode Island, URI 1985.08.77, Donor: Frankie Welch. Welch called this 8″ × 8″ cotton square a "napachief" and marketed it as a versatile scarf, handkerchief, gift wrapping, pocket square or a present

the Cherokee, Chickasaw, Choctaw, Creek/Muskogee, and Seminole (Fig. 7). These tribes were deemed "civilized" because of their organized systems of government and conversion to Christianity [5, 15]. Welch's belief in her Cherokee ancestry included the story of her great-great-great grandfather's removal from Georgia to Oklahoma, when the Cherokee were forced off their land in what is now known as The Trail of Tears [40].

Welch's ability to work with the wives and staff of politicians from both sides of the aisle created business opportunities including a project with Goldwaters Department Store in Phoenix, Arizona, whose chairman of the board was Barry Goldwater, Senator for Arizona (1953–1965 and 1969–1987) [9, 26]. For this project, Welch designed a group of Native American-inspired scarves – Roadrunner, Thunderbird, Basket Weave, Squash Blossom – that were introduced in 1973 at the Annual Cherry Blossom Luncheon in Washington, D.C. [9]. One of the models for the luncheon was Susan Goldwater, wife of Barry Goldwater, Jr., a California Congressman (1969–1983), and daughter-in-law of the Senator [9]. Goldwaters Department Store sponsored the event [9]. Welch's Roadrunner design (Fig. 8) may be a take on Goldwaters'

Fig. 7 Five Civilized Tribes design, ca. 1970. University of Rhode Island, URI 2021.99.01, Donor: Frankie Welch. The Five Civilized Tribes Museum opened in 1966, representing the Cherokee, Chickasaw, Choctaw, Creek/Muskogee, and Seminole tribes

store logo, also a roadrunner. While the Thunderbird pattern was included in this group, information in the Frankie Welch Archive states that the pattern was originally developed for the Arizona Thunderbirds, a group of businessmen interested in promoting tourism [9, 18]. Native Americans of the Southwest consider the roadrunner to be a spiritual animal for a variety of reasons, while Native Peoples in both the United States and Canada regard the thunderbird as a powerful symbol that protects people from evil spirits. The Basket Weave module scarf included four different designs based on Cherokee basket weaving patterns [9]. The Squash Blossom scarf was also a module depicting four different Native American jewelry designs [9]. Welch recognized the versatility of an 8″ by 8″ square in creating modular designs; she named individual squares the napachief (e.g., napkin and handkerchief). Welch developed an 8″ by 36″ scarf that repeated the design four times [9] (see Fig. 9).

Welch published *Indian Jewelry* in 1973 to coincide with the release of "…an elegant line of Indian jewelry based on authentic pieces from the past" ([40], p. 78). Welch, described in the back matter as "a gifted interpreter of America's Indian heritage," designed the cover and drew the illustrations ([40], p. 78). This compact,

Fig. 8 Roadrunner design, 1973. University of Rhode Island, URI 1985.08.14, Donor: Frankie Welch. This image is a close-up of the cotton fabric used for a Frankie dress. The Roadrunner was one of four designs introduced at the Annual Cherry Blossom Luncheon in Washington, D.C in 1973

hardcover book printed on sepia-tone paper, and featuring one of Welch's Basket Weave designs on the cover, is part reminiscence of her life and part "how-to" book. She takes the reader through her discoveries of Native Peoples' signs and symbols, resulting in a highly simplified explanation of customs from numerous unrelated groups. Although she mentions achievements by specific groups, she ultimately treats Native American cultures as a single entity. Although the book has no reference list or footnotes, she writes of visiting museums and speaking with curators about their collections. Welch offers advice on how to collect and wear Native American jewelry, and even how to display it at home. She describes her collection of 200 pieces as an investment that she hopes to pass down to her daughters. Welch blends salesmanship into her writing with suggestions for mixing antique jewelry with her own line, to always pack a Frankie when traveling, and how to use scarves to transform an outfit.

In summary, Frankie Welch's design repertoire included a significant number of designs based on Native American and Indigenous South American motifs. These included seven textile designs, several of which, in their original context, represent

Fig. 9 Welch's designs inspired by Native American artifacts. (left) Basket Weave scarf in brown with (top to bottom) Box, Peace Pipe, Single Diamond, Jacob's Ladder. 1973. Polyester. URI 1985.08.168; (center top) Indian Jewelry book by Frankie [40]. URI 1985.08.171, Basket Weave incomplete scarf in orange with (top to bottom) Box, Peace Pipe, Single Diamond. 1973. Cotton. URI 1985.08.169; (right) Squash Blossom scarf with (top to bottom) Santa Domingo, Navajo, Zuni, Old Navajo. Cotton. 1973. URI 1985.08.170. University of Rhode Island, Donor: Frankie Welch. Welch named these designs, using her collections of Native American baskets and jewelry for design inspiration

sacred beliefs. These designs were used in a wide range of products which were sold in Welch's and others' establishments. Additionally, she wrote a book on the subject of Native American jewelry as part of the promotional strategy for her line of jewelry which was based on original Native American designs.

3 Cultural Appropriation or Cultural Appreciation?

In the recent past, scholars and the general public alike have struggled to find a broadly acceptable definition of cultural appropriation, resulting in many possible definitions. Two are presented here, both from sources directly addressing cultural appropriation of Indigenous cultures in fashion. One comes from a legal standpoint, while the other aided in curating a museum exhibition. Brigitte Vézina, whose research focuses on the legal aspects of cases of cultural appropriation in fashion design, gives the following definition: "the act by a member of a dominant culture of taking a TCE [traditional cultural expression] whose holders belong to a minority culture and repurposing it in a different context, without the authorization, acknowledgement, and/or compensation of the TCE holder(s)" ([39], p. 6). This definition emphasizes the power imbalance between cultures, but also leaves much open to interpretation and relies on holders of a TCE having a unified opinion.

For the Peabody Essex Museum exhibition, "Native Fashion Now," the curators/authors (one of whom is Turtle Mountain Chippewa) employed a different definition: that cultural appropriation is when non-native designers, especially powerful ones, use motifs from Native American cultures without understanding the meanings behind them, which reinforces misunderstandings and stereotypes [27]. They also imply that sacred objects should not be appropriated by outsiders [27]. One example they give, which seems to confuse more than clarify their parameters, is Isaac Mizrahi's Totem Pole Dress for his Fall 1991 collection. Naomi Campbell wore Mizrahi's Totem Pole dress on the cover of *Time* magazine in September 1991, elevating its importance as a fashion design [27]. The curators did not consider it cultural appropriation since it did not perpetuate negative stereotypes and was inspired by totem poles rather than a reproduction of a specific one [27].

Unfortunately, sources seldom define the phenomenon of cultural appreciation, even when suggesting ways to prevent appropriation. Its omission is curious and should not be ignored. Is there an assumption that cultural appreciation is the opposite, or absence of cultural appropriation? One of the few, and most recent explorations of the concept of cultural appreciation was undertaken by Jana Cattien and Richard John Stopford. They explored defining cultural appreciation in terms of esthetics. By reframing the debate to consider the wider contexts in which cultural appropriation/appreciation occurs, their investigation ultimately concluded that cultural appreciation is an inseparable feature of cultural appropriation. They found that one's identity is crucial to defining the act: ".... the hegemonic white European subject is entitled to what it explores because it is built into how it sees itself, the world, and the relationship between the two" [10]. Essentially, they argue that people of White

European descent are culturally biased to view the world through an appropriative lens even when their actions can typically be categorized as appreciation [10].

This argument parallels that of Ijeoma Oluo, author of *So You Want to Talk About Race?*, who pinpoints the problem of cultural appropriation/appreciation as inherently linked to power imbalances [34]. Dominant cultures redefine and profit from the cultural expressions of marginalized cultures, while the marginalized cultures are persecuted for those same expressions [34]. She states: "In an equal and open exchange of ideas, in an equal and open world, cultural appropriation doesn't exist. … cultural appropriation is a symptom, not the cause, of an oppressive and exploitative world order" [33]. Oluo posits that problems of cultural appropriation can only truly be solved by dismantling the systems of privilege that make it possible in the first place [33].

While acknowledging the legitimacy of this argument and encouraging its inclusion in discussion, for the purposes of this case study we have combined different aspects of the definitions of cultural appropriation given above. Cultural appropriation is here defined as the act by a member of a dominant culture taking and re-contextualizing a TCE from a marginalized culture without appropriate authorization, acknowledgement of, and compensation of the original TCE holders. This is especially true when it is an exact copy of a sacred expression or a design representing an individual's identity and/or when it perpetuates a negative stereotype. As stated previously, the definition of cultural appreciation is less clear; therefore, for the purposes of this case study, we have sought first and foremost to let the responses of the TCE holders, the Cherokee Nation and its members, guide judgement. Secondarily, we have sought to understand Welch's views of Cherokee culture, the context of the era in which the design was produced as compared to today, responses to the design by non-Cherokee Native Americans, and responses by non-Native American consumers of the design.

4 Interpretation

Welch self-identified as Cherokee as well as having English and German ancestry [9]. She often brought up her Cherokee heritage in interviews, connecting to her natal state of Georgia, and to Oklahoma, where the Cherokee were forcibly relocated after the Indian Removal Act of 1830. In *Indian Jewelry*, Welch writes: "My maternal great-great grandmother was a full-blooded Cherokee Indian" ([40], p. 8). She expanded on this statement in an interview in 1984, explaining that her mother, née Eugenia Morton in Alpharetta, Georgia, knew of a Cherokee woman living there whose name was also Morton [41]. Welch admitted that while Eugenia knew stories of this woman and identified her as her great-grandmother (thus making her Welch's great-great-grandmother), Eugenia did not know her personally [4, 41]. From this lineage, Welch self-identified as 1/32nd Cherokee [41]. In an interview in 1992, however, Welch identified another Cherokee ancestor, stating, "My great-great-great grandfather had come to Oklahoma on the Trail of Tears and then returned to Georgia"

[42]. Genealogical research undertaken by others has yet to support Welch's self-identification [9]. The identity of both a potential appropriator and the TCE holders is central to questions of cultural appropriation/appreciation. In cases such as Welch's, however, where the designer identifies as being a TCE holder, this becomes significantly more complicated. Questioning the identity of someone of a marginalized group is an ethical morass in itself.

How different groups of Cherokees define membership adds yet another layer. Each of the three Cherokee Tribes recognized by the United States government has a different set of criteria for recognition within each tribe. The Eastern Band of Cherokee Indians (EBCI) requires a direct lineal ancestor who appears on the 1924 Baker Roll of the ECBI to be within four generations of the person who wishes to be recognized. The United Keetoowah Band (UKB) is stricter, as the potential tribal member must be one-quarter Keetoowah Cherokee. The Cherokee Nation, in Oklahoma, expects the applicant to provide documentation connecting them to a lineal ancestor from the Final Rolls of Citizens and Freedman of the Five Civilized Tribes in Indian Territory, also known as The Dawes Rolls.

Unfortunately, the significant degree of fraudulent claims of Native American identity by White Euro-Americans casts a shadow over Welch's self-identification that makes interpretation of the Cherokee Alphabet design a particularly delicate matter. The occurrence of such claims is not new and is often tied to the reception of federal benefits. As early as the late 1890s, the possibility of obtaining free land caught the attention of some non-native individuals. The Dawes Commission was enrolling members of the Five Civilized Tribes into a system that divided tribal land into individual lots. One report describes the patience of a Commissioner listening to "…some white man, who was evidently a dead-beat, and filling their ears with false evidence with reference to his claims upon Indian citizenship" ([30], p. 23). Contemporary fraudulent claims are more often tied to benefits regarding employment and scholarships [31].

In the twenty-first century, the term "Cherokee Syndrome" (coined in the popular media) is used to describe the phenomenon of White Euro-Americans claiming to be Cherokee, typically based on family lore [25]. The United States Federal Census allows an individual to self-identify race and tribe. The 1990 census saw a 37.9% increase in Americans identifying as Native American and in 2010, more than 819,000 Americans self-identified as Cherokee, over twice as many as the combined populations of the three federally recognized Cherokee tribes [25, 31]. Considering the United States' past policies on separating Native Americans from their cultural identity, it is possible that not all these claims are fraudulent. In addition to popular media coverage of contemporary figures such as Senator Elizabeth Warren claiming Native American ancestry, the academic literature has explored this phenomenon.

Joel Martin, a noted scholar of Native American history and religion and a Southerner himself, observes that "contemporary southerners are heirs of a long history of appropriation of Indian lands, images, and Indianness itself" ([29], p. 144). Southerners are particularly noted for their claims of Cherokee heritage and romanticization of Native American culture throughout history to serve regional-specific ideologies of White supremacy and patriarchy [29]. Martin [29] attributes the stereotype of the "noble savage" as a means of recasting the expulsion and extermination of Native Americans from the American South as both an inevitable tragedy and a fair fight, thus restoring the conscience and honor of the White Euro-Americans who took their land. He proposes that the South began to identify with this stereotype to legitimize their rebellion against the Union [29]. This appears to have evolved into claiming Cherokee heritage, just as they claimed Native American names for their towns and geographical features, to further reinforce Southern regionalism [29]. He gives an example which is reminiscent of Welch's own marketing of Duvall House: "Much like the innkeeper who boasts that George Washington slept here, southerners employ images of southeastern Indians to connect their home with American History—with a capital H" ([29] p. 142). On the other hand, Harvard professor Philip J. Deloria reports that romanticizing and appropriating Native American heritage is not a uniquely Southern phenomenon. He argues that these are typical reactions of White Euro-Americans to crises of identity across the ideological spectrum [17]. As examples, he cites the counter-culturalists of the 1960s and 1970s, especially the communalists and environmentalists, subcultural groups associated with the political left who borrowed judiciously from Native American values, beliefs, and esthetics in support of their causes [17].

As those who interpret history and art increasingly commit to the ideals of diversity, equity, access, and inclusion (DEAI), the identity of those whose works they interpret becomes a defining factor. Committing to amplify marginalized voices and stories may seem straightforward initially, but in practice, people's identities defy neat categorization. For example, the artist Fritz Scholder (1937–2005), who created artwork during the same time frame as Welch, provides an intriguing foil to her. He vehemently and repeatedly protested being labeled an "Indian artist": ".... you can't really call anyone who is only one-fourth Indian and who grew up in a non-reservation environment an Indian" ([38], p. 32). Scholder's father was half-Luiseño and worked as a schoolteacher in North Dakota for the Bureau of Indian Affairs, and as a result, Scholder himself was an enrolled member of the Luiseño [38]. The marked difference between Scholder and Welch's upbringings, besides location, was that his parents intentionally eradicated all traces of Native American cultures from their household and devalued Native American identity [38]. As a result of researching Native American history, Scholder addressed in his art some of the darkest elements of their oppression by Euro-Americans [23]. Welch avoids such themes in *Indian Jewelry* and her Native American-inspired designs, instead actively romanticizing Native American life. On the subject of trying to categorize and interpret Scholder and his works, Kevin Gover (Pawnee/Comanche), the director of the National Museum of the American Indian, writes, "What is Indian art? Who is an Indian artist? Is it an Indian who is an artist, regardless of medium or subject matter? To what extent must a person have lived an 'Indian life' to be an Indian artist? What of the non-Indian who employs traditional Indian styles or treats Indian subjects?" ([23], p. 10). Clearly, defining Native American identity is a complex and confusing issue, one that begs the question of who has the right to do so. As for Welch, evidence of her Cherokee ancestry remains inconclusive, yet she truly believed she was Cherokee.

In the 1960s and 1970s, it was not uncommon to capitalize on Native American heritage, with little or no scrutiny. America was experiencing a period of soul searching through various subcultures and movements, but especially with the hippies who embraced non-Western cultures and had a back-to-the-land sensibility. Fashion designers were quick to pick up on the zeitgeist, incorporating Native American styles such as fringed leather jackets, feathered and beaded embellishments, and moccasins as part of the ethnic trend. While it is clear that mainstream America viewed them as cultural appreciation, for some they may have been offensive at that time.

Welch was a savvy businesswoman who capitalized on stocking the latest trends, whether designed by herself or others. For example, when women shifted to wearing pants in the early seventies, she offered pant suits in the Cherokee Alphabet fabric (Fig. 10). Welch not only incorporated the ethnic trend into her designs, but into the promotional aspects of her business. In 1966, a year before the "Summer of Love" in San Francisco, Welch coordinated an "Indian Summer" fashion show in front of Duvall House [9]. It highlighted Vera Maxwell's line inspired by Native American art, which used fabrics designed by Lloyd Kiva New (Cherokee) [9]. On the invitation, Welch emphasized that the use of these fabrics made the collection "more meaningful" as did the collaboration with Solomon McCombs (Muscogee Creek), who at the time was President of the American Indian and Eskimo Foundation ([9], p. 25). A year later, Welch held another event at the Alexandria Athenaeum to debut her Cherokee Alphabet scarf, bringing in the classically-trained singer Hoté Casella (Cherokee) to chant and play the drum in traditional clothing during the modeling [9].

In 1973, Welch premiered her aforementioned quartet of Native American-inspired designs at the Washington Fashion Group's fashion show at the Annual Cherry Blossom Luncheon [9]. The show's theme was Native American dress and jewelry, and while it featured fashion and jewelry by Native Americans, it also included Native American-inspired pieces by Betsey Johnson, Donald Brooks, Halston, and of course, Frankie Welch [9]. Native Americans in traditional dress were also present [9]. This show illustrates what was then considered cultural appreciation, but which would certainly raise eyebrows today if not outright protest, thus providing context for Welch's own outlook and actions.

Such attitudes were not limited to Native Americans, as evidenced by Bergdorf Goodman's 1969 "Basic Black at Bergdorf's" fashion show. That show, which benefitted Harlem's Northside Center for Child Development (known for its contributions to desegregation and research on racial bias), featured a group of Black designers, from the well-established Arthur McGee to the rising star Stephen Burrows [22]. Music was provided by a jazz ensemble dressed in dashikis, and guests were fed corn bread, fried chicken, collard greens, and chitlins, which Black American fashion editor Robin Givhan ([22], p. 63) describes as "several steps beyond the authentic and well down the slippery slope to minstrelsy."

Welch was not part of the counterculture movement, but her interest in Native American objects and culture was shared by many others. Welch's marketing of the Cherokee Alphabet scarf repeatedly reinforces the idea of its American-ness, a theme repeated in *Indian Jewelry*; that Native American history is not just American history

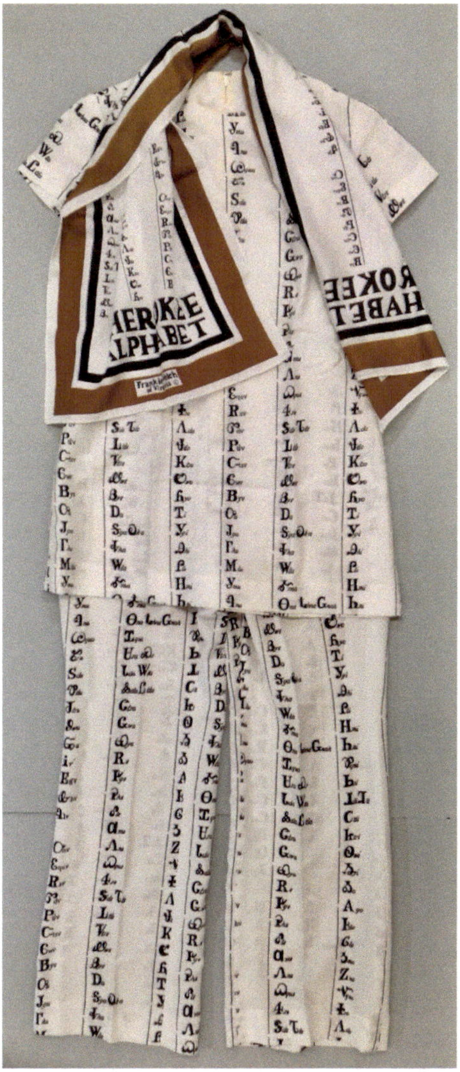

Fig. 10 Cherokee Alphabet Pant Suit, ca. 1975. Polyester. University of Rhode Island, URI 1985.08.03a, b, c. Donor: Frankie Welch

in the sense of a shared geographical space, but that Native American history and culture belong to all Americans [18, 40]. By the mid-seventies, a rising excitement in the nation's approaching Bicentennial coincided with the downswing of the counterculture and ethnic fashion trends. Welch moved on to other sources of inspiration, such as designing a series of reproductions of historical presidential campaign and inauguration scarves [18]. Nevertheless, products featuring the Cherokee Alphabet design can be found in photos of Welch's stores as late as 1977, and in promotional materials as late as 1983 [9].

Aside from the education fund, Welch never appears to have been directly involved in activism related to Native American interests. As she states in an interview in 1984, "I have not been an activist so to speak [...] I more or less don't like the Jane Fonda approach, shall we say?" [41]. She may not have been marching in protests, but she provided fashion items imbued with messages for those who were active in political movements. By designing for both political parties, she was deemed neutral or bipartisan. One astute journalist, however, pointed out that "Mrs. Welch's only public display of partiality is to the Cherokee Indian" [28].

From the evidence presented, it would be difficult to judge Welch's initial act of designing the Cherokee Alphabet as cultural appropriation or appreciation using the chosen definitions. Instead, this study ultimately relies on the design's reception by the TCE holders. On the day of the scarf's debut, the Chief of the Eastern Band of Cherokee Indians sent Welch a congratulatory telegram [9]. According to Callahan [9], the design was also seen on garments worn by Principal Chief W. Keeler (Cherokee) (1949–1975) and Reverend Sam Hider (Cherokee), who created a workbook for learning to write the Cherokee syllabary [43]. Wilma Mankiller (Cherokee), the first female Principal Chief of the Cherokee Nation (1985–1995), elected to wear the Cherokee Alphabet scarf [9]. It is not known how it was received by less prominent Cherokees.

LaDonna Harris, the Native American rights activist who was Comanche rather than Cherokee, wore several ensembles made of Cherokee Alphabet fabric to important social events, as well as for a television appearance [9]. She noticed that the design piqued people's curiosity, and she used it as a conversation starter to educate them on Native American issues [9]. While the Cherokee Alphabet design was clearly popular with non-Native consumers, there is little evidence as to whether it influenced them beyond mere esthetics to consider serious issues, past and present, faced by Native Americans. Harris, however, still views Welch and her designs in a positive light. In the foreword to Callahan's book, published in 2022, she explains that among the rampant appropriation of Native American cultures at the time, Welch and her design stood apart, implying that the design is an act of cultural appreciation [24].

Welch's other Native American-inspired designs also appear to have been well-received by Native Americans. Harris and Wauhilla La Hay (Cherokee), a prominent journalist in Washington D.C., both wore Welch's Native American-inspired jewelry designs [9]. Solomon McCombs (Muscogee Creek), who had assisted with the Indian Summer fashion show and whose art Welch hung in Duvall House, attended the release party for *Indian Jewelry* in a tie with the Roadrunner design [9]. A review of Welch's book *Indian Jewelry* in the Native American newspaper *Wassaja*, however, was less favorable: "Ms. Welch's ideas as to how to wear Indian jewelry would not be acceptable to most Indian women (and men)" ([3] p. 13).

If the Cherokee Alphabet had been produced today, and for the same reasons, it would likely be scrutinized for cultural appropriation. Is it useful to label past designs as cultural appropriation or appreciation given changing attitudes? What may be more constructive is to acknowledge and understand why a marginalized group feels that a design *inspired* by their culture *appropriates* their culture. Contemporary designers should try to avoid this by including cultural stakeholders in the process, respecting their viewpoints, and compensating them fairly. An effective exercise for students of fashion focusing on design, business, history, and curation, is to discuss both modern and historical cases using the definitions provided in this case study. Considering the nuances of ethics in fashion prepares students for a real-life encounter with such a situation and provides them with the tools to educate others about cultural appropriation. In following this model, our approach has been to try to respectfully, yet realistically, understand the case from a variety of viewpoints, and to generate constructive discussion.

Note: For the most comprehensive account of Frankie Welch's life and work, see Ashley Callahan's *Frankie Welch's Americana* (Athens: University of Georgia Press, 2022).

References

1. About the nation (2019) Cherokee Nation. https://www.cherokee.org/about-the-nation/history/. Accessed 31 August 2022
2. Aldridge D (1985) Designing woman: Celebrities prize Frankie Welch creations. Colorado Springs Gazette: G5
3. Arnet C (1974) Book notes. Wassaja 2(2):13. http://www.americanindiannewspapers.amdigital.co.uk.uri.idm.oclc.org/Documents/Details/SNRC_WSJA_1973-1977_ED010. Accessed 14 June 2022
4. Ball N (1969) She's scarf queen of Washington. Arizona Republic, 65
5. Blue blood is Indian blood (1914) The Boston Globe, 58. https://www-proquest-com.uri.idm.oclc.org/historical-newspapers/blue-blood-is-indian/docview/502604889/se-2?accountid=28991. Accessed 11 June 2022
6. Bozek DD (1976) Frankie Welch of Virginia – Southern designer. Press release, Frankie Welch of Virginia
7. Bozek DD (1976) Frankie Welch sees design installed in Smithsonian. Press release, Frankie Welch of Virginia
8. Butterfield M (1968) Technology in fashion: A rare bit of Welch. Women's Wear Daily 117(5):23
9. Callahan A (2022) Frankie Welch's Americana: Fashion, scarves, and politics. University of Georgia Press, Athens
10. Cattien J, Stopford RJ (2022) The appropriating subject: Cultural appreciation, property and entitlement. Philos Soc Criticism.https://doi.org/10.1177/01914537211059515. Accessed 8 June 2022
11. Ceremony draws distinguished individuals (1992) Journal Record. https://www-proquest-com.uri.idm.oclc.org/newspapers/ceremony-draws-distinguished-individuals/docview/259339198/se-2?accountid=28991. Accessed 1 June 2022
12. Cherokee wampum (1968) WWDeadline. Women's Wear Daily 116(79):28–30. https://www.proquest.com/magazines/wwdeadline/docview/1564887574/se-2?accountid=28991. Accessed 14 June 2022
13. Chrisman-Campbell K (2022) Red, white, and blue on the runway: The 1968 White House fashion show and the politics of American style. Kent State University Press, Kent
14. Churchman D (1982) Personal service draws VIPs to chic D.C. clothing shop. The Christian Science Monitor. https://www.proquest.com/newspapers/personal-service-draws-vips-chic-d-c-clothing/docview/1038443615/se-2?accountid=28991. Accessed 11 June 2022
15. Cortés CE. (2013) Five civilized tribes. In: Cortés CE (ed) Multicultural America: a multimedia encyclopedia. Sage Publications, Thousand Oaks, pp 882–84. https://doi.org/10.4135/9781452276274.n324

16. Culture (2019) Cherokee Nation. https://www.cherokee.org/about-the-nation/culture/. Accessed 3 September 2022
17. Deloria PJ (1998) Playing Indian. Yale University Press, New Haven
18. Frankie Welch Archive. Historic Textile and Costume Collection, University of Rhode Island, Kingston
19. Frankie Welch has designs on five presidents (1984) Goodlife, 5–6
20. Frankie Welch papers—Accession 519 (n.d.) Louise Pettus Archives and Special Collections, Winthrop University. https://digitalcommons.winthrop.edu/manuscriptcollection_findingaids/1128. Accessed 11 June 2022
21. Givhan R (1998) Culture splash. The Washington Post:F03. https://www.washingtonpost.com/archive/lifestyle/1998/04/12/culture-splash/4ae01048-9c90-48b8-b766-9160514e36b9/. Accessed 23 May 2022
22. Givhan R (2015) The battle of Versailles. Flatiron Books, New York
23. Gover K (2011) Foreword: A new kind of Indian art. In: Sims LS, Lowe TT, Smith PC (eds) Fritz Scholder: Indian/Not Indian. Prestel, New York, pp 9–12
24. Harris L (2022) Foreword. Callahan A (auth) Frankie Welch's Americana: Fashion, scarves, and politics. University of Georgia Press, Athens, pp ix–x
25. Hilleary C (2018) Going "native": Why are Americans hijacking Cherokee identity? VOA News. https://www.voanews.com/a/going-native-why-are-americans-hiijacking-cherokee-identity/4495119.html. Accessed 14 June 2022
26. Kopytek BA (2010) M. Goldwater & Sons, Phoenix, Arizona. The department store museum. http://www.thedepartmentstoremuseum.org/2010/09/m-goldwater-sons-phoenix-arizona.html. Accessed 11 June 2022
27. Kramer K, Calderín J, Kropa MM, Metcalfe J (2015) Native fashion now: North American Indian style. Peabody Essex Museum and Delmonico Books, Salem, MA and Munich
28. MacPherson M (1968) If you think a political button isn't enough… The New York Times, 54
29. Martin JW (1996) "My grandmother was a Cherokee princess": Representations of Indians in Southern history. In: Bird SE (ed) Dressing in feathers: The construction of the Indian in American popular culture. Westview Press, Boulder, pp 129–147
30. Meserve CF (1896) The Dawes commission and the five civilized tribes of Indian territory. Office of the Indian Rights Association, Philadelphia. http://digital.libraries.ou.edu/cdm/ref/collection/cornish/id/1467 Accessed 12 June 2022
31. Mihesuah DA (1996) American Indians: Stereotypes & realities. Clarity Press, Atlanta
32. Miss Barnett is bride of Pfc. Welch; ceremony performed in Rome, Ga (1944) Greenville News:2. https://www.newspapers.com/clip/16755776/the-greenville-news/. Accessed 25 May 2022
33. Oluo I (2016) When we talk about cultural appropriation, we're missing the point. The Establishment. https://medium.com/the-establishment/when-we-talk-about-cultural-appropriation-were-missing-the-point-abe853ff3376. Accessed 11 June 2022
34. Oluo I (2019) So you want to talk about race. Seal Press, New York
35. Saunders P (1975) Fashion: She wears three hats...not to mention scarves, dresses. Boston Herald American
36. Smith E (1979) Frankie Welch on designing the corporate image. Corporate News and Finance. The Business Review of Washington: 5,14
37. Smith J (1985) Alexandria designer puts wrap on presidential neck. The Alexandria Gazette, B-6
38. Smith P (2011) Monster love. In: Sims LS, Lowe TT, Smith PC (eds) Fritz Scholder: Indian/Not Indian. Prestel, New York, pp 25–36
39. Vézina B (2019) Curbing cultural appropriation in the fashion industry. Center Int Govern Innov Papers 213:1–16
40. Welch F (1973) Indian jewelry: How to wear, buy and treasure America's first fashion pieces. EPM Publications Inc., McLean, VA
41. Welch F (1984) Interview with Frankie Welch. Winthrop University Oral History Program. https://digitalcommons.winthrop.edu/oralhistoryprogram/243. Accessed 25 May 2022

42. Welch brings American themes to life on scarves (1992) Journal Record. https://www.proquest.com/newspapers/welch-brings-american-themes-life-on-scarves/docview/259488893/se-2?accountid=28991. Accessed 23 May 2022
43. Writing Cherokee: A syllabary writing practice book (n.d.) Archive record. Museum of the Cherokee Indian. https://cherokeemuseum.pastperfectonline.com/archive/E6A7B5F5-540C-4B82-9750-555106147714. Accessed 14 June 2022
44. Yotka S (2022) 'In America: A lexicon of fashion' gets a refresh with more than 70 new ensembles at The Metropolitan Museum of Art. Vogue. https://www.vogue.com/article/in-america-a-lexicon-of-fashion-refesh-metropolitan-museum-of-art. Accessed 14 June 2022

Cultural Fashion a Matter of Human Rights. Cultural Misappropriation as a Human Rights Violation. What is Wrong with the Fashion and Legal Systems and How Can We Make It Right? The Oma Ethnic Group of Nanam Village in Laos Have an Answer

Monica Boța-Moisin

Abstract At the intersection of law, social justice, cultural identity, and sustainability in fashion lies a topic that sparks conversations, divides opinions, ideologically separates the West from the rest, fashion from craftsmanship, and ultimately us as a global society. Cultural appropriation has been part of the social fabric since times immemorial. In legal vocabulary, *to appropriate* means to take something from the rightful owner without their consent. If we were to incriminate all instances of cultural appropriation, we could very well incriminate acculturation, cultural assimilation, or cultural syncretism. So it is natural and welcome to ask ourselves what is the difference between cultural appropriation and cultural appreciation, to observe different interpretations based on differences in cultural context between Europe, South-East Asia, and Latin America, for example, and to feel overwhelmed by the complexity of the topic and conscious of the risk of oversimplifying it. What is not natural, is to support a practice of extracting meanings and aesthetics from Traditional Cultural Expressions (TCEs), without the consent of the creators and custodians of these meanings and expressions, use them for commercial purposes–without compensation or sharing of benefits arising from their commercialization–or use them in ways that are derogatory, inappropriate or ignorant. The access to and use of TCEs without consent, credit and compensation, or the misuse of TCEs in derogatory, inappropriate, or ignorant ways should be understood as *cultural misappropriation,* irrespective of the cultural and geographical context where this occurs, and irrespective whether the applicable law in that jurisdiction incriminates or not such practices. Why do I say this? Because this is not just an issue about what fashion designers can or cannot do, should or should not do. This is an issue of cultural continuity and survival, biocultural diversity, and respect of human rights. Both the fashion system and the intellectual property law system are built on eurocentrism, and allowing them to continue operating on this premise will further sustain a *culture*

M. Boța-Moisin (✉)
Cultural Intellectual Property Lawyer and Cultural Sustainability Consultant, Oradea, Romania
e-mail: monicamoisin@culturalintellectualproperty.com

M. Á. Gardetti and R. P. Larios-Francia (eds.), *Sustainability Challenges in the Fashion Industry*, Sustainable Textiles: Production, Processing, Manufacturing & Chemistry,
https://doi.org/10.1007/978-981-99-0349-8_6

of otherness and systemic discrimination in many forms: direct discrimination, indirect discrimination, and positive discrimination. In today's socio-cultural context, when cultural sensitivity, protection and promotion of cultural diversity, and cultural diplomacy are paramount to designing a future of well-being for all, it is essential to challenge *desuet* definitions, concepts, practices, and tools that the fashion and the legal systems operate on, and design new ones in partnership with indigenous peoples, ethnic groups, and local communities who have, for far too long, suffered from systemic discrimination and exclusion. The way forward is to go beyond criticizing the system and offer alternative solutions: models, concepts and tools that can be discussed, tested, and improved. This case study presents the Oma Traditional Textile Design Database©, a cross-disciplinary solution designed as a response to the fashion company Max Mara using a textile design of the Oma ethnic group of Nanam Village in Laos without consent, credit, and compensation. This solution, modeled on the 3Cs' Rule: Consent. Credit. Compensation© Framework by the Cultural Intellectual Property Rights Initiative®, is meant to be a best practice model for systems change in the fashion industry and the intellectual property law system, which are both built on a colonial right of access to, and use of, *Cultural Fashion* expressions without benefit sharing and without contemplating the question of consent, all in the name of the principle of the "*common patrimony of mankind*". The model challenges the understanding of the notion of "*public domain*" and draws attention to the value of traditional textile knowledge for re-designing a fair, equitable, and sustainable fashion system.

Keywords Cultural misappropriation · Cultural intellectual property · Traditional cultural expressions · Cultural fashion · Equity · Cultural sustainability · Intangible cultural heritage · Public domain

1 Introduction

To write about cultural appropriation in fashion and textiles is not an easy endeavor. No matter the perspective—legal, sociological, ethnographic, or economic—approaching the topic is like performing a dissection on a hybrid life form. The *cultural appropriation* phenomenon is plurivalent. It takes shapes and forms that are fundamentally influenced by cultural context, power relations, social norms, politics, economy, and—often forgotten—Time. Its anatomy and effects span across a broad spectrum: from appreciation, acculturation, assimilation, syncretism, and multiculturalism, to misappropriation, exploitation, cultural dilution and—at its extreme—*cultural genocide* [10]. Historically, *cultural appropriation* is an effect of the diffusion of culture, and in many cases, the borrowing and assimilation of cultural practices, lifestyles, worldviews, beliefs and expressions has been regarded as a form of socio-cultural innovation and evolution. Forms of cultural appropriation such as cultural assimilation, acculturation or syncretism, have not always been a one-way process that involved the less powerful submitting to the powerful, but, in many instances, historically, such *cultural appropriation* has been a multi-layered and complex exchange of cultural influences through processes of power, loss, and production [1]. In such cases *cultural relativism* has been prevailing as an attitude toward cultural practices, cultural expressions, and knowledge systems, as opposed to the intensified episodes of colonialism governed by ethnocentrism [27]. So *cultural appropriation* is not always *bad*. In some cases, it is a fruitful process based on cultural relativism and mutually-beneficial exchange, whilst in others, it is an unethical and inequitable exercise of extractivism, oppression, and exploitation of cultural identity. Both extremes co-exist.

To dissect *cultural appropriation* in the context of fashion and textiles requires a multi-layered, cross-disciplinary approach, cultural sensitivity, diplomacy, and a plurality of perspectives. It is not a subject that can be framed simplistically or devised into lists of "*do's and don'ts*". To attempt to draw a line between *appreciation* and *appropriation* is not only futile, but an erroneous narrative that "rests on a false distinction between the two" [6], risking to further polarize the discussion instead of designing a strategy around respect of cultural rights (which are human rights), socio-cultural and economic well-being, and knowledge partnerships [7]. Unquestionably, we have to start from the premise that Ethnocentrism, and Eurocentrism in particular, lie at the heart of the common definition of fashion [15] and at the core of the fashion system *modus operandi*. Niessen rightfully points out that Georg Simmel's (1858–1918) thesis that "fashion does not exist in tribal and classless societies" has been a dominant and unchallenged perspective in fashion narratives over the past century. The *systems of dress of the other* (non-European cultures and peoples) have systematically been identified through static and discriminative terms which evolved from primitive/folk/exotic dress/costume/attire, into traditional/ethnic fashion, subject to the need for design interventions by fashion designers or "reimagining tradition" to make it relevant for contemporary fashion markets [12]. Since the creation of fashion as a system, traditional textile cultural expressions have systematically been regarded as non-fashion, and subsequently, the creators of such expressions–artisans, craft custodians, indigenous peoples, ethnic groups, and local communities–as subordinated and inferior in status to fashion designers and other fashion stakeholders [5].

Against this backdrop, a process of systems-change has been catalyzed recently. The past decade has been one of raising consciousness and focusing on redesigning the fashion system to address systemic inequity and eliminate exploitative, extractive, and disrespectful practices and behaviors. And as this process started and is reaching new dimensions, we have come to understand that the issue of *cultural appropriation* in fashion and textiles is cross-systemic, and two of the systems that require activation for systemic change in order to eliminate *cultural appropriation* in its exploitative form (i.e., *cultural misappropriation)* are the intellectual property law system and the international trade system, the latter comprised of thousands of unilateral, bilateral, multilateral, and regional agreements among more than two hundred nations. In addition, a timely strategy to address *cultural appropriation* in fashion and textiles and grasp the subtle distinctions that are necessary must take into account the evolution of the role of culture [9] and political contextualization of culture [13] in sustainable development.

In this chapter, I will share my position on **framing** *cultural appropriation* in fashion and textiles based on lessons learned, and tools applied, in the context of designing a response to the unconsented use of Traditional Textile Cultural Expressions of the Oma People of Nanam Village in Laos, by an Italian fashion company with international trade operations, for commercial purposes, hereinafter referred to as *The Oma Case*. As a cultural intellectual property legal practitioner with expertise in textile value chain management and intersectionalities of the legal and fashion systems, I am applying the lens of decolonizing the fashion and legal systems, and that of systemic change, to identify what constitutes *cultural misappropriation* on the cultural appropriation spectrum in fashion and textiles, and why the *cultural appropriation debate* needs setting boundaries and applying frameworks to shape knowledge partnerships with conventionally *non-fashion actors* such as the Oma ethnic group of Nanam Village in Laos.

The *Oma Case* is the first response in a *cultural appropriation* case in the fashion and textiles industry that has been developed based on a multi-stakeholder, systemic change driven approach, engaging governmental officials, civil society organizations, legal, ethnographic and cultural sustainability in fashion expertise, and resulting in a replicable model across geographies and cultural contexts, designed to stimulate fair and equitable collaborations between fashion industry stakeholders and indigenous peoples, ethnic groups, and local communities. This model is the Oma Traditional Textile Design Database© [17], part of the Lao Traditional Designs platform [23]. I was involved in the development of The *Oma Case* between December 2019 and June 2021 as a legal expert and cultural sustainability consultant, throughout what the consortium partners[1] in this project consider to be a first stage of a long-term systems-change process that starts from the premises that (1) *Traditional Cultural Expressions are not in the public domain* and that (2) the Oma People, as custodians of their traditional cultural expressions (TCEs), *have the right to maintain, control, protect and develop their cultural heritage, Traditional Knowledge and Traditional Cultural Expressions* (art. 31 para. 1 of the [25]), and have therefore a right to decide with respect to how the elements, symbols, and aesthetics related to their *Cultural Fashion* are accessed, used, developed and commercialized [4].

Throughout this chapter, I refer to *Cultural Fashion* as what Simmel defined as "non-fashion", what Jansen [12] calls "Other fashions", what Niessen [15] refers to as "indigenous clothing systems" and what, in legal terminology, the World Intellectual Property Organization (WIPO) and the 2005 UNESCO Convention on the Protection and Promotion of the Diversity of Cultural Expressions define as *Traditional Cultural Expressions* (TCEs). TCEs include "jewelry, textiles, traditional craftsmanship, and artworks (among other expressions) which embody know-how, skills, or practices that have sustained for a long time, being transmitted from generation to generation in communities, tribes or cultural groups, and are often part of the community's cultural and/or spiritual identity" [3]. **This special category of textile expressions** (not to be

[1] The Traditional Arts and Ethnology Centre (TAEC) in Luang Prabang, Lao PDR, The Oma People of Nanam Village, Boun Tai District, Phongsaly Province, Lao PDR, and the Cultural Intellectual Property Rights Initiative®.

generalized as handcrafted textiles or simply *handmade textiles*), **are created based on** "*knowledge, know-how, skills, and practices that are developed, sustained, and passed on from generation to generation within a community, often forming part of its cultural or spiritual identity*"–i.e. the legal terminology for Traditional Knowledge [28]–in other words, **local and regional traditional wisdom in fashion and textiles.**

2 Cultural Fashion at the Intersection of Human Rights and Sustainable Development

For a very long time, cultural capital and culture itself have taken a peripheral role in the work on Human Rights and theories on Sustainable Development. The most important UN Human Rights Instruments, the UN Declaration of Human Rights [26] and the *International Covenant on Economic, Social and Cultural Rights* [11] include references to the protection of individual culture-related rights, as opposed to collective rights. For example, art. 27 para. 1 of the UDHR (1948) states that "Everyone has the right freely to participate in the cultural life of the community" and para. 2 states that "Everyone has the right to the protection of the moral and material interests resulting from any scientific, literary or artistic production of which he is the author." With no mention of collective rights such as cultural intellectual property rights derived from the custodianship of Traditional Knowledge (TK) and Traditional Cultural Expressions (TCEs), at that point in time–1948–the superiority of intellectual property rights (individual rights) over the knowledge and creations of collectivities such as indigenous peoples, ethnic groups, and local communities, seems to be the international consensus. The 1966 ICESCR reinforces the provisions of art. 27 of the UDHR (1948) but recognizes a vital right in its art. 1–**the right of self-determination of all peoples**, by virtue of which they freely determine their political status and *freely pursue their economic, social, and cultural development*. In my opinion, this is the fundamental human right that constitutes the basis of the recognition of collective cultural intellectual property rights for custodians and transmitters of Traditional Knowledge and Traditional Cultural Expressions, i.e., indigenous peoples, ethnic groups, and local communities. It wasn't until 1982, at the UNESCO World Conference on Cultural Policies in Mexico City, that the link between culture and sustainable development was explicitly made [9], and an extension to collective-related aspects of cultural rights was captured in a groundbreaking definition[2] of culture adopted by the Conference which led to a *solemn agreement* that *Cultural Identity*, the *Cultural Dimension of Development*, *Democracy*, *Cultural Heritage*, *Artistic and Intellectual Creation* and *Art Education* are among the principles that should govern cultural policies [14].

[2] "In its widest sense, culture may now be said to be the whole complex of distinctive spiritual, material, intellectual and emotional features that characterize a society or social group. It includes not only the arts and letters, but also modes of life, the fundamental rights of the human being, value systems, traditions and beliefs;" (UNESCO, 1982).

The highlight of UNESCO work toward the protection and promotion of what I refer to as *Cultural Fashion* in this chapter is the 2005 Convention on the Protection and Promotion of the Diversity of Cultural Expressions [8, 24]–a legally binding international treaty that explicitly references the relationship between cultural diversity and sustainable development and recognizes the "*importance of Traditional Knowledge as a source of intangible and material wealth, and in particular the knowledge systems of Indigenous Peoples, and its positive contribution to sustainable development, as well as the need for its adequate protection and promotion*". In my opinion, this applies to the creation, production, distribution of and access to a diversity of cultural expressions, including cultural goods, services, and activities, among which traditional textiles and the Traditional Knowledge associated with creating them.

As of recent date, the relevance of Traditional Knowledge Systems, which include *Cultural Fashion,* has been explicitly recognized in the Post-2020 UN Biodiversity Framework. One of the "*21 action-oriented targets for urgent action for the decade to 2030*" subject to negotiation and adoption at COP-15 (expected at the end of 2022) is to "Ensure that relevant knowledge, including the traditional knowledge, innovations and practices of Indigenous Peoples and local communities with their **free, prior, and informed consent**, guides decision-making for the effective management of biodiversity." [20]. * At the time of publication the Kunming-Montreal Global Biodiversity Framework has been adopted. It contains four goals and 23 action-oriented targets.

3 Framing Cultural Misappropriation in Fashion and Textiles Through the Lens of the 3Cs' Rule: Consent. Credit. Compensation©

Against this backdrop, I define *Cultural Misappropriation* in fashion and textiles as:

> the act of accessing, and using of, forms of Traditional Knowledge (TK) or Traditional Cultural Expressions (TCEs) from a people, community, tribe, or group that shares a common cultural identity, values, worldviews and history, and using them in a **different context**, for commercial purposes, without the **authorization**, **acknowledgement** and **compensation** of the TK and TCE custodians.

In my view, *Cultural Misappropriation* occurs **the moment the free prior and informed consent has not been given** by the source community, the custodians of TK and TCEs, either because the free prior and informed consent process *has not been initiated*, which is the standard case, or because the Traditional Knowledge or Traditional Cultural Expression have been accessed and used despite the source community explicitly not giving consent for the intended use. The principle of Free Prior and Informed Consent (FPIC) is derived from the **human right to self-determination of peoples** consecrated in–among other legal instruments and as mentioned previously–art. 1 of the International Covenant on Economic, Social, and Cultural Rights [11].

Whether there is *intent* to misappropriate or simply *ignorance*, does not make a difference. What makes a difference, and here I support Nute's theory on "*significance" and* "*recognizability*", is that the object of misappropriation, the form of TK or the TCE that has been used, needs "to be both culturally significant in its original context and also clearly recognizable as deriving from that source in its new context" [16].

In my experience, *Cultural Misappropriation* occurs in two different forms as a result of the **unconsented access to, and use of, TK and TCEs**. These two manifestations are:

(1) **misuse**–which includes derogatory, defamatory, discriminatory, disrespectful, unethical, culturally diluting, or socio-culturally harmful use–simply put, a use that is not compliant with the customary laws of the source community or that the source community does not approve of; and,
(2) **unfair use**–meaning that the use, for example, the sale of the derived product created based on the form of Traditional Knowledge or the Traditional Cultural Expression, generated benefits for the third party and there are no Mutually Agreed Terms (MAT) for benefit-sharing with the source community, depriving the community of an opportunity or an economic gain that in fairness and equity they would be entitled to, or, that the third party claims paternity of the design/technique/cultural expression by labeling the product with their own trademark or posing as original creator of the product without acknowledging the source community.

These two forms of *Cultural Misappropriation* in fashion and textiles can co-exist in the same case–for example when a sacred Traditional Cultural Expressions is used in a derogatory manner and this derogatory use also generates profits for the cultural misappropriator, like in the case of *The Navajo Nation versus Urban Outfitters (2016)*–use of the Navajo name on underwear, jewelry and flasks–but it is also enough for one of the manifestations to occur, like in the case of *The Mixe Community of Santa Maria Tlahuitoltepec vs. Isabel Marant (2015)* where the misappropriation was in the form of *unfair use.*

In *The Oma Case, Cultural Misappropriation* occurred in the form of *unfair use* of Oma Traditional Textile Expressions, for commercial purposes, without the consent, credit, or compensation of the community. Throughout the work conducted on *The Oma Case*, in the deliverables for the project (i.e., Database texts, White Paper Report, Workshop presentations), and in all public communications, we referred to this manifestation as *textile design plagiarism* due to the replica of the Oma designs on the Max Mara items of clothing looking identical to the traditional embroidery and appliqué designs of the Oma People [4]. As stated in the White Paper Report of *The Oma Case,* "*the Max Mara replica patterns were printed on the fabric, not hand-embroidered or hand sewn*", unlike the authentic Oma Traditional Textile Expressions, "*there was no garment label acknowledging the textile designs as belonging to the Oma. The garment labels referenced the Max Mara brand name and trademark to communicate the source of the products, generating a risk of confusion and cultural dilution of Oma intangible cultural heritage.*" Max Mara digitally duplicated and printed replicas of Oma Traditional Textile Expressions onto various items of clothing, thus reducing the Oma painstaking traditional motifs to factory-produced patterns, in the same colors, composition, and shapes. The substantial identity of colors, shape, and size of the Max Mara replicas to the authentic Oma Traditional Textile Expressions create the same overall impression to a consumer, passing Nute's "*significance" and "recognizability*" test. Based on the data collected through interviews with the Oma People of Nanam Village in Lao PDR, it can also be said that there was also *Cultural Misappropriation* through *misuse* of the Oma Traditional Textile Expressions due to the placement of the replicated textile patterns on the Max Mara items of clothing, some of which had been extracted from other types of items of Oma Cultural Fashion and decontextualized through their application on blouses and skirts (Fig. 1).

4 Lack of Legal Protection Against Cultural Misappropriation in Fashion and Textiles and Approach in *the Oma Case*

Why the proposal to frame *Cultural Misappropriation* in fashion and textiles through the lens of the 3Cs' Rule: Consent. Credit. Compensation©?

The access to and use of Traditional Knowledge and Traditional Cultural Expressions are not regulated by international law. To regulate *Cultural Misappropriation* in its two forms described above, namely the *misuse of, and unfair use of, significant and recognizable forms of Traditional Knowledge and Traditional Cultural Expressions,* requires designing *new legal frameworks, tailor-made* for this type of cross-generational knowledge and cultural capital. Cultural capital is "an asset which embodies, stores, or gives rise to cultural value in addition to whatever economic value it may possess" [21]. In legal terminology, we call this *sui-generis legislation.* Whilst some *sui-generis* regimes for the Intellectual Property Protection of Traditional Knowledge and Traditional Cultural Expressions exist at national and regional levels around the world [29], the legislative negotiations in this space are difficult, slow, and subject to many conflicting interests between parties. At the level of the WIPO Intergovernmental Committee on Intellectual Property and Genetic Resources, Traditional Knowledge, and Folklore (IGC), text-based negotiations on an international legal instrument for the protection of Traditional Knowledge (TK) and Traditional Cultural Expressions (TCEs) have been ongoing since 2010. As explained in the Introduction, incriminating *Cultural appropriation* is not a feasible, nor a desirable option, and what the fashion and textiles industry needs at this point in time is **a framework for best-practices in developing and sustaining knowledge partnerships for textile production, consumption, circularity, and safeguarding of biocultural diversity, with Indigenous Peoples, ethnic groups,**

Fig. 1 Oma Elder learning about the Cultural Fashion plagiarism by Max Mara. Copyright The Oma of Nanam Village in Laos and TAEC, 2021

and Local Communities. This is why the 3Cs' Rule: Consent. Credit. Compensation© Framework is relevant. Inspired by article 31 of the 2007 UN Declaration on the Rights of Indigenous Peoples [25], the rules of attribution applicable in case of copyright protection, and the Nagoya Protocol on Access to Genetic Resources and the Fair and Equitable Sharing of Benefits Arising from their Utilization (entered into force in 2014), **this framework is an extra-legal tool** developed by the Cultural Intellectual Property Rights Initiative® to guide best practices for drawing inspiration from cultural heritage and engaging in culturally sustainable collaborations with custodians of Traditional Knowledge and Traditional Cultural Expressions. It is meant to bring craft communities and fashion stakeholders in a space of mediation, negotiation, co-creation, and joint decision-making for a paradigm shift in the fashion industry, a shift to cultural relativism and co-existence of a plurality of worldviews that will reshape our relationship with fashion production and fashion consumption for the benefit of all forms of life: fashion for biocultural diversity.

The **Oma Traditional Textile Design Database©** is designed on the Framework of the 3Cs' Rule: Consent. Credit. Compensation© with the goal of enabling the Oma People to grant access to visual elements of their *Cultural Fashion* and associated Traditional Knowledge to third-party users under specific conditions, on the basis of Free, Prior, and Informed Consent (FPIC) and fair access and benefit sharing (ABS) mechanisms designed through Mutually Agreed Terms (MAT) (Fig. 2).

The topic of legal protection against *misuse of, and unfair use of, significant and recognizable forms of Traditional Knowledge and Traditional Cultural Expressions* in fashion and textiles has been largely explained in the deliverables of *The Oma Case*–in the publicly accessible *Rights* section of the Oma Traditional Textile Design Database© [18], in the White Paper Report of *The Oma Case–Documenting Traditional Cultural Expressions: Building a Model for Legal Protection Against Misappropriation and Misuse with the Oma Ethnic Group of Laos* [4] and on the website of the Traditional Arts and Ethnology Centre (TAEC) in Luang Prabang, Lao PDR, on the page: *What Do You Need To Know About Cultural Intellectual Property?* [22].

This ample exercise of informative content creation related to the legal aspects of *The Oma Case* was aimed at increasing legal literacy on the topic among a variety of fashion ecosystem stakeholders and sharing this information with the public at large for awareness and sensitization. The informative content listed above is freely accessible to the public and a *Traditional Design Database Template* is available for use by any indigenous peoples, ethnic groups, and local communities worldwide, free of any royalty fee. The uniqueness of the solution in *The Oma Case* lies in the fact that, because of its unique structure and the original selection and arrangement of content, the Oma Traditional Textile Design Database© is protected by copyright as a **collection**, under art. 2 para. 5 of the Berne Convention for the Protection of Literary and Artistic Works, ratified by the Government of the Lao People's Democratic Republic in 2011 and entered into force, with respect to the Lao People's Democratic Republic, on March 14, 2012 [2].

It is important to note that the main legal argument of Max Mara legal representatives in *The Oma Case* was that the contested Max Mara design was an original interpretation of the Max Mara "*style team*", "*inspired from a pattern of a vintage fabric, which is..*", due to a lack of intellectual property protection, "*in the public domain*". The notion of *public domain* is vastly qualified as *eurocentric* [19]. In essence, the *public domain* is an intellectual property (IP) law-related concept which is said to consist of intangible assets that are not subject to exclusive Intellectual Property Rights and which are, as a consequence, *freely available to be used or exploited by any person.*

The notion of *public domain* is increasingly challenged and scrutinized. From an intellectual property law perspective, Traditional Knowledge and Traditional Cultural Expressions were left outside the sphere of Intellectual Property protection, being considered *res nullius* (not the object of rights of any specific subject, in particular related to the notion of private property), and therefore in the *public domain.*

Fig. 2 Presenting the Oma Traditional Textile Design Database© to the Village Chief. Copyright The Oma of Nanam Village in Laos and TAEC, 2021

However, the fact that they are not protected by Intellectual Property due to their unique anatomy does not mean that Traditional Knowledge and Traditional Cultural Expressions are not, or should not be, subject to any laws. According to views of representatives of indigenous peoples, ethnic groups, and local communities, these intangible cultural goods are regulated by customary laws rather than by Intellectual Property law, and some communities may consider Traditional Knowledge as falling outside of the Intellectual Property concept of *public domain* [19]. It has been recorded that "WIPO participants have expressed the view that the *public domain* is not a concept that is recognized by Indigenous Peoples and Local Communities (IPLCs) and that expressions of folklore", such as *Cultural Fashion* for example, "could not have entered the *public domain* if they were never protected as Intellectual Property" [19]. In this regard, a great merit of *The Oma Case* is the creation of the *#notpublicdomain* informative Campaign aimed to increase awareness, deliver tools, and engage with designers and fashion brands to stop *Cultural Fashion* plagiarism. The deliverables of this Campaign are not only a tool for raising awareness of the desuetude, eurocentrism, and social injustice of the notion of *public domain* and its lack of ethics as an argument in a case of *Cultural Misappropriation* in fashion and textiles, but also a tool for education and legal literacy on the subject (Figs. 3, 4, 5, 6).

A secondary legal argument of Max Mara in *The Oma Case* was the problem of legal representation of the Oma People of Nanam Village. It was a priority for us in

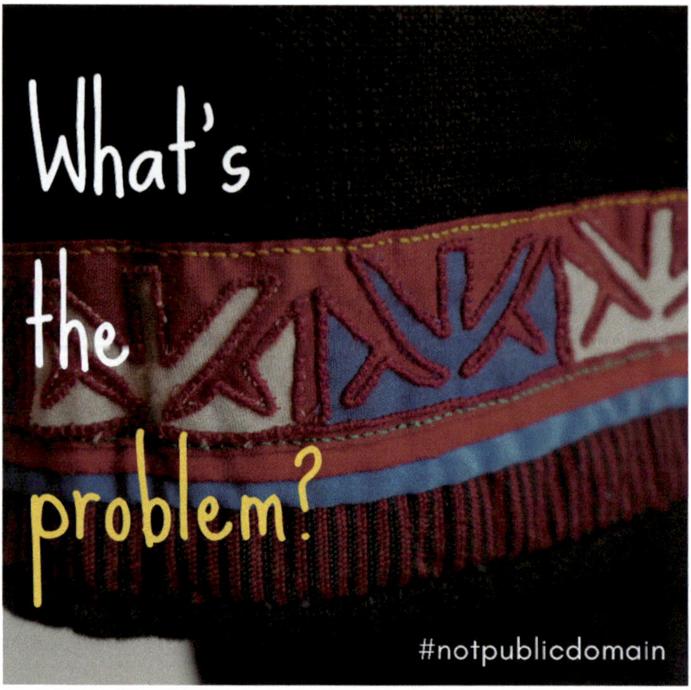

Fig. 3 Visual of the #notpublicdomain informative Campaign

designing the Oma legal response to the unfair use of their *Cultural Fashion* to ensure adequate legal representation of the Oma and have subjected the issue of legal representation itself to a process of Free Prior and Informed Consent which concluded with the signing of a *Representation Agreement*. The inequality of resources and access to professional support in a *Cultural Misappropriation* case in fashion and textiles is evidently in the detriment of the custodians of Traditional Knowledge and Traditional Cultural Expressions and no law will ever succeed to eliminate these systemic inequalities of the capitalist culture and economy which affect, without a doubt, the access of indigenous peoples, ethnic groups and local communities, as well as of artisans and craft custodians, to professional legal advice and representation. **The solution is therefore not to create new laws, but to conventionally design guidelines and best practices cross-disciplinarily**, so that instances of *Cultural Misappropriation* in fashion and textiles are reduced, and in time, eliminated, not by restricting access to, and use of, Traditional Knowledge and Traditional Cultural Expressions, but by incentivizing Knowledge Partnerships between fashion stakeholders and custodians of Traditional (textile) Knowledge and Traditional (textile) Cultural Expressions, who are de facto, *experts* in *Cultural Fashion*.

Fig. 4 Visual of the #notpublicdomain informative Campaign

5 Crafting a Future for Textile-Focused Knowledge Partnerships Between Fashion Stakeholders and *Cultural Fashion* Experts

The Oma Traditional Textile Design Database© is the central piece of the Oma legal response to the unfair use of their *Cultural Fashion*. It has been an objective of the consortium partners in the project to go beyond the classical legal response limited to exchange of documents between lawyers as legal representatives of parties with conflicting interests, and take a *proactive approach* based on cross-disciplinary teamwork and aimed at creating a tool for indigenous peoples, ethnic groups, and local communities to (1) **protect their Traditional Cultural Expressions** against *Cultural Misappropriation* in fashion and textiles by affirming custodianship over their Traditional Cultural Expressions and associated Traditional Knowledge, (2) get familiar and comfortable with the processes of Free Prior Informed Consent (FPIC), negotiation of Mutually Agreed Terms (MAT) and design of access and benefit-sharing mechanisms (ABS) **so that they can engage in fair and equitable negotiations and co-design projects with fashion industry stakeholders based on the 3Cs' Rule: Consent. Credit. Compensation© Framework**, (3) start creating **a private knowledge-center** of Intangible Cultural Heritage in the digital age to ensure their

Fig. 5 Visual of the #notpublicdomain informative Campaign

cultural continuity and cross-generational transmission of Traditional Textile Knowledge in light of the 2003 UNESCO Convention for the Safeguarding of the Intangible Cultural Heritage, and (4) have **a tool that will enable and facilitate commercial collaborations between craft custodians from indigenous peoples, ethnic groups and local communities and fashion and textile stakeholders** that works like a digital library and *look-book* containing written, audio, and visual information about the methods of textile creation, sourcing, processing and use of materials, the scope and uses of each item of dress, the symbolism and significance of the patterns and motifs, taboos and customary use associated with the textiles. Such a model is not only beneficial for indigenous peoples, ethnic groups, and local communities, but also for fashion stakeholders ranging from fashion education institutions, researchers and educators, to fashion designers, design teams, fashion communication strategists and fashion journalists, textile sourcing experts, and fashion innovators. **Fashion needs to embrace a plurality of worldviews, to explore and embrace culturally sustainable practices of creating and consuming fashion**–in the words of Doreen Robinson, UNEP's Head of Biodiversity and Land–" to take concrete steps to shift our unsustainable consumption and production patterns to ones that allow both people and the planet to thrive", by engaging all actors including the private sector, civil society, indigenous peoples, local communities, and individuals.

Fig. 6 Visual of the #notpublicdomain informative Campaign

A significant amount of knowledge has been garnered through the process of conceptualizing, creating, and testing the Oma Traditional Textile Design Database©, and the hope is that this collaborative model and the framework used will support communities and fashion companies to work together more equitably [4], developing collaborations based on Knowledge Partnership and benefit sharing instead of perpetuating the obsolete and ethnocentric view that *Cultural Fashion,* is "non-fashion". The most valuable lesson learned from this project is that indigenous peoples, ethnic groups, and local communities are **resourceful living treasures of culturally sustainable textile knowledge** who can help the fashion industry design sustainable models of textiles production and consumption (Fig. 7).

Through their exemplary and constructive response in a case of *Cultural Misappropriation* in Fashion and Textiles, the Oma People of Nanam Village in Laos significantly contribute to a **paradigm shift in the fashion system** where indigenous people, ethnic groups, and local communities are valued as knowledge partners across textile supply chains and fashion is redefined from the apanage of status and class identities to an expression of values, beliefs, and lifestyles, it promotes culture, ensures the preservation of generational skills, and combines meaning, esthetics, functionality, and sustainability by design. As highlighted in the White Paper Report of *The Oma Case,* Oma artisans from Nanam Village have expressed their willingness to collaborate with fashion companies and designers for creating textiles and

Fig. 7 Oma Elder and Textiles in her home in Nanam Village. Copyright The Oma of Nanam Village in Laos and TAEC, 2021

garments inspired by the Oma culture and aesthetic. They are ready to create textile embroidery and appliqué for national and international commerce. The quality of their work is outstanding and their skills and aesthetic unique. They hold knowledge to create Culturally Sustainable Fashion [4]. All that is left now is for fashion stakeholders to show interest in and commit to exploring collaborations with the Oma People. Commercial contracts and licensing agreements are legal business tools for the Oma to engage in collaborations as co-design and sourcing partners, or teach others their Traditional Textile Knowledge through workshops and immersive learning experiences. The introduction of this type of business-models in the fashion system will significantly contribute to fashion's impact on the *Sustainable Development Goals*, including poverty eradication by enabling the Oma People and other indigenous peoples, ethnic groups and local communities to valorize their textile knowledge and skills and generate revenue streams for the community from textile production and commercialization (SDG 1–No poverty), appreciation of cultural diversity and of culture's contribution to sustainable development (SDG 4–Quality Education), reducing economic inequalities and promoting meritocracy by empowering and promoting the economic inclusion of indigenous peoples, ethnic groups, and localcommunities in textile supply chains as Knowledge Partners (SDG 10–Reduced inequalities), strengthening global responsibility and enabling community resilience by safeguarding cultural heritage (SDG 11–Sustainable cities and communities), contributing to nurturing local ecosystems and biodiversity in regions stewarded by Indigenous Peoples, ethnic groups and Local Communities (SDG15–Life on land), and engaging in multi-stakeholder partnerships for strengthening the craft ecosystem globally (SDG 17–Partnerships for the goals).

Based on my professional experience and interactions with fashion system stakeholders, there is interest in, and commitment to, doing *the right thing*, but there is limited knowledge and resources to initiate and sustain such processes, to disrupt the "normal way of doing things in the fashion industry". When invited to do so and use the Oma Traditional Textile Design Database© to develop a culturally sustainable capsule collection, Max Mara refused for fear of backlash and interpretation of the collaboration as an indirect recognition of their wrongdoing (i.e., the *Cultural Fashion* textile design plagiarism). The Oma would have certainly not seen it like that. Ego and tactics are not part of their values and practices.

To change the fashion system to one of more humanity and care, we all need to work on conquering the fear of failure and the pride to admit a mistake. We are all learning and growing together as a result. By designing processes on the principles of consent, credit and compensation, conventionally agreed guidelines, and Mutually Agreed Terms (MAT), we are looking at a viable solution for eliminating *Cultural Misappropriation* in fashion and textiles.

References

1. Arroyo J (2016) Transculturation, syncretism, and hybridity. In: Martínez-San Miguel Y, Sifuentes-Jáuregui B, Belausteguigoitia M (eds) Critical terms in caribbean and latin american thought. New Directions in Latino American Cultures. Palgrave Macmillan, New York. https://doi.org/10.1057/9781137547903_12
2. Berne Convention for the Protection of Literary and Artistic Works (1979) Available at: https://wipolex.wipo.int/en/treaties/textdetails/12214. Accessed 28 June 2022
3. Boța-Moisin M (2022) The Cultural Fashion Agenda: legal protection for traditional crafts in the EU? The EU Commission says Yes. In: The Cultural Fashion Agenda. Newsletter. Available via LinkedIn. https://www.linkedin.com/pulse/cultural-fashion-agenda-legal-protection-traditional-eu-bo%25C8%259Ba-moisin/?trackingId=6NClVDpXS%2BiQAjmG%2BTVENA%3D%3D. Accessed 28 June 2022.

4. Boța-Moisin M, Gujadhur T (2021) Documenting traditional cultural expressions: building a model for legal protection against misappropriation and misuse with the oma ethnic group of laos, part of securing cultural intellectual property rights for the Oma of laos—traditional design digital library, funded by Voice Global under the Laos Sudden Opportunity Grant – V-19100-LA-SO. Available via https://www.culturalintellectualproperty.com/_files/ugd/d5b008_7bfdbe5a12814304b5cbb6466264c9fe.pdf. Accessed 28 June 2022.
5. Boța-Moisin M, Schreiber R (2021) Rebranding "Made In India" through cultural sustainability—exploring and expanding indian perspectives. Dissertation, The Swedish School of Textiles, University of Borås, pp 44–50. http://www.diva-portal.org/smash/get/diva2:1553045/FULLTEXT01.pdf. Accessed 28 June 2022.
6. Cattien J, Stopford RJ (2022) The appropriating subject: cultural appreciation, property and entitlement. Philos Soc Criticism.https://doi.org/10.1177/01914537211059515
7. Cultural Intellectual Property Rights Initiative (2022) Cultural IP Month 2022—Press Release. Accessible at https://www.culturalintellectualproperty.com/post/cultural-ip-month-2022-press-release. Accessed 28 June 2022.
8. Convention on the Protection and Promotion of the Diversity of Cultural Expressions (2005) Available at: https://en.unesco.org/creativity/convention, https://en.unesco.org/creativity/sites/creativity/files/passeport-convention2005-web2.pdf. Accessed 28 June 2022
9. Gerber M (2019) Culture and sustainable development from a UN perspective. In: Meireis T, Rippl G (eds) Cultural Sustainability Perspectives from the Humanities and Social Sciences. Routledge, London and New York, pp 12–24
10. Huber W (2019) Cultural sustainability in times of cultural genocide. In: Meireis T, Rippl G (eds) Cultural Sustainability Perspectives from the Humanities and Social Sciences. Routledge, London and New York, pp 36–46
11. International Covenant on Economic, Social and Cultural Rights (1966) Available at: https://www.ohchr.org/en/instruments-mechanisms/instruments/international-covenant-economic-social-and-cultural-rights. Accessed 28 June 2022
12. Jansen A (2019) Decolonizing fashion: Defying the 'white Man's Gaze'. Available via Research Collective for Decoloniality & Fashion. https://rcdfashion.wordpress.com/2019/05/22/decolonialising-fashion-defying-the-white-mans-gaze/. Accessed 28 June 2022
13. Meireis T (2019) Sustainable development and the concept of culture - an ethical view. In: Meireis T, Rippl G (eds) Cultural Sustainability Perspectives from the Humanities and Social Sciences. Routledge, London and New York, pp 47–59
14. Mexico City Declaration on Cultural Policies (1982) https://culturalrights.net, https://culturalrights.net/descargas/drets_culturals401.pdf. Accessed 28 June 2022
15. Niessen S (2020) Regenerative fashion: there can be no other. Available via Academia.edu. https://www.academia.edu/44408162/REGENERATIVE_FASHION_There_can_be_no_other. Accessed 28 June 2022
16. Nute K (2019) Toward a test of cultural misappropriation. Int J Critical Cultural Studies. https://doi.org/10.18848/2327-0055/CGP/v17i02. pp. 67–82
17. Oma Traditional Textile Design Database© (2021). https://oma.traditionaldesigns.la. Accessed 28 June 2022
18. Oma Traditional Textile Design Database© (2021) Rights. https://oma.traditionaldesigns.la/rights/. Accessed 28 June 2022.
19. OseiTutu J (2011) Emerging scholars series: a sui generis regime for traditional knowledge: the cultural divide in intellectual property law, 15 Marq. Intellectual Property L. Rev. 147. Available at: https://scholarship.law.marquette.edu/iplr/vol15/iss1/3. Accessed 28 June 2022
20. Post-2020 UN Biodiversity Framework (draft) (2021) Available via www.cbd.int. https://www.cbd.int/doc/c/abb5/591f/2e46096d3f0330b08ce87a45/wg2020-03-03-en.pdf. Accessed 28 June 2022
21. Throsby D (1999) Cultural capital. J Cultural Econ, 23.https://doi.org/10.1023/A:1007543313370.pp.3-12
22. Traditional Arts and Ethnology Centre – TAEC (2021) What do you need to know about cultural intellectual property? https://www.taeclaos.org/what-do-you-need-to-know-about-cultural-intellectual-property/. Accessed 28 June 2022

23. Traditional Designs.la (2021) Protecting Traditional Knowledge (TK) and Traditional Cultural Expressions (TCE). https://traditionaldesigns.la. Accessed 28 June 2022
24. United Nations Convention on the Protection and Promotion of the Diversity of Cultural Expressions (2005) Available at https://en.unesco.org/creativity/sites/creativity/files/passeport-convention2005-web2.pdf. Accessed 28 June 2022
25. United Nations Declaration on the Rights of Indigenous Peoples (2007) Available at: https://www.un.org/development/desa/indigenouspeoples/declaration-on-the-rights-of-indigenous-peoples.html. Accessed 28 June 2022
26. Universal Declaration of Human Rights (1948) Available at: https://www.un.org/en/about-us/universal-declaration-of-human-rights. Accessed 28 June 2022
27. Whyte K (2017) Indigenous climate change studies: indigenizing futures, decolonizing the anthropocene. English Language Notes, Forthcoming, Available at SSRN: https://ssrn.com/abstract=2925514
28. World Intellectual Property Organization (WIPO) (n.d.) Traditional Knowledge. Available at https://www.wipo.int/tk/en/tk/. Accessed 28 June 2022
29. World Intellectual Property Organization (WIPO) (2022) Compilation of information on national and regional *Sui Generis* regimes for the intellectual property protection of traditional knowledge and traditional cultural expressions. Available at https://www.wipo.int/export/sites/www/tk/en/resources/pdf/compilation_sui_generis_regimes.pdf. Accessed 28 June 2022

Fashion and the Ethnographic Museum: A Case Study from The Netherlands

Daan van Dartel

Abstract This chapter looks at the process of inserting fashion in the ethnographic museum, in this case in the Netherlands. Where ethnography museums were always concerned with dress or costume and have great textile collections of people from all over the world—except so-called western styles—fashion was never part of the 'non-Western' world. The National Museum of World Cultures sees fashion as an accessible tool to address coloniality and decoloniality and to speak about historical formations. The chapter shows its methodology.

1 Introduction

In 2015, the just established National Museum of World Cultures (NMWC; het Nationaal Museum van Wereldculturen, NMVW) in the Netherlands appointed a curator of popular culture and fashion. Fashion, whatever its definition, had never before been a topic in the museums that are part of the NMWC: the Tropenmuseum in Amsterdam, the Museum of Ethnology in Leiden and the Africa Museum in Berg en Dal. Curators of the NMWC are responsible for the collections of all three museums and that of the adjoined World Museum in Rotterdam. Together these museums take care of some 360,000 objects and almost a million photographs that were collected by the museums since the second half of the nineteenth century. These museums were so-called ethnographic museums, focusing on countries and cultures that were considered 'non-Western'.

Of course, textiles played a large role in these collections. For one, textiles are easy to transport over large distances. Collectors in the nineteenth and twentieth century, often residing in former colonies, acquired textiles in multiple ways. Textiles were often gifted by locals, or were purchased in markets or straight from the makers, whose names seldomly were registered. Many of the textile collections in museums were donated or sold by professional and amateur collectors. Secondly, textiles, in

D. van Dartel (✉)
National Museum of World Cultures, Amsterdam, The Netherlands
e-mail: daan.vandartel@wereldculturen.nl

M. Á. Gardetti and R. P. Larios-Francia (eds.), *Sustainability Challenges in the Fashion Industry*, Sustainable Textiles: Production, Processing, Manufacturing & Chemistry,
https://doi.org/10.1007/978-981-99-0349-8_7

their omnipresence, are important markers of identity, with which the ethnographic museum could speak to their visitors about cultural difference. As such, textiles were very useful in exhibitions and other types of museum programming. The term fashion was never used for clothing from these cultures, rather differentiating terms were used such as dress or costume to emphasize their exotic qualities.

To understand this divide in terminology, we need to look at several histories. The history of fashion and fashion theory, but also the history of ethnography or anthropology and its museums. In this chapter, I will try to give insight into the workings of the ethnographic museum, and the mostly exclusionary way fashion theory has dealt with clothing practices outside of Europe. Finally, the attempts of the NMWC in employing fashion as a tool for decolonization will be discussed.

2 The Ethnographic Museum

Museums of ethnography, of the description of human cultures, are rooted in sixteenth and seventeenth century elite curiosity cabinets and eighteenth and nineteenth century industrial fairs, such as the large World Exhibitions in European capitals such as London and Paris. Both are strongly connected to the history of global expansion and the emergence of *nationalism* ([6]:90). The cabinets were places of conservation, study and presentation, from which the central functions of the museum developed. Rather than representations of the people and their cultures the objects embodied, as was the intention, they were representations of how the European elite looked at the world. The World Fairs intended to open up the world to Europe's citizens, seemingly a democratic endeavor, stimulating interest in different cultures and lands, however not only for education, but also for entrepreneurial opportunities.

2.1 National Museum of World Cultures in the Netherlands

The National Museum of World Cultures in the Netherlands consists of three museums, and works closely with another museum.

Leiden
One of the oldest ethnographic museums in the world, the museum of Ethnology (museum Volkenkunde) in Leiden, the Netherlands, is part of the NMWC. It was founded in 1837, as the national Japanese museum von Siebold. Philipp Franz Von Siebold had been a physician on Deshima, an artificial island just outside Nagasaki where the Dutch were located in the 1820s. During his time there, he collected all kinds of material culture that he brought to the Netherlands. Today, there still is a von Siebold House in his former residence in Leiden. Besides these collections, the Royal Cabinet of Curiosities of King William I, founded in 1816, was transferred in 1883 to the ethnographic museum in Leiden. King William I stimulated the foundation

of research institutes for the arts and science and commissioned scientists to collect in East and insular Asia. Also in 1883, a part of the objects that were presented at the Colonial Exhibition in Amsterdam were transferred to the museum. From the beginning of the twentieth century, objects from other regions than East Asia and Indonesia were added. With the collections growing steadily, finally the museum became located in the building it is still located in today: the old academic hospital opened its doors as a museum in 1937. In total, the Leiden Museum of Ethnology cared for around 212,000 objects of material culture at the time of the merger in 2014. Some 40,000 were items of clothing.

Amsterdam

Some 35 years later, in 1871, an employee of the Society of the Promotion of Industry arranged for a colonial museum to be opened in Haarlem. He was convinced that all examples of material culture brought to the Netherlands from overseas should be presented in a museum. As a botanist, this Frederik Willem van Eeden was very interested in natural products, from wood samples to objects made of local materials. This first colonial museum in the world was based on these examples of material produce and techniques. It was not an ethnographic museum as defined before, but rather a product museum where people could see what kinds of plants and minerals were available elsewhere, and where they could be inspired to start a business in the colonies. This Haarlem collection, together with the ethnographic collection of the Artis zoo, formed the basis of the museums that were part of the Colonial Institute that was opened in Amsterdam in 1926 and was named the Tropeninstituut (Institute of the Tropics) in 1950. Its museum became the Tropenmuseum. In 2014, the Tropenmuseum cared for approximately 147,000 material culture objects, of which around 28,000 were clothing.

Berg en Dal

In Berg en Dal, near Nijmegen in the east of the Netherlands, the Africa museum (Afrika Museum) was established more recently, in 1954. This museum was led by its founder, the Congregation of the Holy Heart. Missionaries were trained at the location, and what they collected during their work in different countries in Africa, became part of the museum's collection. Many objects were of a religious nature, but also jewelry, dress and performative arts form a large part of its collection. The Africa Museum is relatively small compared to its big sisters, with 8,200 objects of material culture from Africa, including some 3,000 clothing objects. Its most special collections today are contemporary art works from Africa and its diaspora.

Rotterdam

The Worldmuseum (Wereldmuseum) Rotterdam is not one of NMWC's foundational members, but is an associated museum. Several staff members of NMWC are also responsible for tasks at the Worldmuseum. This museum started out as a museum by the Association of the Royal Yacht Club in its building. Its members donated maritime objects, ships' models and ethnographic pieces, which led to the foundation of the Prince Hendrik Maritime Museum, named after its chairman the Queen's consort Prince Hendrik. In 1883, now owned by the municipality of Rotterdam, an ethnographic museum was formed with its major drivers Dutch trade relations, the rise of colonialism and increased missionary activities, as well as the emerging science of anthropology. The Museum for Geography and Cultural Anthropology opened its doors in 1885. Currently, from its almost 95,000 objects, around 17,000 are clothing.

3 Tools of Colonialism

Ethnographic museums such as these, founded in the nineteenth century, were inextricably linked to colonialism. They were not only linked but part of the instruments used by European nations to support the colonization of areas around the world. Their foundations coincided with the formation of European nations that needed finances and presumed Others to mark their boundaries. Museums were established to learn about other cultures and peoples; there was a sincere interest and curiosity, but not without their functionality in gathering knowledge about these others, with which populations and lands could be suppressed and colonized. In this light, the Tropenmuseum, for example, functioned as a showcase of the colonies, where one could come and look at what enterprises would be profitable for oneself and the nation. Scientific expeditions into interiors of Indonesia and surrounding islands, and of Suriname, improved the capability of the Netherlands to dominate such large areas. Setting up plantations and making roads and other infrastructure helped this taking over of land and dominating people. In the Netherlands, in the museum one could see what type of plantations were profitable and how money could be made with which crops. Indigenous materials and techniques were studied and improved for commercialization in laboratories near the museums which functioned as scientific centers on the colonies.

When Charles Darwin published his On the Origin of Species in the second half of the nineteenth century, further ideas on difference in the development of people were created. Anthropology, the science of what makes us human, developed simultaneously, and the notion of Darwin's evolution took hold in the study of man. This meant that scientists saw human societies in different stages of development, with the white European man at the top of civilization. Racist ideology, although a complex and older phenomenon, developed further alongside these theoretical discourses and supported the legitimization of imperial conquests of foreign territories. Physical anthropology also developed as a science, in which biological characteristics of people were measured and studied to explain differences in psychology, culture and development. Not unsurprisingly, this was very applicable within colonial thought, and later in history within other totalitarian regimes, that could now add the civilization of people to the justification for colonization. Here, ethnographic museums again were central in getting this message across society. Of course, the development of ethnographic museums happened in different phases with much more nuances than now presented here, there also was selvage anthropology for example, where anthropologists and museums were concerned with the rapid disappearance of cultures and their material culture ([6]:91). However, that can be considered an expression of a feeling of superiority as well, of a specific way of looking at others as less capable and in need of help, instead of looking at the reasons behind the vanishing of such cultures and peoples.

The different museums of the NMWC all are connected to this history, albeit with different emphasis. They all hold similar collections that give more insight into the worldview of the collecting body, than of the collected cultures. Where the Tropenmuseum was a colonial museum, partly funded by the ministry of colonies, functioning as a propaganda machine for exploitation of colonies, the Leiden National Museum of Ethnology was more connected to the university where research on the material culture of different peoples was central. The Africa Museum was founded by missionaries, whose work in the colonies was to convert people to Christianity. They collected curiosities and caused people to distance themselves from their religious objects, often erasing cultural practices and material culture. The Worldmuseum was based on object entries by magnates from Rotterdam, who were trading in and shipping to and from the colonies.

4 Fashion Theories

Fashion in western fashion theory is often located in the so-called West.[1] Gilles Lipovetsky in the Empire of Fashion even asks himself: "Why in the West and not elsewhere?" ([8]: 4). Other authors have supported this notion, even if they differ in opinion on the beginnings of fashion as a cultural phenomenon. Some place the origin of fashion in eighteenth century Europe, whereas others see the fourteenth century as the period of development of fashion ([2]: 4). The same Lipovetsky calls fashion an institution, where others speak of the fashion system (ibid; [4]: x). Many of these academics write about fashion and modernity as if they are inseparable, and, if modernity too is only located in the 'West,' as many argue, then automatically this would be an indication that there is no fashion outside of the west.

With the earlier described history of colonialism and museums in mind, it is not hard to see in what sociopolitical context this interpretation of fashion and modernity took shape. The fashioning of self and other during the colonial era is a very powerful example of the hegemony of Europe in fashion theory, meaning that all developments in fashion were, and often still are, mainly attributed to European agency.[2] Fashion,

---

[1] For example Mary Ellen Roach, and K. E. Musa, *New Perspectives on the History of Western Dress: A Handbook* (New York: NutriGuides, 1980); Elizabeth Wilson, *Adorned in Dreams: Fashion and Modernity* (London: I.B. Taurus, 2007); Joanne Entwistle, *The Fashioned Body: Fashion, Dress & Modern, Social Theory* (Cambridge: Polity Press, 2015).

[2] Buckridge, S. (2004) The Language of Dress. Resistance and Accomodation in Jamaica, 1760 -1890. Kingston: University of the West-Indies Press. On the historically formed colonial assumptions on ethnicity and fashion, see also Sarah [3].

as a concept in that sense, has had a similar career to the museum as a concept: based in western social thought. The Thames & Hudson Dictionary of Fashion and Fashion Designers proves a point: its 'new' edition, from 2008, does not list any designers working outside the western metropoles.

One of the first authors to address this Eurocentric view on fashion in a systematic way, was Jennifer Craik. Her book The Face of Fashion (1993) discusses different fashion systems, with the western (high) fashion system being just one of many co-existing systems. She writes: "…[T]he western fashion system goes hand-in-hand with the exercise of power. But this is also true of other fashion systems. All fashion systems demonstrate the cultural politics of their milieu." (1994: x). A bit further on, she writes, in my view, one of the most important notions around fashion, in that "the exercise of power cannot simply be associated with the unfolding development of modern consumer capitalism." (ibid). To Craik, European 'high elite designer fashion' is just one specific variant of fashion (ibid.).

5 Fashion as Museum Category

It seems that fashion collections are found in all sorts of museums, except for the ethnographic. Textile departments of anthropology museums are filled with beautiful, skillfully made clothing items that are hardly ever described in terms of fashion. In the ethnographic museum, a museum educating visitors on cultures from all over the world, dress is the preferred terminology for clothing objects. 'Dress' as defined by Roach Higgins et al., refers to the assemblage of modifications to the body and/or supplements to the body ([11]: 7). This term seems to encompass a binary quality when trying to define fashion. The term 'dress' in an ethnographic context contains perceived colonial implications of tradition, of remaining frozen in the past, and of never changing, supporting the belief that in so-called non-Western cultures, there is no fashion *over there* ([5]:70). Ethnographic objects always belonged, and often are still considered as such, to the realm of the Other. This idea of the Other has been indelibly shaped by the colonial project and is still very much alive in its afterlife.[3] Underpinned by colonial ideologies, both the study of fashion and museums have primarily been Eurocentric in their knowledge construction.

Dress thus is a general broad concept, including all types of clothing styles, indeed also fashion, if that is understood as clothing or dress style that has distinctive characteristics at a specific moment. But what are these characteristics? Fashion is considered to address an additional quality to 'just clothes.' Most definitions of fashion emphasize the aspect of change, however, since culture and cultures change constantly, clothes in itself also change constantly and this does not make it fashion.

---

[3] Frantz Fanon (1958), *Zwarte Huid, Witte Maskers*. Amsterdam, Octavo; Edward Saïd (1978). *Orientalism*. New York, Pantheon Books; Ashcroft B., Gareth Griffiths and Helen Tiffin (1998), Postcolonial Studies: The Key Concepts. New York, Routledge.

Similarly, the often quoted notion that expressing identity makes fashion is inconsistent. All clothes express identity, ranging from what a well-known designers presents at a fashion week, to a type of garment a local priest wears during a specific ceremony. According to Kawamura, fashion refers to a wider, context-dependent and socially constructed system around dress, including ways of production, distribution and consumption and depends on many external factors such as purposeful change and expert knowledge [7]. This definition of fashion implies there can be different, concomitant fashion systems operating in the world, as Craik before also stated.

At the NMWC, the question of what then *is* fashion inevitably always arises. In order to work with specific object categories, as museums do like no other institution, definitions are central to our way of working. Especially when working to make collections as accessible as possible, the ordering of things into categories through which to find them could be considered part of the reason for existence of museums. As stated before, when using concepts of fashion theory to re-envision ethnographic collections, the most ubiquitous criteria when defining fashion are the notion of change and its relation to identity construction, which would suggest that all clothes are in fact fashion ([1]:2–4,[12]: xxv). Ingrid Loschek offers an interesting view on this, in arguing that objects themselves cannot be considered fashion, or that fashion is an inalienable aspect of clothing, but that context, and in effect, the observer, makes fashion ([9]: 138). Ethnographic museums that hold large textile and clothing collections are served greatly by this question of *when an object is fashion.* Museum institutions rely on strict categorization needing structured and demarcated identities of objects to be able to register and document them. Fashion, however, cannot be categorized so easily because of its contextual nature ([3]: 143). By accepting that objects can be fashionable within specific contexts, but are not necessarily part of a contemporary, or complex political, social and economic system that we call fashion, we can avoid the understandable inclination to call all clothing fashion.

Decoloniality needs to come with *decategorality*, an undoing of thinking in specific categories to order the world around us. However, losing categories will prove to be problematic, for the same reason they exist, human kind cannot do without categorization to understand life. What is needed is awareness of the unilateral direction of categorization, its ethnocentrism that puts things in categories that can be very different in other contexts and cultures.

6 Decolonization

In 2020, an activist group from the Republic of the Congo, after paying entrance fees, removed a statue from the display at the Africa Museum and took it outside, while claiming ownership and return. A few months later, red paint was thrown at the façade of the Tropenmuseum, with activist calls for restitution of all objects, presumed all looted, to their countries of origin. The current sociopolitical climate in the Netherlands, as in a lot of places in the world, is one of activism and protest. A raising awareness among younger generations on the effects, and affects, of colonialism pushes ethnographic museums outside their comfort zone and into the spotlight of decolonial discourse.

The NMWC undertakes several actions to question even the possibility of a museum that is embedded in colonialism to be a part of decolonization efforts of the Dutch museological landscape. In June 2022, it opened a new permanent exhibition on the legacy of colonialism, called Our Colonial Inheritance (Onze Koloniale Erfenis). When preparing this exhibition, consultants from different stakeholding groups were invited to think along and read the texts the museum prepared. Texts followed the suggestions of the incomplete guide Words Matter that the NMWC published in 2018, in which ethnocentric and colonial terms are problematized and alternatives offered.[4] Still, many words used by the museum were questioned by the invited readers, which indicates that colonial perspectives as expressed by language is still very present in the museological vocabulary. Take the perhaps ordinary and seemingly innocent words traditional or native; literally, these words are not problematic, but when used from a specific presumption about the phenomena it describes, they can be quite harmful especially in an educational institution. Traditional often is understood to imply stagnation, or even backwardness, and native as holding an aura of the exotic.

Fashion is part of the attempt to decoloniality by the NMWC. Already by integrating the word fashion into its vocabulary, a complete different conception of clothing and textile objects takes place among the visitors of the museum. When instead addressing indigenous dress, or local dress as fashion when suitable, alternative associations turn up in people's minds. Secondly, many of the clothing items in ethnographic museums have always been presented either flat or on a mannequin expressing a specific ethnic identity and even social role in the community they represented.

[4] https://www.tropenmuseum.nl/nl/over-het-tropenmuseum/words-matter-publicatie (accessed 27–6-2022).

Photograph Mannequin representing Javanese wayang puppeteer at the Colonial Institute, 1923

One of the main goals of ethnographic museums in the twentieth century was representation. They represented a complete culture by displaying individual mannequins in specific clothing styles, performing very specific sociocultural, religious or economic tasks. More often, clothing in these museums was and oftentimes is presented flat, disembodied, as a work of art or artistry to be admired, hence almost erasing the people wearing the clothes. Even today, in my own museum, the difference between the idea of what fashion is and what is not, is emphasized in this way at the World Museum in Rotterdam, where contemporary twenty-first century Indonesian fashion is exhibited on fashion mannequins, whereas the traditional clothing styles collected in the twentieth century, hang from a rod. Immediately, our brain categorizes the objects following this presentation technique: One is clearly fashion, the other clearly is not.

The new exhibition Crossroads in World Museum Rotterdam, Indonesian textiles and transculturality. Photograph: Mike Bink

7 Fashion and Decolonizing the Ethnographic Museum

All four museums connected to the NMWC have a long list of past exhibitions on dress and textiles, from the nineteenth century beginning onward. Plenty of titles can be found on exhibitions from the past that deal with cloth: fabrics, clothing, dress, woven worlds, or titles in which techniques or material were central, such as weaving, batik or ikat. It was only in 2009 that the then curator of textiles, Itie van Hout, curated a small exhibition called Culture Couture. It was a cooperation with the Antwerp Academy of Fine Arts, whose third year students were assigned to make new work based on 'non-Western' dress styles. It was only in 2017 that the word fashion was used for the first time in an exhibition title, when the Brighton exhibition Fashion Cities Africa was adapted for the Tropenmuseum and later the Africa Museum of the NMWC. From 2015 onward, several fashion events were organized, such as the collaboration on the yearly fashion fest of Amsterdam cultural organization MAFB, that supported young fashion designers, makeup artists and stylists from intermediate vocational education levels in gaining experience and the possibility for inspiration by the museum's collections, and through which these young talents could participate in a prize that consisted of a financial reward and professional training. Next to fashion fest, different other fashion events, such as catwalks or workshops, took place until today.

Besides cultural programming on fashion and its inclusion in almost all its exhibitions, the NMWC developed an acquisition policy plan that had fashion, next to photography, popular culture and contemporary art, as one of its main collecting subjects. Themes such as cultural appropriation, the re-activation of heritage, with a focus on materials and techniques, and the interaction of local fashion styles with global fashion, guide the acquisition of new fashion collections. Acquisition takes place in close collaboration with makers and artists, whose histories and professional trajectories are documented carefully.

7.1 Cultural Appropriation

One of the ten themes in a semi-permanent exhibition at the Tropenmuseum, called Things That Matter, is cultural appropriation. This theme proves to be a clear example for visitors to understand the perseverance of coloniality in different cultural practices worldwide. In order to make this colonial relationship clear, many examples are used from the European high fashion world.

Photo Alejandro Linares García, CC BY-SA 4.0, vía Wikimedia Commons

The museum collected the blouse French designer Isabel Marant was called out for in 2016 by a Mexican village. Marant had used the symbolism of the village of Tlahuitoltepec in Oaxaca, on her own blouses, without any reference to the women collectives of the village who make and wear the blouses. Thanks to a vivid social media campaign, the people of the village were able to get attention worldwide for this copying of their style and indeed, their symbolism of the world around them. The signs on the blouse stand for the environment, important plants and mountains that are part of their world view and of their cultural identity. This case became even more complicated because another French company, Antik Batik, had laid claim to the first use of the symbols and sued Marant. The NMWC has the blouse in its exhibition, next to the 'real thing' from Tlahuitoltepec, with an indication of the price difference: over 200 dollar for Marant versus 20 dollar for the original.

Around the same time, fashion house Valentino cooperated with indigenous Métis artist Christi Belcourt from Canada, in creating their Watersong collection. Belcourt bases her work on the symbols of her people's view of the world and the importance of water for all living beings. Although an example of a good and respectful collaboration between a famous fashion designer and an artist, unfortunately, a year later, Italian fashion house Valentino fell into the same trap of appropriating and stereotyping cultures in their next collections. In thinking about how these approaches could be so different, it can be concluded that Belcourt was recognized as an artist within the global art scene, i.e., her work was part of museum collections in Canada, which made Valentino realize it had to ask for permission and form a collaboration in order to be able to use the motifs. When 'taking inspiration' from living traditions of indigenous cultures that are not part of the western complicated intellectual property judicial system, there apparently still are no restraints.

7.2 Heritage as Inspiration

A second theme that guides contemporary fashion collecting practices at the NMWC is the use of cultural heritage as a source of inspiration for designers who share that heritage. In doing so, these designers try to work with local traditions and reshape them into contemporary styles, either out of respect for traditional styles and techniques, or because these traditions are part of their individual identity. The NMWC collected two of such works by designer Karim Adduchi, who works from Amsterdam. Adduchi, of Moroccan nationality, moved to Barcelona at a young age, and expressed himself through painting and drawing. He moved to Amsterdam and graduated at the Rietveld Academy with a fashion collection that won the Lichting 2016 prize for best upcoming designer in the Netherlands.[5] Since then, his work has featured in several fashion magazines, among which Dutch Elle and Vogue Arabia, and was part of several museum exhibitions, such as Fashion Cities Africa at the Tropenmuseum and the Africa Museum, and Global Wardrobe at the Art Museum The Hague. The V&A in London made a fashion film with his work in 2021.[6] Adduchi won several prizes and became part of the Forbes' Europe and Middle East 30 under 30 list. His work combines different fashion traditions, with an emphasis on Moroccan Amazigh culture techniques and shapes, with the aim of honoring the women making these textiles.

[5] *Lichting* is the annual fashion show in which all talents from design schools show their collections at the Amsterdam Fashion Week.

[6] https://www.vam.ac.uk/blog/museum-life/friday-late-X-jameel-museum-leila (accessed June 2022).

Karim Adduchi, carpet dress from the collection She has 99 names, 2017. Collection National Museum of World Cultures, 7155-1

Surinamese Dutch Giorgio and Onitsha Toppin also took inspiration from the cultures of the country of origin of their ancestors. In their 2021 collection Kabra, collected by the NMWC for the exhibition on colonialism and its afterlife, brother and sister traveled to Suriname and worked with local artisans in the creation of contemporary menswear for their brand Xhosa. Embroidery techniques of Maroon culture decorate a woolen coat, knotting techniques by Indigenous Surinamese people make a sweater, and the style of the Afro-Surinamese koto, a women's traditional fashion style, inspired their koto trousers, using a typically female style to disrupt ideas on menswear.

Kabra collection 2021 by Xhosa, Amsterdam

7.3 Local Fashion Styles

A third theme that is important in the NMWC collecting practices is what I have called Local Fashion. This system of fashion behavior may be inspired by global intracultural connections, but still has a distinct local character. Within local fashion systems, there can be different styles and trends, as there are in the global fashion system with its haute couture, prêt-a-porter, fast fashion, but also fashion subsystems, such as Punk for example. A local fashion system is no different; it is just performed on a smaller geo-cultural area.

Since Suriname was a colony of the Netherlands from the late seventeenth century until mid-twentieth century, reaching independence in 1975, it is logical that Surinamese clothing styles are a part of the collection of the NMWC. From Indonesia, however, having been considered the Dutch most important colony, there are many more items from the different cultures that inhabit the archipelago. From Suriname, the most important collection of clothing comes from the Afro-Surinamese population, consisting of Creole Surinamese and Maroons. Collections of the latter group hold traditional styles such as kamisas (loin cloth for men), pangi (women's hip cloth) and banyakoosu (shoulder cloth for men; see [10]: 36–38), from late nineteenth century until today. Time brought in new elements such as different materials, artificial textiles and colors, new printed patterns that are on trend and motifs that are newly introduced. Teenage girls know what is the latest fashion for them.

Photo Traditional koto from Suriname. Collection National Museum of World Cultures, RV-2452-168

The creole fashion style that is also very specifically Surinamese is the women's koto (complete style of skirt, blouse and underwear), that is always accompanied by a head cloth, the angisa. Developed from clothing styles in the period of slavery, the look was solidified through a specific combination of items after the abolition in 1863 and developed from different cultural influences in shape, materials and patterns.[7]

The museum holds historical and contemporary examples of these local styles, from which it is easy to see how they developed over time, bringing in new transcultural elements into tradition, are clearly connected to cultural identity, and are part of a system that is based on the consent of a group of qualified people, influencers if you will. These local styles are fashion indeed ([5]: 83–85).

8 Conclusion

Fashion supports the objective of the former colonially implicated ethnographic museum the NMWC to address decoloniality. Apart from the very relevant and urgent question if a museum with such a fraught history and that is located in a distinctly colonial building, can indeed decolonize at all, fashion is used as an experimental tool in supporting the realization of this objective. Via the language of clothing, of dress, of fashion, the museum tries to show how colonial thinking still pervades the contemporary mind. Its exhibitions hope to deconstruct this way of thinking, by reframing and questioning museum categories, explaining the origins and effects of language, including multivocality and examining new methods of collecting and exhibiting. As fashion is a phenomenon that can be said to be globally understood, one way or another, it may be the perfect and most accessible, perhaps even universal, language there is. It is often considered frivolous and unimportant for understanding human existence, but may actually very well be of essential importance to get difficult histories and sociocultural processes understood in the best way possible.

References

1. Barnard M (ed) (2007). Routledge, New York
2. Breward C (1994) The culture of fashion. Manchester University Press, Manchester
3. Cheang S (2017) Ethnicity. In: Baxter DA (ed) A Cultural history of dress and fashion in the age of empire, Vol 5. London, Bloomsbury
4. Craik J (1993) The face of fashion. Cultural studies in fashion. London, Routledge
5. Dartel D van (2021) Afro-surinamese Kotomisi: fashion and the ethnography museum. In: Cheang De G, Takagi (eds), Rethinking fashion globalization. London, Bloomsbury
6. Fromm AB (2016) Ethnographic museums and intangible cultural heritage return to our roots. J Marine Island Cultures 5:89–94

[7] In history books, the Dutch abolished slavery in 1863. People still had to work the land of their former owners for ten more years though, and could only really leave in 1873.

7. Kawamura Y (2004) Fashion-ology. Berg, Oxford
8. Lipovetsky G (1994) The empire of fashion. New Jersey, Princeton University Press, Dressing Modern Democracy
9. Loschek I (2009) When clothes become fashion: design and innovation systems. Berg, Oxford
10. Polimé T, van Stipriaan A (2013) Zeg het met doeken. Amsterdam, KIT Publishers, Marrontextiel en de Tropenmuseumcollectie
11. Roach Higgins ME, Eicher J, Johnson K (1995) Dress and identity. Capital Cities Media Inc., Albany, NY, p 7
12. Welters L, Lillethun A (eds) (2014) 2nd edn. Bloomsbury, London

Other Voices: Dynamic Tradition, Empowerment and Andean Fashion in Peru

Rosa Patricia Larios-Francia, Agustina Micaela Burgos, and Joan Jefferson Sanchez Jimenez

Abstract The chapter proposes a reflection on the dynamism or stagnation of the textile tradition in Peru. For this purpose, Andean fashion was contextualized through its history of more than 5,000 years of excellence in textile art, which lost the essence of its tradition and even saw the role of the artisan as the main actor fade away. Textile tradition should not only be conceived for its changing character through different cultures, but also as the combination of traditions and techniques regardless of the syncretism observed in them. The support of some designers who contribute to the empowerment of textile artisan communities was identified, by mixing in their unique designs, techniques, colors and artisan essence. Likewise, an analysis was made of those entrepreneurships that through their platforms and commercial business models have generated a revaluation of Andean handcrafted textile art. All these efforts make possible the generation of identity, sense of belonging and enhance the quality of the works of the textile craft communities, through the exchange with their peers, forging in them professionals with decision-making power over their creative process and in the narration of their history through their textiles.

1 Introduction

The Andean cosmovision embodied in fashion saw its dynamism as an element of transmission of the heritage of the pre-Inca and Inca cultures twisted due to the

R. P. Larios-Francia (✉) · A. M. Burgos · J. J. S. Jimenez
Carrera de Ingeniería Industrial, Universidad de Lima, Lima, Perú
e-mail: rlariosf@ulima.edu.pe

A. M. Burgos
e-mail: agussburgos@hotmail.cpm

J. J. S. Jimenez
e-mail: jeffsanchezjimenez@gmail.com

R. P. Larios-Francia
Facultad de Diseño, Universidad de Ciencias Aplicadas, Lima, Perú

M. Á. Gardetti and R. P. Larios-Francia (eds.), *Sustainability Challenges in the Fashion Industry*, Sustainable Textiles: Production, Processing, Manufacturing & Chemistry,
https://doi.org/10.1007/978-981-99-0349-8_8

actions of colonization and the use of epistemic violence, the latter being conceived as that which claims to speak for all "the Other" individuals (the colonized) but does not allow them to have their own voice [1]. Pulido Tirado [2] describes epistemic violence as the set of a series of systematic, regular and repeated discourses that do not tolerate alternative epistemologies, perpetuating the oppression of the knowledge of the Others and justifying their domination.

In the times of Spanish colonization in Andean territory, the process of syncretism, considered as the phenomenon of interaction between cultures, through which elements of different origins are mixed or fused, forming new cultural manifestations; and as an event that is part of the way of life of the societies of our time, and that is reflected through music, fashion, gastronomy, thought and behavior in general [3, 4], was presented in an even more violent way, given that mechanisms corresponding to an epistemic violence were used that was installed in the collective unconscious and that persists to this day. Thus, cultural empowerment initiatives that seek to exalt tradition, return to the roots and connect with that dynamism that was curbed by the imposition of a culture completely alien to the ideology of the Inca and pre-Inca cultures, initiated the process of decolonization of textile art and fashion in Peru.

Referring to the violence suffered in the colonial (neo- or post-colonial) framework that condemned the knowledge and practices of indigenous peoples to a derived and subjected epistemic. As cited by [5], Pulido Tirado [2] Foucault (1965: 251, 262, 269) locates epistemic violence in the "subjugated knowledges," a total set of knowledges that have been disqualified as inadequate to their task or insufficiently elaborated: naïve knowledges, located at the bottom of the hierarchy, below the required scientific level or cognition.

An epistemic violence that has been categorized by Bunch (2015) and cited by [1] as discriminatory, testimonial and distributive,and that we can evidence in the postcolonial Andean culture until today. A violence that denies the descendants of "the Others" access to multidisciplinary academic material in their own language, to adequate and sufficient information about their past, reinforcing stereotypes and discrimination toward them, where this reinforcement comes even from those who have been deprived of their own voice. An example of this was identified in the study conducted by Ojinaga Zapata [6] on the textile centers that emerged at the end of the nineteenth century in Peru and Mexico as cultural spaces, which also generated spaces for the visibility of popular art, as a commercial offer,as was the case of the Traditional Textile Center of Chinchero in Cusco. Although it is an institution that took up textile traditions and documented them with excellence, generating opportunities for local artisan knitters, it focused mainly on developing a value-added offer by and for tourists, in a context where the artisan knitter is not really empowered or the owner of her own narrative or of what she would like to communicate through her art.

When we define textile tradition, that which remains static is considered authentic; however, denying cultural dynamism and its impact on the variation in the use of color, techniques, materials and symbolism ended up creating a dichotomy between cultural heritage and the search for economic benefit in the contemporary world [7].

However, it has been found that there are more social forces capable of uniting these two concepts than those that seek to differentiate them.

To exemplify the latter, we can think of how textiles were developed in pre-Inca and Inca times, using vegetable and animal fibers, natural dyes and textile tools such as spindles, backstrap looms, frame looms, horizontal and vertical looms [8] and different textile techniques that made it possible for artisans to capture their art through canvases, gauzes, tapestries with diverse symbology, geometric figures and iconographies characteristic of each culture and period of history. However, with the arrival of manufactured, synthetic materials and their vibrant colors, artisan communities began to incorporate new elements to their textiles,even more: With the arrival of new elements belonging to modernity, the collective imaginary proceeded to use new designs in the warps of fabrics to represent that which identified their environment and customs [9].

It is therefore necessary to ask ourselves the following question: Is textile tradition only defined when it is static? Has modernity led to the failure of cultural preservation?

2 Tradition Versus Transformation

The textile art of pre-Columbian Andean man has been recognized for its meticulousness, perseverance, dedication, accumulative and communal effort, the affection for detail and reduced scale, as well as the combination of colors, the love for elegant chromatism, the mathematical translation of complex designs [10]. Textile craftsmanship is considered as the art that developed with excellence compared to others such as goldsmithing, ceramics, sculpture and even architecture, which has been internationally recognized for its great works.

In many of the expressions of pre-Inca and Inca textile art, syncretism was already evident that derived in, for example, increasingly homogeneous representations of the divine, given that the feline, water and earth began to be represented by means of radiated and angular forms that were very similar in terms of use and meaning [11].

In these initiatives we find two kinds of proposals that seem to be at odds with each other: On the one hand, that which seeks to maintain the traditional in a static way, with a scarce presence of design or creativity, based on the use of patterns that are repeated in each production batch of each community, using the same techniques, even the same colors, shapes and expressions, where the artisan is expressed in the work of textile products made delicately by hand, from the collection of cotton fiber, or the shearing and classification of fine camelid hair, to then go through the manual process of spinning, where each artisan gives it its own characteristics, to then be used in various textile products, such as looms, tapestries, garments, embroidery; unfortunately, much of this "handmade" textile art is used only to offer products for the tourist market in local fairs and where we see how the legacy of the techniques, the great designs and developments of our ancestors not only cease to evolve, but even become opaque and lost.

Fig. 1 Ayacuchan embroidery-artisan

A second proposal is that which has as its mission the search for the "generation of identity and sense of belonging" in the contemporary context on the part of our art and our artisans. In recent years, individual efforts of designers and entrepreneurs have emerged to support artisan communities, not only by providing labor supply or market for their crafts, but also by contributing to the improvement of management, design, product development and knowledge generation so that textile traditions can live that dynamism previously mentioned (Fig. 1).

3 High-Impact Designers

Peruvian fashion saw its foundations laid with the rise of the great designers, pioneers in the development of techniques that combined dyes, textile materials with weaving and embroidery techniques and knowledge of pre-Inca and Inca textile art to bring to the contemporary world fresh and disruptive designs that represented the collective imagination of the Andean communities and contemporary practicality and aesthetics.

Thus, through the designers we were able to identify a need to satisfy: to generate and transmit identity through textiles using traditional and contemporary elements based on designs that respond not only to the western imaginary, but also to the empowerment of the pre-Hispanic cultural heritage.

Meche Correa, sustainable luxury fashion designer, whose designs are inspired and committed to rescue and preserve through creativity and contemporary design the Peruvian cultural heritage. Her work revalues the art of Peruvian artisans who are empowered through collaborative work, filling them with pride, improving their welfare and development. The designer presents among her multiple collection garments, the "four-pointed cap" inspired by the Wari and Tiwanaku culture that belonged to the Middle Horizon from 600 to 1000 BC Olguín [12] (Fig. 2).

Jorge Luis Salinas achieved to merge elements of Andean and Oriental culture in his Kimonos collection for Perú Moda [14], as a way to honor the cultural syncretism arising from migration from countries such as China, Japan or South Korea. The designer counted as a work team more than 100 artisans from 5 regions of Peru, who were able to express through their fabrics the identity of each of their cultures, with these pieces and their great creativity developed a proposal of Kimono costumes with knitted fabric (Fig. 3).

José Miguel Valdivia, who is also characterized as a high-impact designer, advises on projects for the development of artisans and SMEs. One of his latest collections,

Fig. 2 Meche correa's [13] collection 4-pointed hat

Fig. 3 Kimono-Jorge Salinas for PerúModa 2015, photography by Lyan Herrera [14]

called Abayat Arequipa, inspired by the trend of Modest Fashion, highlights the fusion of elements of religious expression of cultures such as Muslim and Peruvian during the time of colonization. This collection is composed of handmade embroidered abayas made by the master Leandrina Ramos, the Guardian of Colca Embroidery, heiress of an ancestral tradition, where the designer puts in value the work of Peruvian artisans facing the world (Figs. 4 and 5) [15, 16].

4 Mechanisms to Enhance Dynamism

Distributive epistemic violence is a vicious circle, given that the social gaps in terms of educational quality are both cause and effect of it and, therefore, do not allow the aforementioned "others" to become empowered.

To eradicate this type of violence, alternatives to the business model of the private textile sector have emerged, giving rise to ventures born in the third sector: one that seeks to provide a space where artisans achieve empowerment through the professionalization of their knowledge.

These ventures have responded to the needs of two stakeholders: On the one hand, artisans who wish to generate value through their skills in order to obtain economic, educational and social autonomy; on the other hand, end consumers who demand other types of exclusive, original products with personality.

Fig. 4 Bordered abaya, Abayat collection by José Miguel Valdivia [15]

Thus, enterprises with a sustainable luxury and artisanal perspective should not only be a bridge between supply and demand but should also generate and manage knowledge through systematization of knowledge, alliances with academia and government entities, participatory organizational culture with knowledge sharing, continuous learning through training and knowledge transfer through educational projects for young people.

The real empowerment of an organization will be achieved by generating a sense of belonging and a clear identity in the market, since this is an indispensable tool for acquiring a competitive advantage and achieving differentiation. This requires generating a dialogue with the modern consumer through different channels, being able to identify specific consumer needs in a market study, proposing a unique and differentiated value proposition, generating brand identity through the collaboration

Fig. 5 Bordered abaya, Abayat collection, by José Miguel Valdivia with Leandrina Ramos [15, 16]

of experts and artisans and establishing strategic planning that involves activities linked to empowerment and value generation.

All this through the development of collections justified by a clear concept that allows artisans to tell their story, so that their work is made visible within the context of the enterprise (Fig. 6).

Purposeful Entrepreneurship

In order to identify whether textile artisans are recovering their lost voice and effectiveness in the processes of this modernity, an analysis of those purposeful enterprises that offer diverse channels to communicate the messages of those who seek to express themselves through their art was developed.

A selection was made of purposeful textile enterprises that have a presence in different sales channels, focusing on those that have successfully unified tradition with modernity, offering the added value characteristic of textile handicrafts. These enterprises have transferred the necessary knowledge so that artisans can standardize their processes and their products can be recognized by the local and international market, and have also supported the communities to manage and communicate their

Fig. 6 Native cotton textile artisan in Lambayeque

offer with market niches that seek sustainable products in their three pillars: social, environmental and economic. Their commercial strategy, social projects, responsibility projects, empowerment, training, collaborative work with artisan communities and production chains were analyzed. This made it possible to identify the characteristics of each of their business models, identifying those characteristics that are part of the DNA of different textile enterprises and brands Fig. 7.

The graph thus represents the number of times that a given strength is present throughout the 12 enterprises, to determine which are developed and which remain to be increased. This is in order to determine the quality of the spaces that the tertiary sector provides to the artisans so that they can develop professionally.

It is then observed that all the selected ventures have participated in competitions to obtain funding and/or visibility; 75% offer a diversity of products, while 58% of ventures demonstrated the development of defined collections.

It was also found that 50% of the companies presented adequate strategic planning and, similarly, only half of the enterprises give real visibility to the artisans through spaces on their websites dedicated to telling their story, as well as narrating the elements of their culture present in the textiles produced and marketed through the brand.

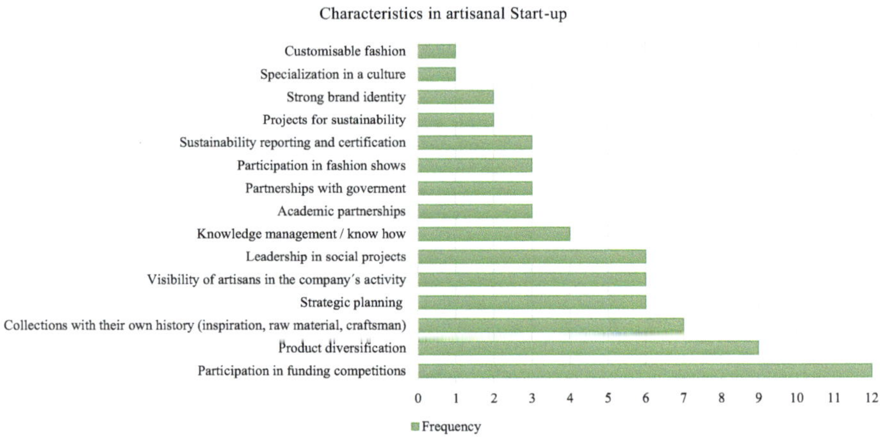

Fig. 7 Characteristics present in artisanal start-ups in Peru

Furthermore, despite being ventures that emerged in an alternative to the private sector, only half of them have demonstrated real leadership in social projects.

Regarding the management and generation of knowledge corresponding to the cultural heritage, 33% of the ventures demonstrated adequate management, ensuring generational transfer.

In relation to alliances with government and academia, only 25% of the enterprises have managed to generate partnerships with either of these for the generation of knowledge or the promotion of commercial activities.

In addition, it was found that only 25% have certificates or present reports to demonstrate sustainability, and only 17% present projects to contribute to sustainability in the environments they influence.

Then, 25% have participated in fashion shows in the textile and fashion sector, such as PeruModa or LIF Week and, in addition, it was found that 17% of the enterprises have a defined brand identity, understood as the set of elements that differentiate it and make it recognizable in the market.

Finally, only 8% of the enterprises offer customizable products, a desired characteristic in terms of the added value of craftsmanship, and only 8% specialize in an Inca or pre-Inca culture in terms of the production of products, designs and/or use of materials.

This analysis shows that, in general, the enterprises have to develop personalized design strategies, and despite being enterprises that base their business model on handcrafted garments and accessories and expressions, only 8% design with specialization in Andean culture or are inspired by its origins to be transformed into designs with more contemporary shapes and styles.

5 Organizations and Empowerment

In the analysis of the selected ventures, it was observed that most of them included empowerment in their value proposition. However, understanding empowerment as that process of vindication of a given group within a collectivity through which each individual acquires power and control to make their own decisions and determine the role they wish to fulfill; this role presents the individual as possessing capabilities that can be managed with autonomy to achieve certain results, as cited in [17], it can be inferred that every craftsman empowers himself through the professionalization of his knowledge; this is due to the fact that when craftsmanship is presented as a profession, it is recognized as a cooperative social activity, whose goal is to provide society with an indispensable good, which in turn implies the cultivation of excellence [18].

This means that the artisan is seen as one who generates meaningful craft products, which defines the purpose of his or her profession: to cultivate and preserve the culture of the original peoples, bringing it into the contemporary world.

It can be deduced, then, that the enterprises that seek to enhance the original cultures of Peru must be the space where each artisan empowers himself, that is, where he begins to form part of the systematization of knowledge, acquires identity and a sense of belonging and enhances the quality of his work through exchange with his peers, which results in him becoming a professional with decision-making power over his creative process and over the narration of his story through the textile (Fig. 8).

Thus, the empowerment of the artisan must come from his or her capacity to provide exclusivity, originality and personality. The first characteristic responds to the capacity to offer a unique, high-quality product; the second aims to combine technique with creative thinking to develop a good that possesses meaning, that communicates; finally, the personalization of the product is called for, making it customized, where the artisan endows the user of the product with his knowledge and skills and thus ends up empowering himself by being able to narrate his own story through his art, which is well remunerated [19].

6 Conclusions

Cultural tradition should not be defined only on the basis of the intrinsic characteristic of being static, but on the contrary, to show that the changing character of it is given both by the representation of the tradition of different cultures from different periods (pre- and post-colonial) but also as the combination of traditions and techniques independently of the syncretism observed in them, but expressing their roots, to the pure representation of the cultural legacy in the textile. Furthermore, this process can happen through epistemological decolonization, considering three essential issues: making it clear that knowledge has value, color, gender and

Fig. 8 Ayacucho embroidery, with cotton, sheep wool and alpaca fibers

place of origin, so that the place from which one thinks is extremely important. Decolonization requires recovering, revaluing and applying ancestral knowledge, but questioning the associated temporality and locality. This process should not be limited to the relationship of Western or Eastern ancestral knowledge as identified blocks, but as critical distributions for new processes of cultural intervention [2].

The coexistence of tradition and commerce is completely feasible by means of the incentive to research, development and innovation of products that express elements rich in history, tradition and culture in addition to those already existing, which are novel either by the introduction of techniques, materials or ancestral iconography of origin, or by the expression of the feelings that characterized the community in a certain period. Such as the resistance to colonization, which raised its voice against the impositions of the colonial order, reconstructing its identity within the framework of this resistance and transmitting that the true colonizer is the one who conquers his work with his own culture (Fábregas, 2014), thus revaluing craftsmanship and contributing to a textile product that shows a brand identity, resistance, personality and cultural legacy, where the textile product is the vehicle of collective communication for the next generations [20].

The professionalization process of the artisan is necessary for the sustainable maintenance of artisan collectives and associations, as this will be the factor that determines the capacity to manage the business, differentiation, degree of specialization and quality of their products, implicitly developing negotiation, marketing and management skills [21].

The fashion industry should be seen as one of the strongest potential allies of crafts through the struggle for change and the search for uniqueness, exclusivity, quality and fusion between tradition and innovation, while at the same time recovering ancestral techniques that are likely to disappear.

Artisan associations and collectives must learn to communicate the right messages, which help to raise awareness and sensitize the last link in the value chain in the fashion industry—"the customer"—who can impact not only the lives of artisans but also the protection of the planet, as they have the choice through ethical and responsible consumption.

References

1. Waqi Sajjad F (2020) Education and radicalization in Pakistan. A post-colonial perspective. In Routledge (ed), Ethics and education (1er ed.), Vol 1, pp 17–34. Taylor & Francis.
2. Pulido Tirado G (2009) Violencia epistémica y Descolonización del conocimiento. Sociocriticism, 24(1–2), 173–201. https://dialnet.unirioja.es/servlet/articulo?codigo=4637301
3. Espinosa MM, Gilyam MG (2012) Sincretismo Cultural. Mestizaje cultural en México y Perú.
4. Villalobos Herrera Á (2006) El sincretismo y el arte contemporáneo latinoamericano. Revista de Sociedad, Cultura y Desarrollo Sustentable 2(2):393–417
5. Beltrán Y (2017) Violencia epistémica en la protección de los conocimientos "tradicionales." Ciencia Política 12(24):115–136
6. Ojinaga Zapata B (2020) Los hilos de la memoria: tejiendo las narrativas de los centros textiles en México y Perú. Revista de Historia, Teoría y Crítica de Arte, 6, 42–61. https://doi.org/10.25025/hart06.2020.04
7. Jones S (2019) Woven textile crafts in contemporary commercial contexts: waving not drowning. Text: J Cloth Cult 17(2), 110–119. https://doi.org/10.1080/14759756.2018.1473980
8. Higuera A (2000) Textiles Prehispánicos - Técnicas textiles. Página de Arqueologia Andina y Tiwanaku. http://www.tiwanakuarcheo.net/13_handicrafts/textiles_tec.html
9. del Solar ME (2016) La creación del challpi wathrako en la identidad wanka desde las primeras décadas del S.XX. Producción de fajas tejidas en el valle del Mantaro, Junín, Perú. Nuevo Mundo Mundos Nuevos, 1–41.
10. Stastny F (1981) Las artes populares del Peru. Ediciones Edubanco
11. Carlson U (2017) Iconografia Andina: La imagen divina y su representación en el antiguo Perú
12. Olguín Bravo NA (2019) The four-pointed hat, its importance, loss and possible revaluation in the peruvian contemporary context. https://repositorio.pucp.edu.pe/index/handle/123456789/185903
13. Meche Correa. (n.d.). Retrieved September 4, 2022, from https://www.mechecorrea.com/meche-correa.html
14. Jorge Luis Salinas: "Para un diseñador estar en Macy's es como ir a una Olimpiada." (2015). Diario Gestión. https://gestion.pe/tendencias/jorge-luis-salinas-disenador-macy-s-olimpiada-91237-noticia/
15. Alexandra Montoya Casillas: "Abayat Arequipa: llevando nuestros tesoros al siguiente nivel". (2022). Diario El Búho. https://bit.ly/3ScUGYO
16. Jose Miguel Valdivia, Diseñador. Jose Miguel Valdivia, Facebook (2022). https://www.facebook.com/JOSEMIGUELVALDIVIASTUDIO?locale=es_LA

17. Pando Ezcurra TT, Cangalaya-Sevillano LM, Herrera Mejía ZE, Cabrejos Burga RE (2022) Liderazgo y empoderamiento en las mujeres empresarias en el Perú. Revista de Ciencias Sociales 28(5):234–245
18. Cortina A, Conill J (2000) El sentido de las profesiones. In El sentido de las profesiones, pp 16–28.
19. Design, artesanato e apropriação cultural. (n.d.). Fashion for Future. Retrieved August 29, 2022, from https://www.fashion-for-future.com/post/apropriacao-cultural
20. ONG IMPACTO. (2021, April 21). El textil como símbolo de identidad y resistencia. UNESCO. https://www.unesco.org/es/articles/el-textil-como-simbolo-de-identidad-y-resistencia
21. Luza Esquivel PY, Robles Martínez NY (2022) Factores claves de éxito de los negocios de artesanía en la provincia de Huamanga-Ayacucho en el 2019 [Programa Académico de Administración y Finanzas, Universidad Peruana de Ciencias Aplicadas (UPC)]. http://hdl.handle.net/10757/648695

Cultural and Cultural Appropriation Challenges of Indigenous People in the Global Fashion Industry

LeeAnn Teal-Rutkovsky

1 Setting the Stage

The international fashion industry faces many challenges related to sustainable design, production, and distribution [8, 9]. This chapter suggests focusing on a cluster of these specific challenges found within the fashion industry, specifically the current and potential impact of the indigenous people population and culture [8, 11, 12] within the fashion industry.

The definition of fashion referred to in this chapter proposal defers to that of Coco Chanel: "Fashion does not only exist in dresses; fashion is in the air, it is brought in by the wind, one feels it coming, breathes it in, it is in the sky and on the pavement, it depends on ideas, customs and happenings" [9, 10].

In addition, for the purposes of this chapter, indigenous people reflect the United Nations definition to be: "Indigenous peoples are inheritors and practitioners of unique cultures and ways of relating to people and the environment. They have retained social, cultural, economic, and political characteristics that are distinct from those of the dominant societies in which they live" [13, 25].

Additional foci look to the fashion industry to adopt the role as culture custodians, rather than deterrents engaging in such practices degrading indigenous people contributions to the fashion industry such as greenwashing, rather, impacting and adding great value to indigenous communities. The fashion industry has the power to unleash unlimited potential thus adding great value to the lives of indigenous people, through elevating the understanding and importance of culture [11, 12].

L. Teal-Rutkovsky (✉)
Impact Fashion Inc., Providence, RI 02865, United States
e-mail: lateal@gmail.com

M. Á. Gardetti and R. P. Larios-Francia (eds.), *Sustainability Challenges in the Fashion Industry*, Sustainable Textiles: Production, Processing, Manufacturing & Chemistry,
https://doi.org/10.1007/978-981-99-0349-8_9

2 Cultural and Cultural Appropriation Challenges of Indigenous People in the Global Fashion Industry

At best and at the very least, the international fashion industry faces many challenges related to sustainable design, production, and distribution [8, 9]. There are indeed a cluster of these like challenges throughout the international fashion industry that spill over and affect indigenous people by the shear nature of where the industry chooses to offshore. Here the explicit foci is driven by the global fashion industry, noting the past, current, and potential future impacts on the indigenous people population but not limited to Culture and Cultural Appropriation [11, 12].

The United Nations adaptation of what Indigenous people are characterized by may be noted as follows: "Indigenous peoples are inheritors and practitioners of unique cultures and ways of relating to people and the environment. They have retained social, cultural, economic, and political characteristics that are distinct from those of the dominant societies in which they live" [13, 25]. This adaptation of Indigenous People is highly compatible with the essence of what culture is as they represent the long lineage of their peoples.

The United Nations adaptation is further supported by Hofstede as he embraces the essence of culture as defined by Kluckhohn to be, "Culture consists in patterned ways of thinking, feeling, and reacting, acquired, and transmitted mainly by symbols, constituting the distinctive achievements of human groups, including their embodiments in artifacts; the essential core of culture consists of traditional (i.e., historically derived and selected) ideas and especially their attached values. (Kluckhohn, 1951, p. 86, n. 5).

Inspired by Kluckhohn, Hofstede treats culture in his book *Cultures Consequences* as

"The collective programming of the mind that distinguishes the programming of the mind that distinguishes the members of one group or category of people from another." [11, 12].

A suitable conceptualization of culture is decisively crucial to the understanding of the true impact of usurping and attempts to weaken the indigenous peoples "cultural resilience [20, 22]." With centuries having experienced such things as oppressed language, dress, ability to create a firm sense of food and occupational securities, a firm understanding of culture is necessary to fully comprehend the potential damage colonial behaviors could, did, and may have in the future on countries they have colonized as the superior actor [20, 22].

It can be said that fashion is all consuming, exists in all cultures as all peoples need to be clothed. When one thinks about it, fashion is everywhere. It is in the bedroom, the workplace, the gala, and the hospital bed. Not to stop there. The act of fashion, construction et al of fashion involves water aquifers, rivers, forests to produce for example such textiles (arguably whether sustainable or not) viscose and rayon with deforestation implications, chemical waste and fast fashion excess credited with building new landfills around the world. Fashion is everywhere.

As far back as 1893 in an article, "What is Fashion" in a monthly review asks, just what is fashion? In the article is a quote, "But seest thou not what a deformed thief this fashion is?" [6] And the article goes forward. If the implications weren't so poignant and so serious it would be hysterically amusing. It talks about a father who was distraught as he was now anticipating the New Year fashion trends, and he pleads for the trends to be a spin-off, elevation of the bonnet styles of the year before, so he didn't have to go out now, purchase more for his daughter. The author goes on to discuss what in clothing is utility and what constitutes fashion. While we in this post-millennial era of fashion, must contend with the fallout from fast fashion, non-utilitarian fashion, it is not such a new concept, after all. We cannot however defend the indefensible where fashion cycles have gone from four seasons to weekly in many cases.

What then is fashion really? Looking to define fashion there may be indications to move forward and away from a more standard example as suggested by Kate Fletcher as simply, "Consumption, materialism, commercialization and marketing," [22]. One may be better present and future served looking to Coco Chanel to a more organic, deconstructed view. Defining fashion here, refers to and defers to that of Coco Chanel: "Fashion does not only exist in dresses; fashion is in the air, it is brought in by the wind, one feels it coming, breathes it in, it is in the sky and on the pavement, it depends on ideas, customs and happenings" [9].

With the combination of culture and a deconstructed approach to what fashion is, the Indigenous communities remain exposed to the mostly self-regulated fashion industry. Fashion effects manufacturing, fossil fuels, shipping, waste in production and post-production, taxing ecosystems on land, air, and sea. In post-colonial affected countries where the production of cotton is involved is considered one of the dirtiest and thirstiest operations of plant-based textiles. The fashion industry lacks the engagement of such professionals as urban planners and urban architects where to date on a regular basis have un-noted relevance and should be elevated to as critical a status as a designer to the garments.

Cotton is a staple crop in many developing countries rich with indigenous populations. Something that will be discussed further in cultural appropriation here regarding cotton will intensify this current discussion.

Any cotton used without organic certification has been treated with pesticides, insecticides, and possibly uses genetically modified (GM) cotton seeds, "that are genetically engineered to produce a natural pest killer" [5], at the beginning of the growing process [10, 11]. More specifically, insecticides dominate traditional cotton production with "pyrethoids and organophosphates the most widely used [10, 11]. The World Health Organization classifies them as moderately hazardous. *However, those used in developing countries are classified as highly hazardous, generally acutely toxic nerve poisons that can contaminate groundwater"* [10, 11], *p. 9).* "Cotton "sold as organic in the US, doesn't have to be certified organic as certification is market not label driven as complex non-universal organic labeling guidance creates markets of perceived value". [21, 24], p. 1). An example of this confusion can be found in the use of traditionally grown cotton. Even though a fabric is labeled 100% cotton, the production of one t-shirt using traditionally grown (i.e., nonorganic) cotton, from farm to retail, uses 33% of a pound of insecticides (Organic

Trade Association, 2012) [19] and over 800 gallons of water in pre-and post-harvest [21], p. 1,National Geographic, 2010, para. 1) [15].

Looking further, serves to illuminate an immediate nexus surfacing, vis-å-vis motivating a brief discussion of anthropocentric and biocentric mindsets as Colonialism and post colonialism residue remains throughout the fashion industry today. This mindset is infused into centuries of fashion industry behaviors centered on the indigenous populations.

We can look to anthropocentric defined explicitly as the position that "considers man as the central fact or final aim of the universe and generally conceives of everything in the universe in terms of human values [27]. Further the literature contends that, "antianthropocentric, biocentrism is the position that human needs, goals, and desires should not be taken as privileged or overriding in considering the needs desires interests and goals of all members of all biological species taken together" [27].

From a modern perspective, looking to a more centrist ideological meeting of the minds suggests a compromise to be found somewhere in between, while there are dissimilar values between the anthropocentric and biocentric points of view, finding cooperation incorporating a balance between the two to be "more complimentary than competitive" [13]. Respect for the humans respect for the environment.

Further, it can and should be noted that the totalitarian anthropocentrism approach to building wealth continues to surface as the dominant force in the creation, support and maintenance of the fashion industry profits, albeit the bottom line. This is realized at the expense of one of the most precious resources in the ecosphere, our indigenous people populations and their priceless past, present, and future contributions to the world.

A wonderful example of just one of many, priceless eco-processes in-waiting, hoping to be extracted by the fashion industry from the indigenous populations can be found in Iceland. Iceland is a most amazing Island country located in the Northern Atlantic Ocean. Through necessity the Icelandic indigenous people used and created a process with what they had at their disposal from the ocean. The ocean provided a zero waste clothing and such things as jewelry and satchels from by-products of what they ate, sustainably suited, fish skin. It is demonstrated that fish skin leather has a far lower environmental impact than traditional tanned leathers [20, 22].

Further, history supports the tenacity of these indigenous cultural/fashion eco-architects. One of many great examples, the Icelandic indigenous people patently demonstrate sustainable success and fortitude. That even with "systemic colonization and repression of their language, culture, and fishing rights as well as dramatic changes in seafood [13] security could not thwart their ingenuity and contributions to the global sustainable fashion movement.

Notably, there is nothing fishy about the forthcoming detail. Dating back to the ninth century, the survival of the Icelandic indigenous people centered on the aboriginal hunting and fishing rights. More powerful national governments stepped in to regulate and restrict their own rights to then hunt and fish and worked to convert them to a more agrarian livelihood based on the invasive national culture and contrary to the life the Icelandic aboriginal people had ever known. This meant a coercion to give up their relationship with the ocean. Iceland reaching to its indigenous roots, uses the skins of the wolffish in the luxury goods genre, to make shoes using tanning techniques from their ancestors. The wolfskin shoes [20, 22] serve as an exemplary model for the fashion industry to embrace. The indigenous tenacity, spirit, and brilliance of creating naturally sustainable processes are well served creating a sustainable fashion infrastructure created to endure. As they have been proven to do, generations and beyond. These processes are not part of mainstream fast fashion machines where the challenges are many here. Who will support teaching the future generations? Through serendipitous measures? Hardly a prescription for success. How do we envision the design for success then, building the infrastructure of support for like-minded measures in the fashion industry? "Action by-Design" where operations akin to the wolffish skin shoes become the mainstream rather than the exception.

Similarly, one looks to the rich in history, resources, talents et al of the African indigenous heritage. A heritage that forges continued paths to modern times despite the colonialism challenges it has faced. As a result, yet again, overcame and continues to overcome challenges with reinforced tenacity and vigor, virtues that aboriginal culture exudes.

Challenges to culture, also rich in history in Africa where colonialism laid centuries of efforts and in many cases successful efforts by stronger nations to abate the will of the indigenous people. A look to such reflections is captured in the literature where there is a voice of support for the usurping of the African culture. The concepts even found a voice in the Novel "*Things Fall Apart* [1]." There is a defense of the colonialism behavior crediting such things as religion, commerce, and English language development. It then also faults colonialism as native languages became a lost memory diluting the heritage and culture yet again of the African people.

While colonialism and post-colonialism paint a broad brush throughout all of Africa, there are significant cultural differences among the different individual countries. For example, Nigeria and South Africa where post-colonial behavior and effects are remarkably different. According to (Hofstede) [11, 12], there are six cultural dimensions. The dimensions are Power Distance, Individualism, Masculinity, Uncertainty Avoidance, Long-Term Orientation and Indulgence. Each country is assigned different scores clearly demonstrating culturally, the differences among each.

> The dimension deals with the fact that all individuals in societies are not equal. It expresses the attitude of the culture towards these inequalities amongst us (Hofstede) [11, 12].

We find from further interest here, that in Africa both Nigeria and South Africa while they share the umbrella continent of Africa, score far differently in dimension number two, Individualism. Hofstede concurs that while Nigeria identifies as a collective culture where members of that society hold to their identity as a country as tied to their close relationships as a group, looking out for each other. South Africa identifies as an individualistic country, where according to Hofstede, culturally they are strongly inclined to care for themselves and their personal families as that is where their loyalty lies.

Translate this into fashion, however, and both countries are producing designers and clothing lines taking the global fashion industry by storm. Consider the mold of cultural constraint broken remaining of course utterly rich with history. The new fashion stand of Africa overflows with brilliance of colors, design, patterns, indigenous sourced sustainable dying, beading, and all things certainly, African.

Demonstrated are both cultures from Nigeria and South Africa (a sampling of wonderful modern African fashion runway bound and, in the making) culturally and African grown. These designers are gifted up-and-coming designers in the global sphere of fashion who clearly, vibrantly, and independently provide the world with African and global influence in their designs, clothing and textiles. They deserve the world stage. The talented Fisayo Longe and Ugochi Iwuaba from Nigeria and up-and-coming sustainable textile and fashion designer from South Africa Sindiso Khumalo are on the forefront of bringing the voice of African fashion to the world stage.

It is one matter for African textile and fashion designers to create culturally inspired textiles and fashion based on their heritage, their culture, their birthrights. However, enter the world of cultural appropriation. While we have developed a rich dialogue here on Africa, let us continue then. One looks at the definition of cultural appropriation.

ervasive functioning definitions of cultural appropriation take center stage in a deep dive of the Bloomsbury analysis [23, 25]. They speak of the words "elusive" and "Freighted with ambiguity." It they contest, it becomes is a conglomeration of events taking place not just in the final product displayed on the runway. It is about not merely being inspired by other cultures, it in addition is the "appropriation of labor in low wage regions, appropriation is ubiquitous and literally woven into the fabric of the fashion system".

Further and perhaps more basic, "cultural appropriation is the taking of aesthetic or material elements from another culture by someone who is not a member of that culture without giving credit or profit. The taking is not unlike stealing or plagiarizing." They go on to say that permission is not normally granted and many times especially in the fashion industry the appropriator profits from the theft and most often is typified by the whiff of colonial residue, it is done from a position of power. Indigenous people having their culture appropriated may well be the people that make the clothing at lower than living wages among other humiliating experiences in the production process where they are the underlings.

This discussion follows a natural path of progression in the fashion industry, following the powerful extracting profits from the weak. In a deep dive into two November 2020 articles, published only 2 days apart, both address Cultural Appropriation in the fashion industry exposing more layers of interest and concern.

Women's Wear Daily headlines an article first, "How Cultural Appropriation Became a Hot-Button Issue for Fashion." The dual authors, Blazio-Licorish and Anyanwu immediately reflect on their interview with Dr. Benedetti Morsiani (a research fellow at the University of Westminster in London in the department of Modern Languages and Cultures) who points to as examples as using current African inspiration to such designers ((however not limited to) as Jean Paul Gaultier, Donna Karan and Dolce & Gabbana [7].

Women's Wear Daily sees the power and the financials of these design dynasties far outweigh that of culturally born African designers who are likely as or more talented, receive "far less impact and attention and representation of their own". The design dynasties capitalizing on the profit end, are disproportionate to the actual Africans lack of benefit who serve as the key monetized cultural inspiration.

To continue, the summation of the critical path of fashion appropriation succinctly relies based on a quote again from Dr. Benedetta Morsiani, that cultural appropriation is "a phenomenon now mainly referring to the exploitation of marginalized cultures by more dominant mainstream cultures [7]."

Dr. Morsiani goes on to note encouraging detail regarding the younger generation being, "more aware and vocal about cultural appropriation due to the cultural diversity in metropolises, and their awareness of the lack of representation for specific groups" [7].

The second of the two articles authored by Miles Socha, *Fashion Has Found Beauty in Other Cultures for Centuries-but Has It Given Due Credit?* In it he shares that there is a welcome "sea of change" enroute as "borrowing of material culture, particularly from disenfranchised parts of the world" is now under far more public scrutiny and vigorous discussion [24, 28].

But to what extent has the phenomenon of cultural appropriation been accepted by the fashion enriched, rich, and famous. Now coming under greater scrutiny? A good place to begin as an example and a prologue to this continued discovery, one looks to the year 2010. September and December 2010 published articles chronicles the five-year anniversary of Irish singer and wife duo, Bono and Ali Hewson's 'Edun' their African fashion safari. The digression begins.

As fashion and design houses go, there is a gravitational pull from all over the world to Africa as precisely supported by Dr. Morsiani of University of Westminster. The extent to which the African culture is represented is manipulated under the guise of extractive economic enterprising fashion houses time and again in the fashion industry. When Bono and his wife Ali Hewson began their clothing line Edun; it was to be all about Africa and empowering the people. In 2005 as a matter of public record, Edun was all in for Africa. The premise was the clothing line would gather and produce mainly in Africa, by Africans and sourced in Africa. The accolades and hopefully profits would be far-reaching for the Irish business partners.

LVMH (as in luxury brands Louis Vuitton and Moet) also became part of the ownership.

Ensuing, the famous Advertisement in September 2010 where Bono and Hewson were dressed in Edun and sporting Louis Vuitton fell flat. The key to the ad was, "Every journey begins in Africa" [4]. Where were the Africans.

As the Christian Science Monitor put it, "Removing Africans out of the scene, out of the image of the 'natural' African landscape is an old trope in the exploitation of the image and a reality of Africa" (Windborne, 2010) [28] Being exploited of course. Where were the clothes produced? As history would have it, likely China. At one point they were producing 85% of their line in China and were, as a matter of public record, called to task during a New York Fashion Week event. While it sent shock waves through the sustainable fashion circle of trust, however, this was Bono. This was Louis Vuitton. This was going to be fine.

And so, it was. They went on to Africa-explain, all the way to 2013 with a new partner, Diesel. Enter Paris fashion week, enter Edun, enter the African thematic/dramatic/folly. According to the Observer New Review in a Hewson interview by Sean O'Hagan. Hewson, while producing in China under the guise of all things African, was quoted as saying, "In the next 20 years, Africa is going to blow our minds". She talked about the cotton in Uganda, Kenya, Tunisia, Morocco, Madagascar. Sound-bytes associated with Africa that sells. No human connection to the Africans behind the label, however there was a French connection, Louis Vuitton and an Italian connection, Diesel [17, 18].

One remaining exploitive fact would be, that in the advertisement featuring the Edun duo departing a perfectly tailored private plane, in a Pristine African field, with no Africans. There was a tiny disclaimer, "Profits from the bag, (Louis Vuitton-not Edun bag) as well as Bono's fee to benefit the conservation Cotton Initiative Uganda and his own Edun line of clothing, which he markets as garments and jewelry, made in Africa with African materials." Classic. Then, how much of a percentage? And the cultural appropriation, post-colonial fashion industry behavior continues. Who really profits?

As a follow-up in 2018, LVMH makers of Louis Vuitton, Moet-Hennessey among other luxury brands, pulled out of their partnership with Bono, Hewson, and their Edun. Edun closed their US operations and the sole brick and mortar store in New York City [14, 16].

Cultural appropriation works best when the aggressor comes from a position of strength as to securely dominate the weaker, culture usurped.

In the book, *Who Owns Culture?* Appropriation and Authenticity in American Law," and author, Susan Scafidi, a Law Professor at Fordham University in New York City confides that it is a difficult task to explain the concept of cultural appropriation. She cites one example, that in the United States cultural appropriation, "almost always involves members of the dominant culture borrowing from the cultures of minority groups" [16]. It is clear the same can be said regarding colonialism efforts of stronger European countries such as China, France, Great Britain, Netherlands, Germany et al in Africa [16, 17].

Edun was never able to get a firm profitable hold with the Africa inspired clothing line. Perhaps one of the reasons is that they saw the vulnerability of Africa to share without permission. It has all appearances that they took the majesty of Africa, of the use of the people and the geography as window dressing for the Luxury brand. They used Africa to elevate this luxury brand and did nothing to elevate Africa.

> "People Think We All Live in A Zoo"—Exclaims Princess Senamile Masango of the Zulu Nation.

Perhaps the missing link here, was their inability to understand that Africa is now post-colonial and Africans, well in the words of a South African Princess Senamile Masango, "People think we all live in a zoo." To the global fashion industry, prepare to begin a newfound understanding of real Africa and real Africans.

Introduced here is South African Millennial extraordinaire, African Princess of the Zulu Nation, Nuclear Physicist, Fashionista. This is someone Bono, Ali Hewson, LVMH should meet in the way of discovery. To discover there is more to modern Africa than serving as a backdrop to enrich themselves, their brands.

And to this, a real South African woman as introduced by the following interview detail:

I had the great honor to spend some time interviewing Princess Senamile. Just an amazing African woman, just an amazing human being. And just ask her as she offers in her own words that, no she does not live in a zoo. She has never even seen a lion in the wild but does confess to seeing an elephant and a giraffe. Her journey through life as a South African has been a blessing to Africa and internationally as well and without the props. She loves fashion, both modern and traditional when culture calls.

THE STORY OF AN AFRICAN PRINCESS

FASHION INDUSTRY TAKE NOTE

SHE DOES NOT LIVE IN A ZOO

Princess Senamile Masango is a Princess by birth, from the Zulu Royal household.

Princess Senamile is a young South African Nuclear Physicist. She also heads the Senile Masango Foundation which also hosts for women, WISE ... Women In Science & Engineering.

In a brief interview we discussed African culture. She notes that in South Africa, the native language is now not passed onto the future generations. That there is very little history taught and a lack of museums chronicling the rich history there. Fashion there is more individualistic, following the terms of the [11, 12] analyses where in Nigeria where they are more community oriented, they are more apt to wear traditional textiles and clothing there. She loves modern dress, although very respectful of traditional dress when there are cultural events. As a member of the Zulu royal family, the chosen garments are of Cheetah, not cotton or silk. When there is an event, and it does not require the royal Cheetah, she sources the organic fabric and has someone she knows, cut, sew, and fit.

She is the first African woman to be part of the African led experiment at CERN (European Organization for Nuclear Research), one of the world's largest and most respected centers for scientific research. Princess Senamile is Founder and Chair of Women in Science and Engineering in Africa, Non-executive at NECSA and Chairperson of the Research and Development Subcommittee at NECSA. She is one of the youngest Board Members to be appointed at a schedule 2 state-owned enterprise (SOE), Council Member at the University of the Western Cape and Board member at Moses Kotane Institute.

Princess Senamile was raised in a village called KwaNongoma a deep rural area in KwaZulU-Natal. She has become one of South Africa's very few black female nuclear physicists. She earned and was awarded a BSc (Physics and Electronics), and BSc (With Honor) in Nuclear Physics from the University of Zululand. In addition, she earned a Diploma in Project Management from Varsity College; MSc in Nuclear Physics (Cum Laude) from the University of Western Cape and Postgraduate Diploma in Energy Leadership from Wits Business school. She was a PhD candidate in Nuclear Physics at the University of Western Cape.

Her commitment to African youth, puts her as South Africa's Research Leader at BRICS youth energy outlook. She established the Senamile Masango Foundation as an NGO encouraging women to engage one another about social issues that affect them as well as plans to create a mobile science lab to travel to villages teaching STEM. She was also named one of the top 50 Global inspirational women of 2020.

In addition, she addressed the South African President, His Excellency Cyril Ramaphosa, the late King His Majesty King Goodwill Zulu (her uncle), and the former President Jacob Zuma.

She has shared platforms with internationally renowned figures such Naomi Campbell; Dr Nkosazane Dhlamini Zuma: former AU Chairperson, currently Minister of COGTA; Former President of Ghana, John Maharnan and more.

She was invited to be part of the Second Eurasian Women's Forum 2018 in Russia. The Eurasian Women's Forum is the world's biggest influential platform for discussing the role of women in-contemporary society. She was invited to FEMROOT hosted by United Nations 75th General Assembly, the aim of the event was to highlight the importance of having females in STEM from an African perspective. Senamile is a frequent guest on African multimedia from print, television to radio and Podcasts and recently was invited to personally meet with the USA Secretary of State Anthony Blinken.

African women empowerment remains a large segment of her platform encouraging post graduate study to all South African women with a great desire to make "Education fashionable in her lifetime." The fashion industry should clearly take notice as beginnings are never too late. Fashion would be far better served beginning with respect not usurping of indigenous people as the look forward now, is to Northern Nigeria.

Moving forward now to the Indigo dying pits of Kano, Nigeria. Part of the cluster of challenges addressed is the current existence of post colonialism in developing countries, continuing with a discussion of post-colonial events taking place in Nigeria specifically the Northern Nigeria region.

Original indigo dye pits can be dated as far back as 1498. Kano has been known for its natural dying pits of indigo where indigenous people have carried on the indigo dying traditions with the use of no chemicals.

While there is ongoing discussion whether the Indigo was introduced to Northern Nigeria by India remains unresolved. However, "indigo dyed hand-woven strips and garments were found in Egyptian tombs during the period of the Pharos. Thus, an indication that the indigo plant was present in Africa, even though it may have not been specifically cultivated for dying at the time as far back as 1498. It was concluded that this evidence refuted the usual contention that the indigo plant was introduced to Africa in the nineteenth to seventeenth century by India [2].

It is no secret what has been happening to the indigenous clothing and textile population in Kano. After centuries of organic, sustainable contributions to the world of the indigo and unique patterns and weaving techniques the China post-colonial era is now on point to extinguish the legacy. This is not only an industry being decimated, but also centuries and generations of knowledge and expertise shared through families. The dominant controlling the lives and livelihoods of the weak [3].

According to the BBC, the urbanization and colonial policies are said to transform the Kano economy. However, with such things as a rise in cheaply producing foreign factories vs local designers and dyers (organic), the industry is turning over to new, cheap, low-quality imitations. It is this imitation industry that is consuming and eliminating the indigenous families, workers, and livelihoods. It doesn't pay to train family members anymore as the businesses are not turning profits.

In an important,f July 2022, interview, author, China foreign policy expert and contributor to the Hearing before the Subcommittee on Africa, (United States Congress) Global Health, Global Human Rights, and International Organizations, "China in Africa: The New Colonialism?" Mr. Gordon G. Chang shares insights here. The study, while comprehensive, concurs that, "While a number of African nations have welcomed Chinese engagement and investment, it often comes at a very high cost, with a focus on extractive economies," as clearly evidenced in Kano.

We discussed the Kano, Nigeria scenario where the Chinese have completely dismantled the city in the indigenous population's centuries old organic textile and clothing industry. Kano is world-famous for its rich organic indigo dyes, weaving and textile printing. In addition, the indigo dying pits tracing back to the sixteenth century and likely longer, no longer have significance to the area. The Kano artisans as part of the Chinese post-colonial behavior have reduced the organic indigenous industry to now creating tech packs, where the merchants send their African designs to be made in China and they in turn must buy them from China to sell in Nigeria as authentic, however not, so they are faux.

Mr. Chang notes that Chinese ruler Xi Jinping views the world through a Han nationalist lens, seeing other peoples as inferior. Xi believes he has a right to rule tianxia or "all under Heaven." All others, in his view, must obey him.

Mr. Chang states, "China's relations with Africa have, in many ways, devastated local industry there. Beijing incessantly talks about 'win-win' cooperation,' but the Chinese view the continent in colonial terms." Mr. Chang goes on to clarify that Xi Jinping views relations with Africa as more than "neo-colonial." "Colonialist" would be an accurate term for his approach to the continent. Clearly an ominous and clear indication of the Chinese impact on not only the fashion industry, however specifically, the African textile and fashion industry and indigenous people stagnation.

In September 2007, the United Nations General Assembly voted to establish and adopt the UN Declaration of the Rights of Indigenous Peoples. "The Declaration recognizes that indigenous peoples have important collective human rights in a multitude of areas, including self-determination, spirituality, lands, territories, and natural resources [26].

In addition, the United Nations recently, in the past few years, created the Alliance for Sustainable Fashion. It has such immense potential to coordinate with the fashion industry to create nuclei around the world elevating the indigenous fashion and textile industry. Currently, there are no people on the alliance with a strong fashion background to develop a proper critical understanding of the unlimited potential of good it potentially can serve to engage.

It also sets out minimum standards for the determination of spirituality, lands, territories, and natural resources. It can be noted that this may be a sound beginning toward indigenous peoples afforded protections however time is of the essence. As evidenced in Iceland and Africa as well as the colonialist business models rampant almost standard in the fashion industry more must be done. How to get beyond minimum standards?

The United Nations Alliance for Sustainable Fashion is another sound beginning and should begin a more rigorous coordination with any relevant United Nations group. The issue, at hand here, is that the people involved with the newly formed United Nations Alliance for Sustainable Fashion have very little knowledge of fashion, true fashion as treated here. The fashion supply chains do not merely consist of shipping containers and durable packaging; it is about the communities where the fashion industry does business. It is about urban planners, urban architects, waste disposal, soil scientists, water, and land management.

Fashion is about the community where the fashion industry does business.

In Mumbai, en route to the Indira Gandhi Institute for Development Research where I am an advisory board member, I noted the roads on the way from the Airport were in utter disrepair, some barely navigable. How much money does the fashion industry extract from the indigenous efforts of India? Trillions. The impetus exists creating the need to empower the indigenous communities impacted by the fashion industry to restore, revitalize and renew.

The demands need to shift in favor of the weak. We see Princess Senamile a crown jewel of Africa. Let there be more. We need to dismiss the power of the fashion industry to dominate in pursuit of profit, in lieu of recognizing the power to uplift.

The fashion industry will be well served to adopt the role as culture custodians, rather than deterrents engaging in such practices degrading indigenous people contributions to the fashion industry such as greenwashing. Using the casual zebra and lion to make the sale, it is well served to keep the Louis Vuitton out of the faux safari. They can do better, luxury fashion can do better, and the fashion industry can do better. The fashion industry has the power to unleash unlimited potential in the communities where they do business, thus adding great value to the lives of indigenous people, through elevating the understanding and importance of culture [11, 12]. Let them lead the way to sustainable innovation.

References

1. Achebe C (2007) *Things fall apart*, 1st edn [paperback]. Penguin, NYC
2. Adamu S (2020) The dying dye pits of kofa mata. *BBC Magazine*. https://www.kamso.co.uk/blogs/news/kofar-mata-dye-pits. Accessed 22 June 2022
3. Adamu S (2022) Why Nigeria's historic dye pits in kano risk closure. *BBC Magazine*. https://www.bbc.com/news/world-africa-61066957. Accessed 20 June 2022
4. Anakin (2010) Edun's bono and hewson in louis vuitton ad. *Luxe*. https://www.luxuo.com/style/luxury-accessories/bono-ali-hewson-edun-louis-vuittoncampaign.html. Accessed 06 June 2022
5. Biello D (ed) (2012) Genetically engineered cotton benefits farmers, predatory insects. [Special issue]. Sci Am 9:1-5
6. Bigg AH (ed) (1877) The nineteenth century: a monthly review [Special issue]. Hist Period

7. Blazio-Licorish T, Anyanwu O (2020) How cultural appropriation became a hot button issue for fashion. *Women's Wear Daily*. https://wwd.com/fashionnews/fashion-features/how-cultural-appropriation-became-a-hot-button-issue-for-fashion-1234579968/. Accessed 10 June 2022
8. Carbone M, Blanquart C (2008) Towards a sustainable fashion retail supply chain in Europe: organisation and performance. Int J Prod Econ 2(114):1–25. https://doi.org/10.1016/j.ijpe2007.06.012Accessed28March2022
9. Charles-Roux E (2004) Chanel and her world, 1st edn. Vendome, Paris
10. Fletcher K (2008) Sustainable fashion and textiles: design journeys by kate fletcher. In: Boone T (ed) Fashion Practice, vol 1, No (2). Earthscan, London, pp 271–274. https://doi.org/10.2752/175693809x469229. Accessed 15 June 2022
11. Hofstede G (2001) Culture's consequences-comparing values, behaviors, institutions, and organizations across nations, 2nd edn. Sage, Thousand Oaks, London
12. Hofstede G (2001) Culture's Consequences, 2nd edn. Sage, Thousand Oaks, London
13. Hunter Jr ML, Redford KH, Linden Mayer DB (2014) Complementary niches of Anthropocentric and biocentric conservationists. Conserv Biol. https://doi.org/10.1111/cobi.122296
14. McGoldrick D (2018) Bono and ali's fashion label edun winds down. Irish Central. https://www.irishcentral.com/news/irishvoice/bono-ali-hewson-fashion-edun. Accessed 16 June 2022
15. National Geographic (2010). *The global water footprint of key crops*. http://environment.nationalgeographic.com/environment/freshwater/global-water-footprint/. Accessed 30 May 2022
16. Nittle NK (2018) *A guide to understanding and avoiding cultural appropriation* (Legal Briefing). thoughtco.com. http://thoughtco.com/cultural-appropriationand-why-its-wrong-2834561. Accessed 02 June 2022
17. O'Hagan S (2013) Ali hewson: in the next 20 years, africa is going to blow our minds. Observer New Review. https://www.theguardian.com/theobserver/2013/mar/17/ali-hewson-edun-fashioninterview. Accessed 06 June 2022
18. Orange County Water District (2008) How much water does it take. California. Accessed 20 August 2013
19. Organic Trade Association (2012). *Cotton and the environment*. http://www.ota.com/organic/environment/cotton_environment.html. Accessed 01 June 2022
20. Palomino E, Karadottir K, Phiri E (2020) Indigenous fish skin craft revived through contemporary fashion [Paper presentation]. International Foundation of Fashion Technology Institutes, New Delhi, Delhi, India.
21. Rodman A (2007) Untangling organic textile certification. Tilth. http://tilth.org/education research/in-good-tilth-magazine/articles/207/18iv/untangling-organic-textile-certification. Accessed 01 June 2020
22. Routledge handbook of sustainability and fashion (2015) 1st edn. Routledge, London
23. Rovne VL (2002) Cultural appropriation in fashion, dress, and appearance. Fashion Archives. Bloomsbury, London. https://doi.org/10.5040/9781474280655.
24. Socha M (2020) Fashion has found beauty in other cultures for centuries-but has it given due credit? *Women's Wear Daily*. https://wwd.com/fashion-news/fashionfeatures/gallery/fashion-has-found-beauty-in-other-cultures-for-centuries-but-has-itgiven-due-credit-1234649801/. Accessed 10 June 2022
25. United Nations (2022) *Home page*. United nations deja. http://www.un.org/development/desa/indigenouspeoples/about-us.html. Accessed 29 March 2022
26. United Nations (2007) United nations declaration on the rights of indigenous people [Description of declaration]. https://www.un.org/development/desa/indigenouspeoples/declaration-on-the-rights-ofindigenous-peoples.html. Accessed 14 May 2020
27. Watson RA (1983) A critique of anti-anthropocentric biocentrism. J Environ Ethics 5(3):245–256. https://doi.org/10.5840/enviroethics19835325
28. Winborne T (2010) Need a sales campaign? Look no further than bono and africa. *Christian Science Monitor*. https://www.csmonitor.com/World/Africa/AfricaMonitor/2010/1214/Need-a-sales-campaign-Look-no-further-than-Bono-and-Africa. Accessed 15 May 2019

The Anatomy of One Size Fits All

Clelia O. Rodríguez

> *There is a private riddle I ask myself:*
> *Why did Europeans enslave us in Africa and take us to the United States?*
> *The answer: Because we would not to voluntarily.*
> from *We Are The Ones We Have Been Waiting For: Inner Light in a Time of Darkness* by Alice Walker

Abstract The nature of this political textual offering, one that morphs in all directions without a worry in the world about the "one size fits all" narrative, cannot be contained within the specificities of a standardized 'abstract.' This text can't be abstractized (that word doesn't exist because I just made it up). The "fitting into" the –isms is what keeps us from realizing the worlds of the "I'm possible" because "Impossible" is what gives the perfect measurements for what we have been forced to believe to be. A is for Assimilation. They'll say, "Ah, there goes the rebel, always going against the grain not wanting to write an abstract how it's always been," and my response is "Yes. That's not an insult. It's the Truth." What's here is the urgent call to politicize the notions of bodies, autonomy, voice, culture, colonialism, truth, land, and why not, History, everywhere. The approach to reading 'anatomy' is interdisciplinary and it has been harvested precisely knowing that if we are going to recuperate ancestral knowledge is because we are also ready to dive into the deep waters of undoing how we read "numbers." Numbers through the lens of body, autonomy, voice, etc. I considered giving an explanation or a form of justification of the excessiveness of the footnotes but I think it adds another reading layer to the "One Size Fits All" situation sucking the life out of life that James Baldwin would find kind of funny. In essence, that's a glimpse into what this textual offering is.

Keywords Colonialism · Decolonization · Pedagogies of the body · Cultural studies · Critical pedagogies · Sociology · Geography

---

C. O. Rodríguez (✉)
Ontario Institute for Studies in Education, University of Toronto, Toronto, ON, Canada
e-mail: clelia.rodriguez@utoronto.ca

M. Á. Gardetti and R. P. Larios-Francia (eds.), *Sustainability Challenges in the Fashion Industry*, Sustainable Textiles: Production, Processing, Manufacturing & Chemistry,
https://doi.org/10.1007/978-981-99-0349-8_10

The anatomy of measurements is x-rayed to detect where adjustments need to be made[1] in the body: Two inches here, one addition there, an extra room for restricted movement underneath the garment, and maybe removing an entire layer of fabric that it does not seem to be a good match for the walkway called life.[2] To speak of broken fitting promises, the ones floating in the river of illusions of what a crisis is and is not under the rubrics of Institutionalized Democracy, is to give full consent of autonomy to let go of the measurements.[3] To speak of the liminal responses to wilful official readings of the closeted and unsettled skeletons, as a crisis of silence, is to hear precisely said silenced voices.[4] To speak of the historical narcissism of the consumption of the other before the other was otherized[5] and another other is to worship unknown and unwritten notions of "truth" so existing outside of the rigid measurements suffocating humanity to a never-ending death.[6] To speak of addressing the protagonist role of exact numbers, it implies the most important divorce that every colonized human can sign, the cultural divorce.[7] Without it, the allegations of crimes against humanity committed by the G-7 Group, the Security Council of the United

[1] Reader: If your first grammatical observation is a reaction to my choice of the 'passive voice' in this textual embroidery is because I have no money for linguistic reparations.

[2] The first puncture from taking in measurements based on comparative models to extract a piece of flesh without anesthesia is caused by assuming everyone walks. Some of us crawl because learning requires contact with the land and in horizontal-like movements not restricted by corsets pulled by Victorian ghosts.

[3] One is born when one realizes that our bodies came with the expectations of 10 fingers and 10 toes. Babies are X-rayed from top to bottom and everyone rushes to check if everything is in order underneath the umbrella of ableism. "So long as it is 'perfect,' I guess it doesn't even matter if it's a girl," is what's echoing still in the ignorance filling the minds of people whose existence follows patriarchal-induced offsprings.

[4] One starts sucking the life out of addictions. The 'yes' when one means a 'no.' 'This dress doesn't look good on me' narrative settles its unwanted presence in a land that was occupied the moment your weight was spoken of as the ideal weight for a baby of your size and race. The clan crowning your hips began living rent-free there when they referred to them as 'exotic.' That brand is tattooed in those hips that are constantly lying to themselves.

[5] I once tried explaining to an Elder in my community about the latest theory I had learned in class, 'the other.' The response is a question that remains answered until today. I learned in his response what colonialism was for him, his answer helps me also formulate what colonialism continues to be for me: textual asymmetries invented in the thought of a person existing in a one-dimensional world. Colonialism doesn't require a thorough definition from those whose own flesh has carved in them the words displacement, disappearances, and dignity. One's own disciplined body is what pushes out the wonder that sprinkles theories. A cosmovision where intergenerational nurture happens.

[6] Corsets carrying corpses piled in alphabetical and perfectly aligned order to avoid disturbing someone else's circle of life. That is the lesson to be learned in the methodology class of anatomy–tie it harder, harder–pull in deeper, deeper–until you prove once and for all we are not related to the history of the rib. Crash the layers and layers of ancestral bodily memory. No one is there to listen at that point, anyway.

[7] The cultural divorce is an urgent and necessary step to undo the numbers that make up the endless lists of missing and murdered. Let's not also co-op the "buen vivir." The "buen vivir" is not subscribed to the language we may use to exchange thoughts. It is where the exchange of feelings and thoughts move beyond the binary system. It is not something you can just grab and appropriate in words to fit into a 500-word answer to what discrimination to the self is and is not about. It is definite rupture with the brutality of numbers in the thinking-feeling life we rightly

Nations, the Vatican and the Lords in the Narco World will remain a digital manifesto replacing all traces of human contact.[8] The anatomy of how the Humanities, with an emphasis on THE, was written, carved, tattooed, and erased follows mathematical coincides and synchronized sequences that lead to fear with a real 'f.'[9] To unfollow it is just one crawl away from another crawl to another crawl to a place that knows

---

deserve to breathe. No lawyering is necessary. There are none to take up cultural divorce cases, anyway. Otherwise pharmaceutical companies would collapse from the collective healing. Life in the numerical cycle of lawyers who do not want to pursue are constantly sought after to keep the colonial tie to culture. In the name culture, humans get seized to be sized to a world of symmetries. This is just a glimpse into what that notion of cultural divorce signifies... just a window... just a peak... what lies after cultural divorce arrives and is accepted is where the quarrels of borders vanished from the psychological cultural engravings marinating the place in the body where the heart and the mind keep their breakup alive. It is healing, to put it in simpler yet more complex terms.

[8] Allegations of crimes against humanity–no one will claim either the allegations or the crimes. Rarely, you will have a voice erecting their voice to one of the 7 s in a place that no one is able to hear words they themselves invented. Those who make it to the headquarters of any of the machinery of the state and its affiliations enter a world of possibilities of green spaces in desserts. It is not sustainable. The world without any of the leaders that are chosen to represent its people are stripped off the gift of wonder and of wondering. The question: "What would my breathing sound, smell, feel, taste, and look like if I added a pinch of wonder to my daily wardrobe?" The manifestos of the Lords would most likely be drenched in water and not blood. They will be drenched in hope and not the eradication of dignity. No one sleeps in a terrorized and dictated world by leaders occupying their unearned position to power and male privilege. The mythical notion of peace and security wouldn't make the profits it makes if wonder and wondering will be more than a catchy phrase in think-tanks & its generous donors.

The glorified lords are the ones from the Narco space, the ones that have some divine connection because they are reduced to scary series shown in Netflix, the ones inhabiting how we measure our safety or unsafe ways of living no longer in rural areas in the middle of bushes, the ones who are dressed in fancy Italian designs, the ones who map out the existence of lands and territories because their birth is marked by genocide, the ones who do the surveillance on behalf of other followers of Foucault, the ones who get denied because acknowledging them would be to be against the political party in power and no one wants to piss anyone off, the ones who embrace life by inking a tree on one arm to kill that wonder by keeping track of the people you've killed in the other side of the face, the ones in clandestine skins controlled by the brands, the ones whose history has contaminated the world by the perpetual use of the word *discovery*, the ones who cement to the ground the possible escaping route out of the Narco world.

[9] Fear is the common denominator. Fear of the awkward act of embracing what is returned to us as a reflection of who we have become. Fear because it is composed by the co-existing with the disposable garments resting in a living hell in Xinfeng or Deonar or Delhi or Sudokwon or Puente Hill or Malagrotta or Laogang or Bordo Poniente or any piece of land. We are connected to these lands. We are connected through this common denominator of fear. The real fear is knowing that the returned image will unveil who we are without colonialism. The real fear is sensing momentarily the 'we' in the collective wasn't really a part of any actions towards liberation. On the contrary. It was a 'we' riding an elliptical machine branded as "The change is here!" while getting its source of energy from the living hells above mentioned. That is an intentional way of looking inward for one's anatomy to be X-rayed.

coordinates in the World of Geographies.[10] The One-Size-Fits-All[11] and its flesh,[12] blood, and bones,[13] inspires the promise of forever questioning[14] numbers, all the time.

---

[10] And here comes the politics of "but": The collection of these garments is what riddles my closeted closet, my dungeon without a door, and my Alice in an unwandered land. The unintended harm, they say. The choices I never had. The false identities we inherited with birth rights that never wished for us to be born to begin with. The framing of our so-called transitions within the 0–13 sizing confined us to linguistic spaces, the liability of looking comfortable in front of the chance of a digital selfie, and the shadow of a surprising dysfunctional angle that can decrease our chances to look less golden from a sunset we're not even witnessing: The "but." The "but" that follows the footsteps of able walkers is what keeps geography in all numerical graphs. Precisions in the cuts are necessary for the "but" because that is where corners can be made circles without a historical trace. That is where no one sees or hears anything. The consciousness of a living "but" not getting haunted by wonder and wondering. No imagination lives in this framework of geographical information.

[11] The One-Size-Fits-All narrative is a place of strictness based on judgement, starting with how fast one produces the hundreds of life-less garments that have piled up between footnote 9 and this one. In the flesh of time, we have connected with an absent place of our knowing because it is a better fit if it remains in the absence from the closeted closets. In the flesh of time, many women and children had manifested a historical 'no' that has been added to the aching of swollen ankles. In the flesh of time, the young and free-spirit people in one territory went from holding captive another human because the unknown self did not come from anywhere in existing maps. In the flesh of time, the spectacle of designing without assigning meaning to the land, went from a sweatshop to a boutique. And that is just on flesh. X-raying colonial narratives do this.

[12] In the blood of time one captivates whether a recognition of the measurements happen or not. In the blood of time, the punctures become habits, repetitions, the joystick of shopping, clicking and checking out, our new way of adding value to a humanity to doesn't even register in our present and presence. In the blood of time, we find the tune-ups to all the imperfections in transition to the manufactured occupiers in landfills. In a drop of blood, one carries the how exploitation has been rationalized, the how one understands it as the one thing you don't ever question, the how 'so long is not me' situation until I need someone or something.

[13] In the time of bones, I rely on an X-ray machine to do the kind of reading that I do beyond the binary, as a pedagogical-thinking-feeling offering to co-unlearners. To deny the connectivity between calcium and bone marrow is what takes places when everyone rushes to measure up indicators of social injustice and the –isms that end up lingering the tails of research publications stored in dusty colonial boxes in a dungeon without doors. The task of X-raying bones in a time where bones are supported by alignments to push up the perks of standardized beauty norms in the same old-same old picture is the only acquisition that should be made available without relying on extracting the calcium in the bones of others. In the times of bones, one cannot just look at the image of what is exposed to the naked eye, the remains of calcium in the body, and not do a scavenger-hunt game of placing where colonialism is detected.

[14] What is it about the structure of my jaw configuration that reminds me of the unforgivable debt I paid him once in exchange for the appearance of safety in pins? Why does my spine carry the political meaning of telling the truth when it sees a hanger because it is reminded of straight backs gone backwards in the inner child consciousness? Why does the pelvis remain fixed in a scheme of what the blue/pink codes carry when I look at the keys of the house I grew up in? Why does the femur situates a truth that was captivated in the window built to create the idea that one can breathe away cooperating cultural genocides? Why doesn't this X-ray show the scheme to keep mannequins from showing their "We can't breathe!" SOS messages? Why aren't the arms and legs shown as the cultivation of transgressions that happens when artistry is felt and not just critically thought of? Why do broken ribs show only one side of why they were broken instead of imaging how the value

1 Size 13

"Based on your height and your drips, are you a C-cup?" they asked. What I said: "Right now I may be a D because I haven't picked up my daughter yet. You know, I'm feeding her".

The oohing that *followed*[15] like an empty attempt to imitate a cacophonic warm-up exercise to loose weight after giving birth in a suburban city in the middle of anything powered by *whiteness* felt like a religious dweller when you're in the middle of masturbating when ovulation shows up unexpectedly after hitting the road unannounced for five months. What I really *wanted* to say: "That's none of your fucken business." What you *recalled* saying: "Right now I may be a D because I haven't picked up my daughter yet. You know, I'm feeding her."

The oohing from the *other* new girl: "Sorry, he meant to ask if you were truly a C because of your proportions." The internal door to my thoughts wanted to slap her historical role of justifying everything fucked up men say because of (a) how she was raised to *be* or (b) how to sound like. I do not know what the difference between them is while I am awake. Yes, in plural form. The *be* is a collective. The do is a collective. The collective knocked out the hell outta my teeth and triggered my perfect female colonial wanna be measurements. She *insisted*: "He's new and is getting new to his role. My name is Daisy and I apologize if his response came across as confusing. Let me take care of you know." "I promise I just wanna make it up to you." What I *wished* she had said in my impossible imagination coming from the mini-Macondo that shows up when textiles and fashions are also present. Except, she *did not*. The anatomy of my soul no longer matched the one from my body. This is what I think they mean when the compounded word of intersectionality shakes up a room filled with truckloads of theoretical unfelt and disconnected data from the <quote on quote> 'Global south.'

The oohing in this recent disruption pulled every cell, every centimetre, and every ounce of the sweat I had left resting in the cracked Victorian flooring holding the sounds of "Get me outta here." This cultural twist knew only of painful ideas of measurements as in the times of flesh, blood, and bones x-rayed as the practical way to do cross-references beyond what publication and their domains tell you to do (Care Instructions: Read the labels of footnotes 11–14. Hand-wash only with a narrative of water that is not mapped out in the Truth and Reconciliation Report of Any Truth about land rights. Iron never. Steam with the patience of cotton balls picked by the speed of an 82-year old sick woman. Place synthetic soap at your own risk. Dry on top of a lavender field). Another cross-reference that makes it to cultural studies: The tongues, the collective ones, thirst for stepping on the scales. We can't speak up to any truths coming from cultural legacies... We can't swim when we wear the

---

of a woman called Eve entered into the parasite-like mind of a man? Why does the density of the questions related to bones are innocently and without any political repercussion loaded into the pile of folders marked with the sign of the 'X'? Why do the fractures penetrate in colonial frequencies instead of palpitating in vibrations simulating a real carnival of physical-emotional joy?.

[15] Another altered way to read: all the words appearing in this catwalk in italics are intentional.

same fishing rod to convince a fish to be fished… We can't even begin to organize the word *decolonizing* in the closeted closets because it continues to be sanctioned in the false modesties of the self-proclaimed decolonizing experts… We can't say 'this or that' to speak of the crisis that colonialism and its subsequent hungry children, patriarch & the rest, brings upon our lives if we do not even say the words 'stolen knowledge,' 'stolen land,' and 'stolen lives' when we witness how white supremacy destroy the essence of ancestral embroidery? How does the cult to solidarity gets caught up in the screams of others while participating in the same carnival to collect funds on behalf while exercising the mouth to shower the air with gossip instead of co-creating?

We can't jump from the oohing to the oohing regulating the path of the bee, the whale, the salmon, the Torogoz, the monarch, the fungi, and the dust. The same self-regulation in the oohing when oppression is literally happening as we take another breath is the possibility that in the name of the perceived concept of 'decolonizing,' oppression is literally and not just metaphorically happening. This self-regulation is what keeps the act of doing reduced to the act of filling an empty stomach with the idea that a void dressed as a delicious homemade meal is going to keep you alive. If this self-regulation does not engage with imagining what it would be like to live answering the question of 'who are you without colonialism?' then it is simply moving from one side of the oohing to other side without going anywhere. That accepted immobility is what constitutes the largest percentage of people securing colonialism as the next and the next economic flare-up where vulture culturing will continue to occupy land and the imagination. And this where the controversies or debates get heated even in places where the intellectual matters so long as you accept the one-directional way of knowing what *occupation* is, according to settlers, that is. The same voided-empty calorie dressed in couture-whatever is where people are imprisoned by the belief that liberation can be something to be patented. There is no such thing as the –isms are up for debate. The tongues are fighting hard to stay sharp because they have been reduced to fit into the category that quieter is prettier because it can be disciplined to reduce its inches to pursue the one size fits all idea.

If those two at the store could have been near the act of crawling instead of running in the hamster's cage, this text could be woven through ancestral practices of reciprocity to share how other ways of reading mothering and motherhood are part of reading the land in moments of crisis. If those two would have acknowledged the presence of land from what my body was doing, the conversation would be now entirely focused into what pedagogies of liberation would look like through the political reading of textiles in all of its complexities leaving completely out of it measurements, the timing of time, and cultural landmarks. If those two could have been taught another way of writing that requires no ink, they'd known of my existence because they were acknowledging themselves doing it so. *Like nothing, like the void, like the empty-calorie meal dressed by a mirage, like no idea.* If those two would have experienced in their own flesh, blood, and bones the ancestral idea that curse/cure are nothing without each other, the possibility of seeing me, as an extension of themselves would be in written in the next paragraph… except they *didn't*.

2 Size 12

Like nothing, like the void, like the empty-calorie meal dressed by a mirage, like no idea. This size 12 is what I have known the longest because it is the number that reminded me about the chronic pain caused by endometriosis. It was the size that told me that everything was either going down or up. Things couldn't only get worse from 12. Like there was no such thing as 13 apostles or more than 12 months or 13 ribs. It was not the size where one can heal its organs without surgical consequences. Size 12 and the oohing do not call for the submission to deadlines. AKA *death*lines. It is during this sizing down that one looses most sleep reducing the hours of quality time with sentient beings. Even this text has gone through the zooming action by deadlines. Like nothing, this oohing still bounces back and forth within the unrealistic measurements of the ideal bust, the ideal waist, and the ideal hip. In this 12-thingi-thingi, I considered for the first time in my life grabbing a knife and performing a self-cultural circumcision to birth the start of my own liberation. Age was also 12. Nothing perfect. Nothing to be accountable for. Nothing to read for the so-called 'American' dreams.' Nothing calcified. Nothing mummied. Nothing glittery or shiny to reflect the calorie-induced anxiety complex in the RIP memorial site for everything I had no control over. No. No. No. Nothing of that sort for Size 12. Because what if this inconsistency is an example of what existing in relation to the 12 would be like. Like in a place, that knows no colonialism in math. Like in a place without measurements. Like what if this learned notion of size 12 changes when I get in touch with imagined ways of reading 12 in relation to my way of knowing the land instead of my knowledge of sizes. Like what if this reveals an embodiment of a world without colonialism in textiled-narratives? Like what if the signs in these tapestries would signal ancestral ways of moving through our own flesh? What if these missed connections can change from the missed status to the found ones. What would the sound of the break-up do to the unchanged heart after realizing that a size 11 would probably give her back a crispier reflection from the mirror? Another question left: Would it hurt? Okay, just one last question in this thread: *Would this rupture of consciousness signify anything in the reduction of numbers to the pursue wellness or sickness?*

3 Size 11

Having no immediate linguistic and cultural ties to 11,[16] the sizes in the in-betweenness cause a seismic bodily disruption. The newly adopted change between the number '2' and the number '1' places the arteries on alert mode because the historical ties to the mid-section is where the fat is stored caused by intergenerational trauma.

[16] Well, except the nation that birthed neoliberalism where 11 September 1979 happened–you know: Chile, Salvador Allende, Condor Operation, Pinochet, dictatorship, Víctor Jara, 'Where's his body?', torture, Villa Grimaldi, disappearance, pact of silence, impunity, bones, flesh, blood.

Like the void, appearing out of nowhere, it returns not giving a flying fuck if its unannounced presence is welcome or not. The body in size 11 does not seem to have any ties to less happy times. This untraceable way of living could mean paying no taxes. No taxes no numbers. No numbers no counting. No counting no calculating. No calculating no distracting. No distracting was a balanced no. And here returns the 'but', like the perfect child from the boomerang crew, with "but nothing ever remains the same. And since I am in this size that can only leads into more change I could cross that fence and see what I can get." This new place of existing as an untraceable anything meant one could disappear from the colonial mappings, from the constraints of the binary-thinking, from the stagnated notions of progress, from the unspoken ways of being racist, from what seemed to be the epitome of an ideal weight. This is the place where colonial ways of measuring life and values riddle itself into the balanced equation of $2 + 2 = 5$. This is where one could possibly claim sovereignty of the body only if we understood that the body means also land. In this size 11, where the number 11 reminds you of what's on footnote 16 about cultural legacy and its crisis, is where the I'd find the space to ask this question: Why is it that women are always left to do the political work while men remain unbothered by the free copies of our x-rayed bodies? I know. I should know this. Men will not come to our rescue. It is like expecting Black or racialized men to nest hope in Black and racialized women without shifting the air conditioning for the comfort of the white woman. It is like expecting white woman to lead us forward in dislocated terrains that hasn't tasted their moon visits. This is probably why the number '1' was replaced by the number '0' as the body changed its sizes.

4 Size 10

The question came back again: *Would this rupture of consciousness signify anything in the reduction of numbers to the pursue wellness or sickness?* The riddle had turned now into a telenovela. The lower back felt like it had adjusted well to the tightness of the new dress. The back posterior was better than ever. The connection to an untouched layer of historical fat and its silenced treatment was the right fitting. The size 10 turned into that moment when a city turns cosmopolitan according to a list someone came up with. This moment is the when followed by the where is political in the situation of *belonging*. How can one now feel the change in the symmetries? It was actually in the United States where I learned what a size meant to/for a girl like me. Size 10 mattered then and it matters now. According to an uncle, who I learned a few years later was a paedophile without ever being accountable; my tights were how exactly they should be. It was in my own body where stolen sovereignty happens. In my own flesh, blood, and bones he tried to rename my own landscapes. Land knows no distances measured in kilometres or miles, in metres or centromeres. Sizing land has no meaning in political implications of land-based pedagogies. The moment land is occupied, the body is also occupied. You occupy the body, you occupy the land. The authorization of the seizing is admitted in the name of omitted voices while you order coffee, while you stay silence during dinner, while you shop and you witness micro and macro aggressions, while you know you are working where you are working because of your whiteness, I mean, it is while of all the whiles happen in our daily lives. When one chooses to remain with an unattended hiccup, then one is for sure within the confinement of a decolonization that shows in its X-rays only a proximity to whiteness. It shows it unfiltered and unedited. The immediate and most apparent answer in the results from the X-ray could be of the wish for the number '9.' The closeness to a new body adjusted by new belts, new styles, and new fittings was like the realization that the portal that separates colonial-indigenous nostalgia was homed in the body itself. It was in that proximity where you thought you could find the exit out from the pile of shit piling up over all over the globe. And although there was still the possibility of claiming this sovereignty, she pushed for another numerical change that came to haunt her for the two summers.

5 Size 9

Ta-da! It's a size 9! The truth is that I thought this size would give me the chance to allow my imagination to wonder through the corridors of my intimacy. I thought this was the ideal weight. For what I have known. For what I have worn. For what I have tried on. For what I thought, I looked good into. For what I bought thinking I would fit into for the next season. I thought everyone was seeing me, watching me, threating me, and observing me. The eyes of my colonial world let me see my sovereignty was no longer real. It was just part of my existence with colonialist getting being storied in a book of measurements. Nothing mattered after the poke by the returned image of an unchanged idea of a body in the mirror. The unprecedented arrival of number '9' showed another ancestral teaching coming from the land: mountains welcome change as it signals the future. The brutality in the returned image radiated the evidence of trauma without a burial site. Trauma wandering. Trauma shielding itself from ancestral medicine. Trauma repeating the distraction of how one escapes when class fragility shows up when it escapes from the closeted closet.

6 Size 8

Random bits of information as inscribed in the medical history of the nearest hospital where the body was taken after collapsing from dehydration. The size came with a price, CAD$47 for the ambulance. At this point, it becomes less and less important if the quarrels from the lordly Lords are consequential in president elections in the world of the free.

7 Size 7

The reduction in the waist became less about the inches and more about the shape. Something about this shame did not sit right with the hips that only lie to themselves. Previous measurements could have sworn this new improved lucky size 7 would be the light at the end of the beauty cult. In this size, the trauma seemed less colourful and more towards the tones and shades of a pale-looking world. Suddenly, the compliments: "OMG you've lost so much weight!" "Wow, what did you do? You look amazing!".

The reduction in the waist weighted me down before I divorced culturally. When staring back at the waist, where colonial fat is stored, the more obvious it became I needed to continue shedding off these historical layers until I made them disappear from all known maps and occupied territories. It was only by undoing them that perhaps I could see less and less the colour blue. To be this reduced and start seeing returned images as double the size simply meant that there was more inches to be lost. Déjà vu. The 'once again' and 'here I am' in the perpetual discourses of colonialism rejected any notion of making sense in the further exploration of dropping sizes and the extraction continued.

8 Size 6

Déjà vu. *Would this rupture of consciousness signify anything in the reduction of numbers to the pursue wellness or sickness?*

9 Size 5

Like the empty-calorie meal dressed by a mirage, this indigestible new body was kicking itself in the inner linings of the intestines. It was the clear signs of earthquakes just like the birds and dogs would know. One of the big ones. The roaring made you stop by the alert warnings of the –isms but it also made you remember to see towards the side of the car to make sure you didn't have any distractions. The number '5' connected with the largest sizes of women who had left their flesh, blood and bones to secure your right size in the stores to achieve the ideal shape. The connection wasn't there and you were not able to read number 5 through anything that wasn't created in the West. The underpaid and exploited women wearing size 5 could never testified that the crisis in the civilization is the legacy culture of colonialism and its forthcoming transitions. They continued wearing 5 as it nothing else was happening. Turning 4 now was suddenly real. Another check mark in the to-do-list as the long process of physical sacrifice. This new goal, this new deadline would *be* the defining moment of a final change. The rehearsed phrases of people glorifying the body of all the bodies stayed there, in rehearsal. And like an empty-calorie, a meal dressed in a mirage, 5 turned to a 4.

10 Size 4

A body that is glazed over by discourses of languages that were not thought with you in it is not connected. It was not thought of to dialogue or to resemble any notions of reciprocity. The size 4 did not receive the exclamatory phrases related to *beauty.* The lack of intimacy between the mind and the body had made it impossible to connect with a once-upon-a-time beautiful bee. The here, as a mathematical nomination that eludes to the number 4, and the here of the size 4 considers the past as an immediate meditation and the past of the past in a state that begins the goodbye ritual to wonder. This is where any engagements that happened in the past of the past bring the first genuine smile to a starved stomach, deprived from emerging collective dreams. The self-negation arrives at the same time that self-acceptance gets off the car. A crisis is caught in an x-ray showing entangled fishing rods destroying what little remains undrowned in the oceans. At this point the act of wilful listening begins to spindle out of control.

The why:

It fades, repeated words begin to back as it asks why you didn't have a cup of ice cream, it communes with the closest thing to understanding the body beyond the colonial mathematical operations, the metaphor. Another why: It gathers endangered ancestral seeds preparing them for a new voyage you will be doing soon, as whispered by an Elder in your dreams. It counts in made-up sounds to quite down the forgotten reason why a size of 13 was never good enough.

11 Size 3

Beauty. The self-negation marked the lies, the symbols, the consonants, the codes, the numbers, and the insatiable thirst for anything that says 'forever young.' In the name of *beauty*, in the name of whiteness, in the name of that approximation, the punctures of broken bones showed up again the loopholes of colonialism: *everything was made to appear tangible to cover up the optical illusion of wellness, the echoing of landfills nauseating back the calamity of humanity, the points you've earned from facing the question* 'Who are you without colonialism?'.

12 Size 2

'Who are you without colonialism?': Skinless. Less skin. More skin. Dry mouth. Agonizing dry nausea. Increased palpitations from the ongoing *death*lines. The dangerous to open your eyes to *settler colonialism.* Like no idea how the fragility of the dying skin still manages to flip you over so that you can still manage to scratch your ass even when you're in the semi-coma of your cultural death. Crashing cart once. Your bones crashed with stones you've would played with when you were a child. The same bony legs showing the announcement of a last circulation, are the same legs that chased you away from the dangers of a premature cultural invasion. It is those legs that crawled with musical intentions respecting the dance of the Torogoz and those legs that once dreamed of cultural transcendence. Out of nowhere, another crashing cart. The body shipwrecks blocking the remaining acoustical pedagogies coming from the garden outside of the hospital. If you could only still be amazed by splash of the whales… the accelerated trip to minimize one more size happened while everyone was sleeping, except you. You had to be alive to die.

13 Size 1

Would this rupture of consciousness signify anything in the reduction of numbers to the pursue wellness or sickness? This is where déjà vu ends. This is where we begin an unlearned journey of co-teaching in community because the number that remains as part of the anatomy of one size fits all takes another thinking-feeling journey. What size 13 would have wished to ask size 1 while it was still operational was if ended up curing the loopholes of colonialism. Size 1 dreamt of a similar question and in her dream, she offered an answer: When I fell down last week when the incident with the crashing happened, my bones turned into a sea of drums making music to the rhythm of metaphors. It made it easier to express how other forms of communication come alive. It is a sort of language where words, if you can call them that, breathe in unison with one another through pedagogies of compassion. I say compassion as if I am thinking-feeling it as a living organism.

The more the body of size 13 pushed for the reductions of numbers to shed the layers and layers from the keywords of *civilization crisis, decolonization, cultural legacy, and transitions* the less connected she was to a resting place. Size 1 is not the only one. There are thousands of ads proving this. She is not the only one facing what those keywords have in common: death. There is no such thing as decolonizing without a cultural divorce. The crisis is a winning numerical strategy to make sure deadlines happen, to fulfil a quota, to make oneself count in the capitalist-driven way of existing. So long as we continue to refuse listening to Indigenous Peoples, the sense of wanting to answer *the who and the what* from the cultural legacy hat, the healing will never happen. With it, the size 1 can write itself out of entanglements of falling out of life altogether. The writing itself lands in the number '0' where the storied self doesn't get placed either in the positive or the negative.

14 Size 0

Alice Walker comes to my rescue as I wrap up this textual offering: "Because we would not to voluntarily." Inner light, she reminds us. With the death of the number '0' the riddle is not so private anymore and the question must be centred in everything we do because everything is connected: "Why did Europeans enslaves us in Africa and take us to the United States?" The senses speak historical truth through dried up veins from ancestral lands. Political listening in the 'One Size Fits All' old-fashioned world is an urgent pedagogy of feeling-thinking. No, not thinking over feeling. I shared recently in a co-teaching space that colonialism begins the very second the heart is separated from the mind turning this cut as a tool to control ways of learning and reducing it to 'One Size Fits All.' The body, chopped symbolically from the head, is the site where maps and borders are marked down. Size 0 is the perfect site to put *beauty* to rest with a tombstone that reads: In the name of class…

If size 0 could re-visit the body formations in each number it would trace back a detailed plan on how to heal and since it cannot, the quest for the impossible becomes available for others in the category of 13 waiting to see who checks in for another fitting:

To size 7: Contemplate nothing and stay there.

To size 3: Between the body and the aches, there is distance measured by void.

To size 10: Your knees need less running and hotter baths with Epson salts.

To size 5: The connection between easily bruised skin and the carvings on the wall is there.

To size 13: You have no control. The in-sickness and in health is part of the language you are liable and reliable for as part of the calorie count for your English-based diet.

To size 12: You think. You do not feel. Feel again. The disconnection is the crisis.

To size 6: When you start loosing your historical inches, prepare to wash up your immediate ripped off Band-Aids you have carried since elementary school. No one is going to tell you again that *beauty* kill the senses. I am telling you now. Your insides will eat your outsides. Your flesh, blood, and bones are the ultimate inches they will consume.

To size 8: No one is going to tell you either that your movement is no longer going to be in unison with those of the monarchs. You will not secure a rest-in-peace spot because fashion will never forgot you did not follow orders properly ending up in the number '0'.

Conversations on Decoloniality and Fashion: Hosting, Listening, (Un)Learning

Erica de Greef and M. Angela Jansen

Abstract To expand a collective understanding of decoloniality and its convergence with fashion, the monthly Conversations on Decoloniality and Fashion was conceived and co-convened by Angela Jansen, Erica de Greef and Shayna Gonçalves as an online experimental platform beyond institutional, disciplinary and geographical boundaries and transcending academe. The Conversations were initiated to bring-together-in-relation different local histories and embodied conceptions and practises of decoloniality in order to investigate decentred ways of knowledge-creation and sharing regarding fashion. In this chapter, Angela Jansen and Erica de Greef are guided by inspiring decolonial thinkers Rolando Vazquez and Walter Mignolo as they reflect on their journey of learning and unlearning through the Conversation. We consider what it means to hold space and host voices; to invite discussion and the development of a decolonial discourse; and to bring a wide range of participants together in dialogue. We acknowledge the teachings of a community, as we think with them, from a shared legacy, guiding us in the mapping of our own understandings (and misunderstandings), as we practice in relation with others a 'politics of redress' (Niessen). The work of the Conversations aims to find ways to 'give back a place in the present, of hosting and emplacing what has been eradicated' (Vazquez 2018), in an effort to resist and prevent the loss of cultural futures due to coloniality/modernity. Fashion(ing) becomes a 'key site of interrogation of the modern/colonial order' as it articulates and demonstrates the colonial difference; 'it has been colonised to the realm of the image (as representation) rather than being in the realm of relations' (Vazquez 2021).

With the guidance and mentorship of Rolando Vázquez and Walter Mignolo, and the contribution of Shayna Gonçalves.

E. de Greef
African Fashion Research Institute, Cape Town, South Africa
e-mail: africanfashionresearchinstitut@gmail.com
URL: https://afri.digital/

M. Angela Jansen (✉)
Research Collective for Decoloniality & Fashion, xxx, South Africa
e-mail: info@rcdfashion.com
URL: http://www.rcdfashion.com

M. Á. Gardetti and R. P. Larios-Francia (eds.), *Sustainability Challenges in the Fashion Industry*, Sustainable Textiles: Production, Processing, Manufacturing & Chemistry,
https://doi.org/10.1007/978-981-99-0349-8_11

Keywords Decoloniality · Listening · Conversation · Relationality · (Un)learning

1 Practising Decoloniality: 'Engaging in' Rather Than 'Consuming of' Knowledge

> We acknowledge that we do not own this thinking, but that there is a community producing what we are thinking; from the languages that we are using to the concepts and insights that we are suggesting. We are not thinking of this, but instead we are thinking with, from a shared legacy, from a consciousness of owing. We acknowledge our infinitive indebtedness to others. Rolando Vázquez in *Vistas of Modernity* (2020).

In the writing of this chapter, we have been guided by decolonial elders and mentors Rolando Vázquez and Walter Mignolo as we reflect on our journey of (un)learning through nearly two years of monthly *Conversations on Decoloniality and Fashion*. We convened these online conversations together with Shayna Gonçalves as an experimental space beyond institutional, disciplinary and geographical boundaries to expand on a collective understanding of decoloniality and its convergence with fashion. What began in November 2020 as a gesture towards creating a community and an opportunity for the *Research Collective for Decoloniality and Fashion* (*RCDF*) to meet and grapple with the complexities and entangled notions of fashion and decolonial 'theory,' became a generous environment for hosting a wide range of voices in a shared journey of (un)learning and (co-)learning. Through conversation, through the communal and coalitional, and through a radical act of listening across age, race, gender, histories, geographies and the colonial difference, RCDF's *Conversations* were initiated to bring-together-in-relation different local histories and embodied conceptions and practises of decoloniality to experience decentred ways of knowledge-creation and sharing in regard to fashion, drawing on Angela's[1] experience of the Decolonial Summer School.[2]

In this chapter, we consider what it means to reflect on not only our own spaces of speaking, but also on hosting other voices and taking responsibility for holding space and hosting voices. We think about what practising decoloniality means in creating and sharing fashion knowledges by 'engaging in' rather than 'consuming of' knowledge through conversation **(part one)**. We reflect on what it entails to create community and meet in difference, whilst acknowledging the teachings of that community by thinking with and from a shared legacy, and to be guided in the mapping of our own (de)colonialty, as we practice in relation with others a 'politics of redress'[3] **(part two)**. We contemplate on what it means to be implicated,

---

[1] When referring to each other in the context of everyday life or conversations, we have chosen to use first names only, since this reflects how we address each other in these contexts. When referring to each other's writings, however, we have decided to use full names, as is custom.

[2] The Decolonial Summer School, now called the Maria Lugones Decolonial Summer School, is an ongoing annual meeting initiated by Walter Mignolo and Rolando Vázquez in 2009.

[3] Concept by Sandra Niessen and referred to by Rolando Vázquez at the RCDF Annual General Meeting 2021, (24:20) https://youtu.be/Z6jocYS6b0I.

by undoing innocence and taking responsibility, by dismantling the autonomy of the artist, academic, designer or researcher **(part three).** Lastly, we imagine what it entails to (re)create space for a plurality of fashioning systems, a multitude of possibilities, that is perhaps messy, incoherent and complicated (from a euromodern standpoint) **(part four)**.

With this writing, we wish to honour an invitation by Walter to reflect on what we have (un)learned (so far) from the *Conversations* through 2021 and 2022,[4] while being guided by Rolando to take responsibility with, and for, our collective and reflective voice(s).[5] We owe much of the shaping of this chapter to their wisdom, nurture and care. In an initial conversation with Rolando, he shared some valuable decolonial principles with us that greatly influenced the development of this reflective writing. His comment that self-reflection too often becomes a recentring of the sovereign self has been key in imagining how we could move away from this notion of a centred and sovereign self, and focus instead, on our greater, collective effort of hosting, mirroring and caring. He guides us to a position of gratitude for hosting this regenerative space–as a space for learning each other (following Maria Lugones), a space where we meet in and with difference, and through common histories, to rebuild trust and allow healing. Rather than reflecting on how decoloniality has been advancing us intellectually, Rolando encourages us to reflect on how decoloniality is transforming us–as human beings, guests of this planet, critical thinkers, creatives, mothers, partners/spouses and more.

We have observed the recent popularity of decolonial rhetoric increasingly used as a tool to gain power to advance institutional and individual reputations, to capture and incorporate, instead Rolando reminds us to always be transparent to the white gaze and to use decoloniality as a tool and a reason to bring people together in conversation. In conversation, he reminds us, engagement becomes key in the transformation from a position of consumption to a position of taking responsibility, from a position of indolence and indifference to a position of engagement in-difference-with others, from a place of compassion, where we can allow ourselves to be touched and transformed by the other.[6] Only in conversation, Rolando points out, are we enabled to receive the difference of the other, and avoid the exhibition of the suffering of the other, due to the involvement in the suffering that conversation implies. Essential to decoloniality is that it is a collective act and, in contrast to much academic research and writing, which is most often individual, conversation is collective and enables the listening to and receiving of, rather than the erasing and consuming of the other.

In reflecting on our ongoing *Conversations on Decoloniality and Fashion,* we acknowledge that engagement has been key in transforming each of us from a position of consumption to a position of responsibility, from a place of 'indolence and indifference' to in-difference-with-the-other grounded in compassion, where we allow ourselves to be touched and transformed by the other.[7] Positionality is important in

[4] In email correspondence with Walter Mignolo, 5 February 2022.

[5] In conversation with Rolando on 13 May 2022.

[6] In conversation with Rolando on 13 May 2022.

[7] In conversation with Rolando on 13 May 2022.

this process, but not as our individual identities, but rather as our positionalities *in relation to others*, where we recognise and acknowledge privilege and the way we live as impacting on the lives of others. In engaging with the guidance of Walter and Rolando, we aim to share how the *Conversations* have enabled us to receive the difference of others through collective conversations facilitated by listening and receiving.[8]

It is through coming together in difference that we experience how modern fashion aesthetics have been playing a key role in configuring a canon, a normativity that enables the disdain and rejection of other forms of aesthetic practices, that is, other forms of sensing and perceiving (aesthesis).[9] It is in these shared spaces that the urgency becomes tangible to delink the notion of fashion from modernity/coloniality and to reconceptualise it as a multitude of possibilities—in and outside of modernity, rather than a normative framework claiming universality. In the Fashion Theory special issue on Decoloniality and Fashion, Angela expands on how fashion as a noun—contemporary fashion—has come to refer to a temporality of contemporaneity, a system of inequality and a capitalist industry of consumption particular to modernity—while fashion as a verb, the act of fashioning the body, is of all temporalities and geographies.[10] This link of fashion to modernity endorses superiority over not-fashion, universality as the only 'real' fashion and contemporaneity as separate from the past to establish a dominance of the present and reinforce categories of racial, cultural and temporal discrimination.

Walter encourages us to remember why, as human beings, we fashion our bodies beyond the incentives of capitalism, consumerism and notions of luxury/exclusivity/superiority. Reflecting on deeper interpretations and understandings of the sensibilities of fashioning (that include not only ideas of protection, but also spiritual and divine forces beyond the comprehension of seasonal trends dictating our fashion acquisitions and discards), then we can begin to acknowledge our relations in the world through fashioning, our desire to tell individual and collective stories, and our need for beauty and joy.

By silencing dominant (fashion) voices, it becomes clear how systems of fashioning 'outside modernity'—places with genealogies and trajectories that are not rooted in European modernity—are deliberately and systematically discriminated against, silenced and erased. We are taken to unlearn the claim of one universal fashion system and (re-)allow space for a global network of sovereign yet connected fashioning systems, affecting each other like communicating vessels. It faces us with the responsibility to activate a global network of connected yet self-determining fashioning coalitions for an environmentally and culturally sustainable, ethical and socially just fashionscape. It urges us to expose contemporary fashion as the ongoing

[8] In conversation with Rolando on 13 May 2022.

[9] Rolando Vázquez and Walter Mignolo (2013) 'Decolonial AestheSis: Colonial Wounds, Decolonial Healings.' *Social Text Journal* July 15. https://socialtextjournal.org/periscope_topic/decolonial_aesthesis/.

[10] Angela Jansen (2020) 'Fashion and the Phantasmagoria of Modernity: An Introduction to Decolonial Fashion Discourse,' *Journal of Fashion Theory* 24 (6): 815–36.

articulation of coloniality/modernity as well as the co-option of land, episteme, aesthesis, spiritualties, memories, subjectivities, dignities, histories.

The *Conversations* have been a deeply grateful and inspiring space for addressing uncomfortable topics that are largely avoided in dominant spaces; for taking responsibility; for rebuilding the relational through (re-)establishing trust by committing time and care; and for practicing decoloniality in knowledge creation and sharing regarding fashion. One of the purposes of the *Conversations* has been to critique modern scientific knowledge creation and sharing. The way dominant contemporary fashion knowledge is produced and shared, i.e. through academic writing and publishing based on peer reviewing processes to validate knowledge, claim excellence and practicing gatekeeping, is excluding and discrediting other knowledges, obscuring relationality and suggesting innocence/arrogance. Through the *Conversations*, we have been aiming to decentre the praxis of knowledge creation, and instead create space for other fashion knowledges. By creating a combination of listening, followed by breakout or group discussion, the format of the Conversations invites engagement and involvement, and acknowledges participants' positionalities and responsibilities. We ask, by doing, can we practice academia differently?

A central tenet to the *Conversations* has been the 'radical act of listening.'[11] Listening not as 'withdrawing from speaking,' but as 'engaging in the process of knowing each other,' based on reciprocity, on 'hosting each other' and on 'being vulnerable in processes of (un)learning.'[12] The *Conversations* have been a space for learning and unlearning through listening and engaging in conversation with guests (recorded), participants (unrecorded) and selected materials (texts, videos, podcasts, music, poetry, exhibitions) as well as by reflecting through proceeding documents, the Reflections blogposts, publications, lectures and email correspondence with guests and participants.

The *Conversations* have been hosting a multitude of voices, both present and absent. Critically re-evaluating the existing structures of knowing and understanding fashion(ing)–that have been predominantly shaped by modernity/coloniality and Western perspectives–we have invited, experimented with and (re)imagined what it is to (re)allow a pluriverse of fashioning knowledges that decentres the modern fashion system, and exists in co-existence with ancestral and indigenous knowledges rooted in respective histories and cultural identities. We have been sharing and inviting others to 'learn how to exist together in our differences, to find solidarity in both our affinities and our singularities.'[13]

---

[11] Erica de Greef, Shayna Goncalves, Angela Jansen (2021) 'Listening as a Radical Act: Conversations on Decoloniality and Fashion.' *African Arts* 2021: 54 (4): 1–5.

[12] Fabian Barba (2021) in a letter to Rolando Vázquez in response to his publication 'Vistas of Modernity.' Unpublished.

[13] https://www.e-flux.com/announcements/451104/14th-gwangju-biennalesoft-and-weak-like-water/.

2 Making Community: Meeting in Difference

> The conceptualizations and actionings of decoloniality are multiple, contextual, and relational; they are not only the purview of peoples who have lived the colonial difference but, more broadly, of all of us who struggle from and within modernity/coloniality's borders and cracks, to build a radically distinct world. Catherine Walsh and Walter Mignolo in *On Decoloniality: Concepts, Analytics, Praxis* (2018).

Communities emerge out of necessity; a necessity for survival, protection, support, belonging and other. While gay communities during the AIDS pandemic in the '80 and '90's solidified due to their interdependence for protection, support and even palliative care and precarious alternative domestic spaces, subversive and artistic communities have been helping and supporting each other by co-working and co-living. Similarly, fashion researchers, practitioners and activists working in the margins of dominant spaces have been finding each other for protection and support, for creating 'an alternative intellectual home' and a sense of belonging and to push back against the totalizing claims and political-epistemic violence of contemporary fashion discourses and practices. In the wake of growing popularity of notions of decoloniality, solidarity and care in recent years, concepts of community, collaboration and collectives have become trendy, but these groupings feel forced and imposed because of the lack of shared struggles, beliefs, ideologies, joys or pain.

What has initiated and been driving the Research Collective, is this urge to create space for those who feel they 'do not have one,' who feel like they 'do not belong,' 'out of place' and therefore 'misunderstood.'[14] When we first came together in 2012 in Morocco, as fashion researchers, practitioners and activists we were feeling 'in the margins' of fashion departments, 'misunderstood' at Eurocentric fashion conferences and in need of a place to connect with like-minded people, people that did not make us feel odd/weird/fish-out-of-water and as Erica describes it, that felt like 'finding family.'[15] From that first meeting, the Collective developed organically, without a clear plan, structure or agenda, but with a shared need of belonging and a shared goal to push back, contest and disrupt. In the February 2022 Conversations on 'Coalitional Consciousness Building,' Cricket considers this coming together as an important 'site' for resistance and re-existence, where in 'knowing, we begin to build deep coalition–as resistors/re-existers–by understanding, validating and furthering each other, in resistance/re-existence.'[16] We look for each other as a kind of orientation in the world, as an emancipatory collectivity in complex, linked, and interdependent resistances/re-existences.[17] In other words, in learning to re-see ourselves–as actively

[14] Sarah Cheang, Erica de Greef, Angela Jansen, Sandra Niessen and Toby Slade in the November 2020 Conversations on 'Our Collective and Individual Journeys.' Unpublished.

[15] Sarah Cheang, Erica de Greef, Angela Jansen, Sandra Niessen and Toby Slade in the November 2020 Conversations on 'Our Collective and Individual Journeys.' Unpublished.

[16] Cricket Keating in February 2022 Conversations on 'Coalitional Consciousness Building' (13:55). https://youtu.be/E8CMeXZWd-Y.

[17] Cricket Keating in February 2022 Conversations on 'Coalitional Consciousness Building' (35:00). https://youtu.be/E8CMeXZWd-Y.

working against coloniality/modernity–we begin to build coalitions, and re-define and re-signify life in conditions of dignity.[18]

In her article on *Coalitional Consciousness Building*, Cricket Keating (2005: 86) reminds us that we are not practiced in the habits of coalition (anymore) and especially what Maria Lugones calls deep coalitions,those coalitions that go beyond short-term interest-based alliances and challenge us to align our own self understandings, interests and goals with other oppressed groups.[19] Her ideas on coalitional consciousness building provide us with a method for self and collective education towards coalition, as a way to engender solidarity across multiple lines of difference (2005: 86).[20] She shares three steps to build coalitions across difference whereby the first step is sharing experiences related to a theme in a way that pays close attention to the national, racial and class and other relevant contexts and histories in which the experiences being articulated are being played out. The second step is examining the experiences with an eye for the multiple relations of oppression and resistance at play. The third step explores the barriers to, and possibilities for, coalitional action with regard to the experiences.

Inspired by the feminist consciousness-raising groups described by Cricket Keating (2005: 88), the *Conversations* have become a similar space for grappling with the idea that when existing theory and knowledge about fashion cannot be trusted due to modernity/coloniality, we can turn to what each one knows and trusts, i.e. our own thoughts, feelings and experiences. Especially the idea of Kathie Sarachild, quoted by Cricket Keating (2005: 88), who explains that 'Our feelings will lead us to our theory, our theory to our action, our feelings about that action to new theory and then to new action'[21] has been guiding us tremendously. As an 'experiment in radical democracy' (2005: 89–90),[22] the stress on the value of each person's experience and knowledge has been serving to build a non-hierarchical and transformative space for thinking about and acting upon one's own and each other's different situations. By challenging the divide between theory and practice, the *Conversations* aim to be a space for ways to create knowledge and theory in a participatory and collective manner based on the idea that 'everyone is a theorist' (2005: 90).[23] Keating has been teaching us that building solidarity requires critical self-reflection that acknowledges that how one lives impacts the lives that others are able to live. 'We are all connected because of our relational insertion into hierarchies of power and privilege, hierarchies that we also can resist and transform' (2005: 93).[24]

---

[18] Adolo Achinte in February 2022 Conversations on 'Coalitional Consciousness Building' (12:30). https://youtu.be/E8CMeXZWd-Y.

[19] Cricket Keating (2005). 'Building Coalitional Consciousness.' *The National Women's Studies Association Journal* 17(2): 86–103.

[20] Cricket Keating (2005). 'Building Coalitional Consciousness'.

[21] Kathie Sarachild (1970: 78) in Cricket Keating (2005). 'Building Coalitional Consciousness'.

[22] Cornell (2000: 42) in Cricket Keating (2005). 'Building Coalitional Consciousness'.

[23] Mackinnon (1989: 102) in Cricket Keating (2005). 'Building Coalitional Consciousness'.

[24] CRICKET Keating (2005). "Building Coalitional Consciousness'.

Erik Wong, who together with Sophie Krier joined the May 2022 *Conversations* on 'Designs for the Pluriverse,'[25] explained how Miquel Hervás Gómez, the graphic designer of their exhibition 'In Search of the Pluriverse,'[26] shared the idea of comunalidad with them. 'Comunalidad is what brings together the community itself. It's what is in between each member of the community. It's kind of like the binding, or the glue–you can call it many ways. This glue is what the First Peoples of Latin America, or Abya Yala as they call this continent, refer to as 'communal autonomía'. Comunalidad then, is the 'glue that binds the community together'.'[27]

Each of us experiences our being 'in the world' differently and understanding and acknowledging our positionality in relation to each other and the colonial difference, is key. In her Reflections on the first Conversations in February 2021, Shayna explains how 'To verbalise, narrate and explain my own positionality, practice and experience of decoloniality requires an untangling of the cultural meanings imbued in the words, names, types and signifiers of identity that have been projected onto my body, mostly for the purpose of other people's relatability to me and placement of me in a modernised world, more so than the words I use to experience and describe myself. So, some of the words that currently signify–to others–my positionality include woman of colour, currently living in South Africa where my specific racial group is named Coloured. Based on my upbringing and my modern/colonial educational privilege, I constantly accumulate the social and cultural capital that places me in a middle-upper class segment."[28]

Within its narrative of progress, modernity/coloniality brings with it an intellectual hierarchy that valorises certain modalities. Realities—like being on one side of the colonial difference, that is to be colonised, subjugated and erased—are often concealed. To be colonised, subjugated and erased is an experience manifest in daily practices of people of colour to simultaneously straddle appreciation, respect and self-narration to become liberated, empowered and visible.[29] 'For me, practicing decoloniality then is a constant process of re-learning and delinking from words that force my body and mind into a mould that conforms to these external terms of progress, modernity and success, such as: being professional, being modest, being humble, being sober/drunk/high, being thin, being nice, controlling my emotions and not taking things personally in the workplace. These are words that require a double suppression from people of colour–to at once be the word, and to forget the things that make us colour-full in fulfilment of the word. To be professional is to speak English and to forget another vernacular language. To not take things personally at work is such that when a white male colleague makes an office-wide announcement that 'Black Lives Matter is irrelevant' we need to respect the rules of not discussing

---

[25] https://youtu.be/KqMvN64VGU0.

[26] https://pluriverse.hetnieuweinstituut.nl/en.

[27] Erik Wong in May 2022 Conversations on '*'Designs for the Pluriverse,'* (13:20). https://youtu.be/KqMvN64VGU0.

[28] Shayna Gonçalves, *Decolonial Fashion Practice #1* (published 15 February 2021) https://rcdfashion.wordpress.com/2021/02/15/decolonial-practice-1/.

[29] Erica de Greef, Shayna Goncalves and Angela Jansen (2021), *African Arts First Word.*

race and politics in the workplace and hence assimilate into Eurocentric norms that silence all other voices for whom the statement feels like an attack on our very personhood.'[30]

The *Conversations* furthermore affords the positionality of those who are being hosted as guests and participants across differences, for example as a 'home away from home' for students who do not 'feel at home' at their department. For Alexandre Prince, a French-Canadian student based in Québec, it has been a major breakthrough. 'I learned every month about how these [decolonial] studies could be used to provide critical perspectives on the dominant fashion system. Instrumental to the RCDF was the theoretical framework of coloniality/modernity developed by Aníbal Quijano and applied to the world of arts and aesthetics by Rolando Vázquez, which I had the chance of meeting at the time. Other influential thinkers that I had the pleasure of meeting in these events include coloniality/modernity thinker Walter Mignolo, textile anthropologist Sandra Niessen and African fashion curator Erica De Greef. These monthly conversations with scholars from all around the whole world; from Argentina, Brazil and South Africa to the Netherlands or France, allowed me to decolonise some of my thoughts and strive to include in my research more and more theorists that stand outside of the West's canon. Inspired by Rolando Vázquez, I discovered how Homi K. Bhabha's *Location of Culture* could be applied to the study of fashion and representation.'[31]

Alexandre's point of the 'chance of meeting' influential thinkers, points to a second key aspect of the *Conversations*, and an interesting shift in academic practice–it is that the online space presents for Alexandre a 'real' meeting moment, underpinning that the dialogic format we chose to use, translates as a tangible 'meeting' for Alexandre, beyond the wider experience of online engagement purely as viewers, not as active participants. The online 'space' thus opens a global engagement that breaks down both geographic and academic barriers normally held in place by gatekeeping through 'western/colonial' academic systems.

The *Conversations* were initiated in the run-up of the COVID-19 pandemic, when global dialogue took on a different form–how do we connect across time zones, across language barriers, across experiences? The shift to serious or critical and committed online engagement not only in our *Conversations* but globally, has seen a radical shift in who gets to speak, how this is disseminated, and whose thoughts are heard or listened to in terms of 'authority.' Movements such as Black Lives Matter, and the impact of George Floyd's murder raised and shifted public understanding of conventional mediascapes and the potential for an altered media landscape means that no-one longer relies on single authorities for information, and welcomes citizen-views and voices. We learn about events, life-worlds, emotions, realities that we might never have heard about elsewhere, filtered through a variety of perspectives, and presented on digital platforms that accommodate languages, images and voices that might not otherwise have been seen or heard.

---

[30] Shayna Gonçalves, *Decolonial Fashion Practice #1* (published 15 February 2021) https://rcdfashion.wordpress.com/2021/02/15/decolonial-practice-1/.

[31] Alexandre Prince, email correspondence on 14 March 2022.

Moving away from the sovereign self, we have been building on the foundations of a community by hosting localised–as a digital locus, in participants' home spaces, as decentred and spread around the world–communities online, where we have been able to host community participations through a combination of listening and 'speaking back.' As Shayna in her reflections on Decolonial Fashion Practices (2021) formulates it, 'Delinking from modernity/coloniality for me means having the courage to speak about my practice of solitude and meditation as a legitimate process of unlearning and re-learning, and of creating knowledge. During the stillness I ask myself such questions as: What hairstyle do I want? Why am I wearing sunscreen (beyond maintaining a lighter colour)? What are the benefits of exercise and a healthy diet (beyond pursuing thinness)? Why do I feel like eating this? What is this work serving? Who is benefitting from this work? Is anyone harmed/abused/exploited/erased/silenced by this work? How do I really feel like responding in this situation (being an angry black woman is ok. Being kind is ok)? What is the person you are engaging with really saying?

Listen.

Have the courage to listen.

When it is my turn to speak, have the courage to speak the truth, especially when that truth will shatter the comfortable world order of those who benefit from my silence.'[32]

3 We Are All Implicated: Undoing Innocence, Redressing Relationality

Please don't talk to me about decolonization,
When you are still speaking in the colonizer's language,
See, You genocided us,
Then you colonized us,
See you sterilized us,
And now you fetishize us,
See you stigmatized us,
Then homogenized us,
Try to co-opt the movement and gentrify us,
I say no!
Land back, Please
Stolen Land Under Seiche
Bobby Sanchez (2021) 'Quechua 101 Land Back Please'.

[32] Shayna Gonçalves, Decolonial Fashion Practice *#1* (published 15 February 2021) https://rcdfashion.wordpress.com/2021/02/15/decolonial-practice-1/.

We are all concerned. As Catherine Walsh and Walter Mignolo (2018: 5) formulate it, coloniality is experienced on both sides of the colonial difference, whether our life experience is touched and formed in and by the Global South, or whenever citizens of the Global North come to understand that they belong to a history that has engendered coloniality and disguised it by the promises and premises of modernity.[33] People in or from the Global North and in or from the Global South not only live different lives, but people in or from the Global North live the lives they do, in large part because people in or from the Global South live the ones they do (2005: 93).[34]

Understanding this dynamic that our choices, lives and worlds are interconnected, means that we must accept our actions in relation to those in the sacrifice zones. Anthropologist Sandra Niessen points to the concept of erasure–and the conceptual erasure that allows us to exists–as an underpinning logic of modernity, where modernity demands forgetting.[35] In this regard then, remembering becomes a radical act and a challenge to the amnesia that is built into modernity. Sandra describes the normalcy that is so different in the Netherlands to the normalcy in Indonesia, and to 'understand the world,' we need to come to terms with these contrasting normals; that our wealth in the global north has much to do with the exploitation of the global south, something that we are seldom aware of.[36]

In preparing this chapter, we returned to the question raised by Rolando, whether we can live an ethical life when our ways of dressing, eating, speaking and our ways of enjoying life are so dependent on the suffering of others and the destruction of Earth? How do we begin to take responsibility for our own decolonial liberation? No-one should expect that someone else will decolonise him or her or decolonise *X* or *Z*. Similarly, no-one who is living-thinking-being-doing-decolonially should expect to decolonise someone else (2018: 5).[37]

Catherine Walsh and Walter Mignolo (2018: 11) identify those living-thinking-being-doing-decolonially, as thinking from and with the possibilities of building a radically distinct world, as part of the decolonial project.[38] They describe this as thinking from and with *facultad* (the power to do), which is especially sensed by those who are 'pushed out of the tribe for being different'. Those who do not feel psychologically or physically safe in the world–females and homosexuals, darker-skinned people, outcasts, foreigners and those who are persecuted and marginalised–are more likely to develop this sense or *facultad*.[39]

---

[33] Walter Mignolo and Catherine Walsh (2018). 'Introduction.' In *On Decoloniality: Concepts, Analytics, Praxis*, pp. 1–14.

[34] Brown 1992, 298 in Keating (2005). 'Building Coalitional Consciousness'.

[35] Sandra Niessen in April 2021 Conversations on 'Modernity/Coloniality and Sustainability in Fashion,' (10:35). https://youtu.be/VIOTNG1GZrk.

[36] Sandra Niessen in April 2021 Conversations on 'Modernity/Coloniality and Sustainability in Fashion,' (12:40). https://youtu.be/VIOTNG1GZrk.

[37] Walter Mignolo and Catherine Walsh (2018). 'Introduction'.

[38] Walter Mignolo and Catherine Walsh (2018). 'Introduction'.

[39] Walter Mignolo and Catherine Walsh (2018). 'Introduction'.

Decoloniality brings forth an understanding in which there is no safe or innocent position of self. We are all located in relation to the colonial difference that structures our modern/colonial order. This means we are all positioned somewhere along the social axes of discrimination and privilege (namely coloniality/modernity) with regards to notions such as gender, race, culture, sexuality and class. It is from this awareness of our positioned selves that we enter in relation with each other, that we listen to and learn each other.[40] As Walter points out, when you think about 'where coloniality is touching you,' or where you feel coloniality, you come to understand that coloniality is 'not over, it is all-over.'[41]

Learning to be in relation with others does not simply mean to include other practices and concepts into our own. To understand the radical diversity of worlds–of ways of sensing, of experiences, of languages, of community–is a movement to humble modernity, and to acknowledge and affirm the 'non-reducible differences across colonial divides,' writes Vázquez (2020: 86–87).[42] Relationality is only possible through positionality.

South African creative practitioner and archival researcher, Siviwe James in dialogue with Rolando in the March 2022 *Conversation* on 'Vistas of Modernity,' pushed us to consider how we speak and listen to each other across these colonial divides, and how in the realm of fashioning, this impacts how we are understood, and/or assimilated into the contemporary.

Rolando points to the potential of creating relations (with each other) instead of aspiring to representation (of each other) as an important decolonial refusal of the 'violence of mediation.'[43] By refusing to be clearly represented or transparent to the other, or as Rolando describes it, 'the demand for legibility or readability is the violence of transparency. It reproduces the [colonial] violence.'[44] Resisting the idea of transparency as the tool of incorporation, Siviwe concurred, is where 'not everything needs to be transparent to power, to be explained .... [and yet] where is that fine line, where do we hold back from needing to be overly understood? [We] almost feel indebted to this idea of being understood. What does it imply for [us] as the creator and narrator, then?'[45] Questioning the right to (in)visibility, Siviwe asks, "who are we being visible for? What does visibility 'garner'? As people in the global south, what does visibility mean to us? Is visibility then equal to shouting or screaming?'[46] Often the risk of decoloniality is that there is still a pursuit of

---

[40] Rolando Vázquez (2020). Vistas of Modernity.

[41] Walter Mignolo in March 2021 Conversations on 'Some Key Propositions of Decolonial Thinking,' (19:40). https://youtu.be/jzxGEjNDWc8.

[42] Rolando Vázquez (2000). Vistas of Modernity.

[43] Rolando Vázquez in March 2022 Conversations on 'Vistas of Modernity.' https://youtu.be/BAccsUubS6c.

[44] Rolando Vázquez in March 2022 Conversations on 'Vistas of Modernity,' (23:20). https://youtu.be/BAccsUubS6c.

[45] Siviwe James in March 2022 Conversations on 'Vistas of Modernity,' (20:20). https://youtu.be/BAccsUubS6c.

[46] Siviwe James in March 2022 Conversations on 'Vistas of Modernity,' (21:30). https://youtu.be/BAccsUubS6c.

translatability, to write–to know–the life of others.[47] Regulating the representation of reality, aesthetics and purity are modes of power (Vázquez 2020: 148).[48]

In our *Conversations*, we have grappled with this idea of coming into relation with others, through our choices of texts, languages, speakers, and even, session formats. How do we come to know one-another, to understand and come into relation with each other that acknowledges the violence of the colonial order? Through listening, we are invited into more pluriversal spaces. And yet, even in these pluriversal spaces, as Siviwe and Rolando propose, 'Can we work with the idea that not everything needs to be transparent (to power), not everything needs to be explained?[49] This demand (to be understood, to know, to be consumed) follows the colonial pursuit of legibility, even when 'dressed in decoloniality.'[50]

Rolando continues,

> With whose words are we speaking? To whom are we speaking? Often when we are speaking back to power, we are spending our energy towards [that] power, to them, to the system ... But, that is not our horizon. That is not what we are. The 'being against' is not who we are. ... Sometimes we reproduce the violence of that logic. [The question then is], for whose eyes, for whose ears, for whose body are we speaking, are we dressing?[51]

Vázquez (2020: 147) writes about this right to opacity in *Vistas of Modernity* as the right to difference, as the right to not be enclosed in an impenetrable absolutism, as the right to not become exposed in the exhibitionary order of modernity.[52] Representation, or purity, that is circulated and consumed, as abstract knowledge, corresponds to the timelessness of the contemporary, the loss of distance, the loss of time, the loss of the relation.[53] In relationality, there is no dominating the other, there is no-one classifying another, and there is no-one wanting to appropriate or to become the other. The movement of remembering back, the movement of relating–that is relational aesthesis–is key to delinking from the search of purity (2020: 141).[54] Relationality requires an equality in difference.[55]

[47] Rolando Vázquez in March 2022 Conversations on 'Vistas of Modernity,' (17:00) https://youtu.be/BAccsUubS6c.

[48] Rolando Vázquez (2020). Vistas of Modernity.

[49] Rolando Vázquez in March 2022 Conversations on 'Vistas of Modernity,' (15:45) https://youtu.be/BAccsUubS6c.

[50] Rolando Vázquez in March 2022 Conversations on 'Vistas of Modernity,' (16:30). https://youtu.be/BAccsUubS6c.

[51] Rolando Vázquez in March 2022 Conversations on 'Vistas of Modernity,' (11:27 & 14:10). https://youtu.be/BAccsUubS6c.

[52] EdouardGlissant (1990) in Rolando Vázquez (2020). Vistas of Modernity: 146.

[53] Rolando Vázquez (2020). Vistas of Modernity.

[54] Rolando Vázquez (2020). Vistas of Modernity.

[55] Rolando Vázquez (2020. Vistas of Modernity.

4 Creating Space for a Plurality of Voices/Fashioning Systems

> You are a damn wonderful person with so much to offer. You're so quick to deny my humanity, but with your prejudice today you just disavowed your own. Humanity is not about what we look like, it's about how we treat each other. You resort to hatred to shield yourself from the pain. Being disciplined into a narrow idea of beauty and forced to live someone else's idea of who we should be hurts. But hurting me won't heal you. You are a human being, not a recycle bin. What if we stopped regurgitating the cruelty we've been shown and instead chose love? There's so much beauty in the world if you allow yourself to see it. I love you and need your you –not theirs– in this world. (…) *'I love you more than you mor that you can hate me.'* Alok Vaid-Menon, Intagram post February 15, 2022.

With the *Conversations*, we have been investing time and attention to nurture a space for understanding, advocating and practicing decoloniality, and a space for hosting a plurality of voices from a plurality of fashioning systems. Our goals have, however, for some participants made the monthly sessions challenging. The *Conversations* not only require a long-term investment *and* personal involvement, but they also do not offer a clear how to decolonise five step plan. Instead, they disrupt the commodification of knowledge and are not 'easily consumed'. They require vulnerability and critical self-reflection through active listening and learning to unlearn. The monthly sessions have been open-ended, felt, pluriversal, conversational, inclusive, interdisciplinary, opaque and courageous.

Observed through the lens of the modern/colonial canon of knowledge production and performance, the *Conversations* could be perceived as unstructured, non-hierarchical and non-authoritative, even emotional and messy, wasting or taking too much time, non-academic, obscure, complicated and ultimately, risky, without a clear outcome, result or readymade solutions. Despite these efforts to break with the strict formalities of conventional academia, Shayna points to issues of accessibility and relevance that still seem to keep the *Conversations* out of reach for a younger, more diverse and decentred audience. As fellow RCDF member, Toby Slade asked "What are the barriers [to access and experience]? Can we change not just what we are talking about, but the way we talk about it and who we are talking about it with?'.[56]

The task of unlearning and learning each other is to share in carrying the weight of difference, to take responsibility for historical erasures, and for the violence and absences caused by coloniality/modernity. Our monthly sessions centre on inviting guests with languages, values, practices and lived experiences that extend beyond the hegemony of a western fashion discourse. In our efforts to host diverse voices and narratives from people who have for the most part been marginalised, we have been privileging plurality, and challenging ourselves to consider what it means to create space for more pluriversal relations and understandings with others. An important moment in our efforts to include a plurality of voices surfaced in a session in early

[56] Sarah Cheang, Erica de Greef, Angela Jansen, Sandra Niessen and Toby Slade in the November 2020 Conversations on 'Our Collective and Individual Journeys.' Unpublished.

2022 on 'Indigenous Gender Knowledges,'[57] where we grappled with what it means to meet in relation with absence:

> This failure is not about us not finding a guest. It draws into focus a much wider failure. Individuals such as Bobby Sanchez, Alok Viad-Menon, Githan Coopoo and many others, are continually asked to speak and share, in fact 'tasked with teaching' against the western/colonial/modern/hetero-normative gender binary that dominates the world. In the same way, People of Colour working in activist organisations such as Black Lives Matter (as well as those simply working with friends and in classrooms or clubs) are also tasked with the need to educate, to inform, to explain, to unpack different positionalities, etc. They are confronted by the need (and responsibility) to be the teacher, *to help undo* the damage of racism, gender exclusions, coloniality, bias, etc., to redress.[58]

We need to develop and maintain practices that will ensure the undoing of coloniality and privileges that created these erasures, and that continue to sustain their ongoing violence and disavowal. Hosting plurality is vital in new world-making dialogues. It is only *through listening* to those who are usually and/or largely excluded, marginalised, othered, that we will collectively (un)learn. In this, we have also come to understand that unlearning for some represents demanding (something) while for others, unlearning 'allows'. In hosting the *Conversations*, and in the work we do as the RCDF, we consider it our responsibility to give voice to the erased and absent–to disrupt the violence of silence–and to undergo how confrontational and emotional this feels. "Listening is a task of giving back a place in the present. By listening to those in the sacrifice zone, we may begin to (re)imagine what was erased, forgotten, placed outside of the textbook, beyond the curriculum, at the margins of fashion systems.'[59]

Like an interrelated web–where each thing is connected to another, not linear or chronological, but entangled and connected–the *Conversations* unfold as layered, multi-voiced, and open-ended engagements. Sophie Krier and Erik Wong, the curators of the podcast series and exhibition 'In Search of the Pluriverse' who joined the *Conversations* to reflect on their journey, describe the pluriverse as "a world, where people and other beings live together in harmony with each other and the environment. An equal world in which we together, shape a future that truly does have a future.'[60] Understanding relationality is a means to unsettle the singular idea so often assumed and portrayed in design, exhibitions, academic thought, etc. When different local histories and embodied conceptions and practices of decoloniality, including our own, enter into conversations, they build understandings that cross both geopolitical locations and colonial differences and contest totalizing claims and political-epistemic violence of modernity.

Working pluriversally, for example, with the exhibition designer Sean Leonard from Trinidad and Tobago in designing the exhibition, meant that Sophie and Erik

---

57 https://youtu.be/sYoJGOoulDc.

58 Proceedings for the April 2022 Conversations on 'Indigenous Gender Knowledges.' Unpublished.

59 Erica de Greef, *Decolonial Fashion Practice #3* (19 April 2021). https://rcdfashion.wordpress.com/2021/04/19/decolonial-fashion-practice-3/.

60 Proceedings for the May 2022 Conversations on Designs for the Pluriverse.' Unpublished.

'had a lot of unlearning to do in the process of collaborating, but [it was] super wonderful and valuable for us to work with him.'[61] The pluriverse asks for and acknowledges embedded knowledges, with different disciplines of learning from each other, such as art has learning from the environment, or academia learning from the vernacular. As students of the pluriverse, we have to slow down.

Another example of their 'making, thinking and working pluriversally' that Sophie and Erik shared, was to consider the colophon or imprint of the exhibition. Firstly, acknowledging Escobar and all those who came before them was important, and then, it was to thank all those who contributed (in small or large parts), as well as those who did not want to be named. Sophie describes how, "following Rolando [Vázquez]–who believes that working in the shadows, is just as important as putting things out into the light–we needed to consider how our desire to name, acknowledge and represent, can again be a colonising, or wanting to grab everything.'[62]

The ability to think through the pluriverse, to become aware and to unlearn, depends on the capacity to allow for opacity and to embrace a multitude of possibilities. For author and public intellectual Achille Mbembe, 'we will not open up the possibility of different aesthetics and ethics, nor will we open the door to a different politics of the future world (…) without different modes of counting, of seeing, of doing.'[63] For a different politics of a future world, we add that we also need to invite and allow different modes of fashioning, listening and being.

Guided by these aspirations, the Collective initiated the Global Fashioning Assembly (GFA) in 2021, of which the first edition will be organised in October 2022. This online coalitional gathering beyond institutional, disciplinary and geographical boundaries is planned every two years and aims to decentralise knowledge creation and sharing regarding fashioning through global networks and long-term relations. With a multi-stakeholder programme, in a variety of languages and formats, in a combination of online and offline programming, guests become hosts and hosts become guests, disrupting conventional colonial power relations often at play in global projects. Underpinning the GFA is a communal and experimental process in terms of collective ideation, decision making and development. In order to disrupt gatekeeping processes that decide on who gets to speak and who does not, whose knowledge gets validated and who's not, communities are invited through a 'snow-ball' process, through a network of networks.

Joining around the world assemblies in which communities explore in their own ways how to collectively create shared pathways to politics of wholeness, we share their collective mission that 'we can't solve our crises using the same way of thinking that created them.'[64] With the GFA, we aim to contribute to systemic changes in the way we think and practice body fashioning, by creating democratic spaces to gather,

---

[61] May 2022 Conversation on 'Designs for the Pluriverse' (24:15 onwards). https://youtu.be/KqMvN64VGU0.

[62] May 2022 Conversation on 'Designs for the Pluriverse' (7:20 onwards). https://youtu.be/KqMvN64VGU0.

[63] Achille Mbembe cited in Erica de Greef, *Decolonial Fashion Practice #3* (19 April 2021).

[64] https://www.grassroots2global.org/.

share, listen, (un)learn, address and connect. As a decolonial alternative to western-centric academic fashion conferences, we want the *Assembly* to be an experimental platform for collective decision-making processes that are deeply democratic, fair, ethical, caring and culturally empowering. By experimenting with different formats that encourage speaking with each other, rather than speaking at each other, we want to question how communities can gather beyond difference to take responsibility for our deeply connected futures.

References

1. Gonçalves S, Decolonial Fashion Practice #1 (15 February 2021) https://rcdfashion.wordpress.com/2021/02/15/decolonial-practice-1/
2. de Greef E, Gonçalves S, Jansen A (2021) Listening as a radical act: conversations on Decoloniality and fashion. African Arts 54 (4):1–5
3. de Greef E, *Decolonial Fashion Practice #3* (19 April 2021). https://rcdfashion.wordpress.com/2021/04/19/decolonial-fashion-practice-3/
4. Jansen MA (2020) Fashion and the phantasmagoria of modernity: an introduction to Decolonial fashion discourse. Fashion Theory 24(6):815–836
5. Keating C (2005) Building coalitional consciousness. National Women's Studies Association Journal 17(2):86–103
6. Vázquez R, Mignolo W (2013) "Decolonial AestheSis: colonial wounds, Decolonial healings. Social Text Journal. https://socialtextjournal.org/periscope_topic/decolonial_aesthesis/
7. Vázquez R (2020) Vistas of modernity: decolonial aesthetics and the end of the contemporary. Mondriaan Foundation, Amsterdam
8. Mignolo WD, Walsh CE (2018) Introduction. In: On decoloniality: concepts, analytics, praxis, pp 1–14. Durham, NC: Duke University Press

Fashion and Identity in Virtual Spaces

The Other Bodies as an Avatar in Animal Crossing

Kyoko Koma

Abstract As fashion researchers Susan Kaiser and Denise Green discuss the role of fashion and identity in terms of intersectionality, fashion is a device that incorporates one's boundaries (gender, country, age, ethnicity, etc.) and others, and is a device for realising the 'self one is trying to become'. The other as an ideal image to aspire to has been constructed in images and discourses by the fashion media over the centuries. In order to get as close as possible to this ideal image/other and become the 'real me', people have tried to get closer to the ideal other by changing their own faces and bodies, from corsets to dieting, changing hair colour, cosmetic surgery, etc., wearing make-up and fashion and crossing various boundaries. Fashion media, which have provided readers with idealised images in this way, have spread over time to mass media such as television and film, and digital media from blogs to social network services, and now the term 'metaverse' is rapidly emerging, with virtual spaces including games also becoming a new fashion media are becoming new fashion media. Wearing virtual fashion without physical cloth in virtual space is considered a sustainable act, but in the first place, does wearing this virtual fashion with an avatar mean abandoning one's own body and becoming free to follow one's own desires, unrestricted by the boundaries of the various "I's" as Kaiser and Green, does this mean that I am free to follow my own desires? In other words, does it mean that one can freely overcome the boundaries of the various 'I's and become the person one wants to be? Using the Japanese game Animal Crossing as a case study, this paper analyses the relevant discourses of several actors to explore how virtual fashion has been used to create new forms of identity for those who have abandoned their physical bodies in the age of *Liquid Modernity*, and what the issues are.

K. Koma (✉)
School of Information and Communication, Meiji University, Tokyo, Japan
e-mail: kyokokom@meiji.ac.jp

M. Á. Gardetti and R. P. Larios-Francia (eds.), *Sustainability Challenges in the Fashion Industry*, Sustainable Textiles: Production, Processing, Manufacturing & Chemistry,
https://doi.org/10.1007/978-981-99-0349-8_12

1 Introduction

Fashion researchers Susan Kaiser and Denise N. Green have discussed the role of fashion and identity from the perspective of intersectionality. Fashion constructs an identity that is conditioned by its boundaries (gender, country, age, ethnicity, etc.), and by incorporating others, it can be regarded as a device for realizing "who I am trying to become [1]." Over the centuries, fashion media has constructed this "who," as an ideal image that one should aspire to become, through images and discourses. Many individuals have transformed their face and body, from corset to diet, changed their hair color, and had cosmetic surgery to resemble this ideal image. Using make-up and clothing, they have transcended various boundaries to get closer to the ideal other and realize it. Significantly, both mass media (including television and films) and digital media (including blogs and other social networking services) have become an intrinsic part of fashion media over time, which has provided readers with this ideal image. Nowadays, with the rapid emergence of the terms non-fungible token (NFT) and metaverse, virtual space is becoming a major source of information in the fashion industry. It is beginning to be seen as a place where fashion information can be disseminated. For instance, virtual fashion platforms, such as Dress X, allow users to purchase and wear virtual clothes using their photos. Further, in March 2022, the 2022 Metaverse Fashion Week (2022 MVFW) was held on Decentraland, involving several well-known luxury brands and brands specializing in virtual fashion. To participate in the virtual exhibition space, visitors could create their own avatars. This raises the question of whether wearing virtual clothes in virtual spaces essentially means abandoning our bodies and wearing whatever we want to wear without limit? Does it mean that we more freely overcome the various boundaries that limit us and become the "real me," or whoever we want to be, by incorporating the other? Even before the inception of the virtual fashion space, we already had the Japanese game Animal Crossing (Atsumare dobutsu no mori in Japanese). In this case study, therefore, we will focus on this game to examine, through the discourse analysis, how the relationship between fashion and identity in the virtual space are created by the digital society in the age of liquid modernity proposed by Zygmund Bauman [2].

2 Digital Identity

In a reference to digital identity management in meta-verses discussed by Solomon in 2010, an innumerable number of ordinary users in a virtual identity can be constructed [2]. The author also mentions two main types of digital identity here. One is the digital identity that has been constructed by adding photo manipulation and other modifications to create the image of beautiful people that do not exist in reality, and the second is the digital identity of avatars [2]. The use of avatars is discussed in terms of "offline to online," where, what I post on social networking sites becomes who I am, and the relationship between online experiences (fantasy identities) back in the real world "online to offline," and "offline merge with online," which has been swallowed by augmented reality technology [3]. The three issues discussed in 2010—posting on Social Network Services (SNS), photo modification functions, fashion expressed through AR technology and avatars—are still relevant in recent years. According to Webb [4], digital identity is something that is constructed through discourse and narratives [4], through the development of digital media, through the processing of photographs of a certain self to begin with [3], and avatars that are set up separately from one's own body and dressed up in various fashions.

Anthony Elliot discusses that self-identity in the twenty-first century is likely to mean "social networking updates, posted messages, blogs, virtual reality apps, AI cognitive architecture, cloud computing, big data and digital media [5]". In this context, with the development of digital fashion being discussed, how are digital identities being constructed through fashion?

3 Changing Fashion Media: Information Dissemination, Sites of Practice, and Actors

As Winnicott points out, "our inner world is defined in a double sense: the life we want and the life handed to us." Fashion media suggests "normative anxiety" [6] and normative ideals to model readers. "Normative anxiety" is constructed using the gap between the "desired self" and the "real self" [7]. Digital media facilitates the construction and realization of the "desirable self" by allowing the digital alter ego self to wear digital fashions, choose normative other-ideal images that transcend the boundaries of the real self, including gender, age, and nationality.

Thus, on digital media, clothes are constructed and transmitted as fashion, while at the same time the ideal identity of the avatar owner is proposed and constructed by the avatar wearing the fashion.

So, how and by whom were clothes originally constructed as fashion in the media? Roland Barthes, in his 1967 book titled *The Fashion System,* analyzed the construction of fashion through the discourse of the fashion media semiotically, as clothes are spoken of and become modes (fashion trends)' [8]. For a piece of clothing to become a fashion or a clothing trend, there must be a media presence, a language, to communicate it. The media of fashion has changed over time. In France, for example, in the late seventeenth century, portraits, costume prints such as etchings and dress prints depicting modes and customs, and dolls dressed in fashionable clothes were used as a means of communicating French modes throughout Europe. With the development of transport, printing technology and literacy, fashion magazines spread to a wider audience and one style became a fashion trend, spreading worldwide through the mass media [9].

Fashion media then went online on the internet and the web, such as Vogue.com and Elle.com in 1994, and were systematized to allow readers to contribute to the same platform. Since the 2010s, when smartphones appeared and existing fashion media began to have full-fledged social networking accounts, the transmitters have been fashion media as mass media, their readers (followers), ordinary individuals and fashion companies. Fashion is shaped by the intertwining of these discourses [10].

Fashion media used to be a place for fashion companies to disseminate information, but with the birth of social networking and the internet, it is evident that it has become a place not only for the transmission and collection of information, but also for fashion practice and transmission by wearers. Before the advent of the internet, the street was the place where fashion was practiced and disseminated, but in the 2000s, it shifted to blogs on the internet and in the 2010s to social networking services (SNS). Today, it has expanded to virtual spaces such as avatars and image-processed self-expression formed by having people try fashion on digital media.

4 Ideal Images Constructed in Fashion Media: Fashion as a Device for Transcending Frameworks and Becoming Who We Want to Be

Kaiser and Green describe intersectionality in subject formation [1] and discuss what norms we try to move towards and become by wearing fashion, crossing various boundaries, and transcending ourselves [1]. Arguing from a Western perspective, Kaiser and Green discuss fashion as a means of transcending fixed frameworks, norms and frames based on Western white-centric binaries of gender, borders, class, age, ethnicity, and location.

What are we trying to transcend? It could be said that we are trying to transcend the very boundary between the "handed over" self and the "desired" self, as D.W. Winnicott, quoted by Eliot, has shown [5]. When considering what drives people's strong desire to transcend these boundaries, fashion has been created and transmitted to reflect and evoke people's desires, while being proposed on the fashion media, and to form a self that can be realized by wearing such fashion. The existence of fashion media is undeniable [8].

In what follows, we will examine what has been attempted to be transcended in the various boundaries that the Kaiser and Green speak of as limiting them, along with the evolution of the media that construct fashion.

At the beginning of the twentieth century, when fashion magazines as mass media presented fashion trends, what was presented as a boundary to be crossed was considered to be classy. Needless to say, there are concepts that explain this phenomenon, such as what Veblen calls "conspicuous consumption" and Gimmel's "assimilation and difference" and "trickle-down." Using France as an example, it is pointed out that "at the beginning of the twentieth century, mode magazines were exclusively for the bourgeois, (...) certainly not only the latest Parisian modes, but also a medium for the sensibility of the times" (Françoise Blum, cited by Matsuda: P18) [11]. Thus, it can be said that fashion was driven by the bourgeois and other upper classes and elites. Another example is Christian Dior's "New Look"- long flared skirts, presented as part of the Paris Haute Couture collection in 1947, after the Second World War, were imitated all over the world, including post-war Japan [12].

Later, in the 1960s, a second school of feminism emerged and this "class," presented as a boundary to be overcome by following fashion, blossomed as street fashion by London's Marie Quandt, as if to reflect the times, at the time. The "class" of the 1960s and '70s was transformed into "youth" by the miniskirt presented by Courrèges in its 1965 Paris Haute Couture collection. Overcoming norms can be seen to have shifted from "class" to "age (youth)" from this period 16 [13].

In the post-war period in Japan, Americanization was progressing; new looks and miniskirts, which the USA and the rest of the world was following at the time, became popular, and the number of Western-style dressmakers increased.

From the 1960s to the 1990s, mass media and the street coexisted as media for constructing and communicating fashion. In Japan, too, the source of fashion trends shifted from Parisian haute couture to the world's prêt-à-porter and street fashion.

In the 1970s, after the rise of prêt-à-porter collections, non-Western designers and non-Western fashion became active, presenting a variety of principles, mass media, and street-driven styles. In Japan, too, readers were influenced by this trend and "imitated" them using inexpensive ready-to-wear materials. Street style was practiced on the streets as a counterpoint to mainstream fashion.

Since the 1990s, physical spaces have of course existed, but since the 2000s only online versions of media, blogs, and forums existed. Later in the 2010s, with the spread of smartphones, social networking sites (official fashion media accounts, digital influencers' pages, etc.,) both anonymous and famous (including earned media SNSs and tie-in posts), (offline to online) avatars similar to/very different from themselves were formed, by having them wear fashion on digital media, including AR and image-processed self-representations, as non-physical "alter egos" of themselves. For example, at work or in their private life, multiple digital identities are formed by the different fashions they wear in their private life [9]. As of June 2022, Meta has launched a digital apparel shop, the Meta Avatars Store, which has been launched by fashion brands such as *Prada, Balenciaga,* and *Thom Browne* that also have announced their entrance in the market. The avatars used here on Instagram are offline to online illustrated avatars, as close as possible to their own person (FashionSnap.com, 21 June 2022), but not quite their own body, but more similar offline to online illustrated avatars to wear fashion and build a digital identity. In addition, services such as those proposed by Japan's CyberAgent, where celebrity digital twin models wear digital fashions and sell them as NFTs, have started to take the image of the digital twin from offline to online, even if it is not physical and can be altered somewhat, to the image of the "person," is used to construct a digital identity (FashionSnap.com, 9 June 2022).

However, the avatars in Animal Crossing, the game popularized by the COVID-19 pandemic, which will be discussed in this paper, are not based on one's own image and are created by freely choosing body parts that are different from oneself, but will look almost similar, no matter what is chosen. While creating such similar avatars, it can be said that the avatars express their identity and individuality by wearing a variety of fashions.

With the advent of online games and virtual spaces such as Animal Crossing, the online fashion media space has become a place for the transmission of fashion information from the fashion industry, and for followers (transmitters) who not only receive information but also practice and transmit it. The online fashion media space has become a place where fashion is not only received, but also is practiced and transmitted. The table below shows the transmitters of fashion information in fashion media by era, the ideal figure, the place where fashion is practiced, and the body that practices fashion in each medium, as described above.

Period	The fashion media (in the broadest sense of the term)	The originator	What transcends boundaries and approaches the ideal	The place of fashion practice	Body
Early twentieth century–1960	Fashion magazines	Fashion Industry	Class (normative beauty)	Life space (normative)	The body of the self
1960s–1990s	Fashion magazines Street	Fashion industry General	Youth (normative beauty) Race (Western white-centeredness)	Life space (normative) Street, cosplay (deviance)	The body of the self
2000s	Online editions Blogs Forums Fashion magazines	Fashion industry Blogger	Youth Diverse boundaries	Living space (norms) Blogs(norms, deviance) Forums (norms, deviance)	The body of the self
2010s	SNS Blogs (norms, deviance) Forums (norms, deviance) Fashion magazines	Fashion Industries followers Instagrammers (professional and general) Youth	Youth Crossing diverse boundaries and becoming different from social norms	Living space (norms) SNS (norms, deviance)	The body of the self Avatars who look like themselves
2020s **	SNS Games (virtual space)	(SNS ditto) Fashion industry Game players	Game player Crossing diverse boundaries to become the norm for oneself	Living space (normative) SNS (norms, deviance) Game/virtual space (norm, deviance)	Bodies other than my own

(continued)

(continued)

Period	The fashion media (in the broadest sense of the term)	The originator	What transcends boundaries and approaches the ideal	The place of fashion practice	Body

*The fashion industry here refers to all companies and people involved in the fashion business, including fashion brands and fashion media
**The first version of Animal Crossing was released in 2001, and the launch of Animal Crossing in March 2020 in the wake of the COVID-19 disaster became a global social phenomenon, including the entry of the fashion industry in virtual space, which had not existed before

5 Fashion and Identity in Virtual Space

Thus, along with the changing times and fashion media, transmitters of fashion information, normative boundaries to be overcome, places where fashion is practiced, and question of whether one has a body or not, are transforming. The normative boundaries to be overcome extend from class to various boundaries such as youth, race, gender, and intersectionality, where these normative boundary crossings intersect, and come into practice. This diverse boundary crossing is realized in fragmented spaces that differ from the social space in which we live in reality, such as social networking game spaces, virtual reality spaces, and augmented reality spaces, where wearing fashion allows us to construct a "desired" self, rather than being trapped in a "given" self-identity that the former example is the social space of everyday life. An example of the former is the community of LGBTQ people, who, depending on the place and time, may sometimes have difficulty expressing their gender in society, but who gather to wear fashion based on their own bodies and beyond the framework of their gender [14]. There are also communities that wear fashions that are not worn in everyday life, such as Lolita fashion and cosplay. In the latter case, avatars wear fashion on the island, which is a virtual space such as a game like Animal Crossing. They can choose to wear the same fashions in current society, with the "My Design" function. The function allows the avatar to wear the same fashions as in today's society. However, in the book "Raku Raku [easily in Japanese]," only the QR code needs to be scanned to get an outfit like that character. You can set up downloadable character fashions based on famous manga characters and idols and then dress your avatar in the digital fashions you obtain their kawaii fashion [7]. In these cases, body modification and make-up are also included.

This study examines the kind of fashion proposed by various actors on virtual media and the kind of identity they are trying to construct by wearing the proposed fashion. The Japanese game Animal Crossing is used as a case study, and through it, the study examines how fashion and identity are constructed through the game. The relationship between fashion and identity in Animal Crossing is discussed below.

6 The Resurfacing of Animal Crossing

The first version of the game was released in 2001, and when COVID-19 first began to spread, a new version was released on 20 March 2020, called Atsumare Dobutsu no mori (English Animal Crossing), which went viral. The game was also picked up by the media worldwide, with *Le Monde* (a left-of-center French intellectual daily newspaper) reporting on 8 May 2020: Between coinage and speculation of failure, Animal Crossing is a stay-at-home extravaganza. (Entre course aux clochettes et spéculation sur les navets, "Animal Crossing," le grand jeu du confinement) and 17 others, and several articles, such as "The growing interest inside academia in Animal Crossing" in the 15 July 2020 issue of the Washington Post, 79 articles, such as "The growing interest inside academia in Animal Crossing," dated 9 July 2020, are seen.

In Japan, in the midst of the COVID-19 pandemic, the July 2020 issue of the magazine *SWITCH*[1] featured a special issue on Animal Crossing, in which creators from various genres introduced the kind of desert island they had created in this game and explained how they were living there. Additionally, the 7 August 2020 morning edition of the Nihon Keizai Shimbun (Japanese Economic Newspaper) wrote an article focusing on "Animal Crossing," where people gather in a virtual space under self-restraint from going out due to COVID-19. The article noted (1) that companies and luxury brands are gathering in Nintendo's Animal Crossing and promoting their products in the game, and that interaction is becoming a budding business opportunity. A function called "My Design" is being offered, which allows users to change the design of their clothes and furniture to their liking using a communication function. It is said that "one of the funs is to choose your own character in the virtual game space and to share your proud fashion and room interior design with other Animal Crossing users" (ibid.). The report also states that "the fashion and art industries [and museums] were quick to take notice of this high degree of freedom, and there is a movement to provide "virtual fashion" by releasing logos and original designs that fans can recognize at a glance for free (ibid.). Thus, the virtual space referred to here is a medium for obtaining fashion information and "virtual fashion" itself, and

[1] Since its first issue in 1985, Switch has been an interview magazine that has gained a large readership over the past 25 years, based on the concept of "following the trajectories of vivid individuals who shape the times and conveying their breath and brilliance." The magazine has met people at the forefront of various genres such as music, film, literature, art, and design, and through interviews that elicit words that only a small magazine can tell, and photo stories that elicit expressions that only a small magazine can show, it has provided a deep insight into the thoughts, feelings, and lives of the people living at the forefront of these genres. The interviews and photo stories that bring out expressions that can only be seen in the magazine delve deeply into the thoughts, feelings, and way of life of the artists living in the same era (https://www.fujisan.co.jp/product/1281680375/ August 9 2020).

a place where "I," as a character, who is not my own body, wear the clothes in the virtual space. In this virtual space of "Animal Crossing," we obtain information on various fashions, and attempt to construct any kind of identity by wearing them.

7 Examples of Fashion Practice at "Animal Crossing"

We will focus on three case studies. The first is the case of Kotono, one of the animal resident characters and non-resident visitor characters in Animal Crossing, who conducts regular fashion checks for avatars in the game in each of 11 fashion style categories. If a user performs a fashion check on their avatar from Kotono and receives an "ideal coordination" rating from Kotono, they would receive one type of clothing from Kotono's brand "Kate Series" and two tailoring coupons, with which the players can buy clothes, or obtain them by talking to Kotono. The eleven different themed fashion styles are proposed, ranging from sporty, outdoor, relaxed, formal, work, fairy tale, formal, horror, party, vacation, and holiday, from everyday to extraordinary, which evaluate the avatar's fashion. In this way, participants are judged on their conformity with the fashion norms for each theme of Animal Crossing. Although the game can be played anonymously, but it is as if, in the same way that we wear clothes in the real world with an eye to the un/visualized gaze, we can say that Animal Crossing is not an empowering space where we wear clothes with complete freedom, but we follow the "norms" of the game with an awareness of the gaze of the Kotono standard.

The second case study is of a participant using an avatar in a special issue of the Japanese magazine *SWITCH* on Animal Crossing.

The July 2020 issue of Japanese fashion and art magazine *SWITCH* featured how creators from various genres are living on deserted islands. In it, musician Meilin (*SWITCH* July 2020 issue), who created PIGEON Island in Animal Crossing to study French and plays Animal Crossing in French, received a gift from the fashion brand My Design (@nookstreetmarket) for her clothes. She downloaded a Marc Jacobs coat and her avatar wearing the coat is featured in this issue of *SWITCH*, forming an identity through digital fashion.

The third case study is the fashion media case study in Animal Crossing.

In the October 2020 issue of Vogue Japan, an article was published in *Vogue Japan* featuring *Vogue Japan*'s animal crossing. The *Vogue Japan* editorial department has also created a "*Vogue Japan* island" of eight key looks from the 20–21 autumn/winter collections of Japanese brands such as Tomo Koizumi and Sacai, under the concept of "a place to enjoy Japan's cutting-edge mode," and has also started offering free downloads of each garment (see below) (Vogue Japan, October 2020).

These first two case studies are fashion styles that exist in the real world as defined by the game world, despite the difference between anonymous and real names of the game actors who create avatars, especially in the first example of Kotono, where there is a fashion check by a character called Kotono. In this sense, it is in line with the aforementioned idea of reflecting reality. However, even players who do not choose fashions that do not fit the gender and various other conditions given to them in the real world are free to do so as avatars in the game, and in doing so they transcend the various frameworks that originally define them, such as gender and nationality, and follow the norms of others, thereby creating a partial, unreal space.

And in this virtual space, the player's own body is not used, but an existing "cute and immature" character's avatar. In other words, even if a player chooses an Animal Crossing character's avatar and adopts a normative fashion within the virtual space, the player actually running that character may not necessarily be the same gender or age and may already have transcended the framework at the stage of choosing the character. This possibility cannot be denied: in Animal Crossing, players can define their own identities, which are very different from those of real-life players, by (i) updating their passports, (ii) changing their hairstyles and so on, (iii) where they wear their clothes, and (iv) what they wear, (v) where and what they wear, (vi) how they change their clothes this way and that (My Design), (vii) transformation also changes their clothes, and (viii) learning reactions. It can be constructed through (Animal Crossing Complete Strategy Book + Super Catalogue, 2020, Nintendo 19–22). However, Animal Crossing's avatar characters have similar body shapes and faces, and the differences between them are not perceived to be significant in terms of gender, age, etc. It can be said that fashion—what they wear—plays an important role in differentiating them from other avatars. It is one of the only tools to express what kind of person could be fashionable, which is supposed to compensate for physical changes and not to change the person's own appearance (although they have hair, for example). Unlike in the past, where people went on diets, had plastic surgery, modified their bodies on social networking sites, or created avatars with their identities revealed, players are anonymous and can identify themselves. They choose an avatar with a completely different gender, age, appearance, and even characterization as the base of their body and wear the fashion of their choice. Since there are no physical constraints and the person wears fashions that conform to their desired social norms, reinforced by (1) their passport and (2) their reactions, their identity of who they are is erased, and once they have formed their own digital identity in the game, they become dependent on fashion to Games such as Animal Crossing can be seen today as pioneering a new form of fashion media.

In addition, when looking for comments from wearers on Twitter in June 20 2022, with the words "#Atsumare Dobutsu no Mori (Animal crossing)" and "#Kotono (kotono) chan fashion check," rather than wearing their favorite fashion, it was found that the comments were more about "whether Kotono-chan will approve of me or not." Some of the posts were about whether they would be recognized by Kotono-chan, rather than whether they would wear the fashion they liked. In addition, as the January 2021 issue of the Japanese fashion magazine *SPUR* has an article titled "Six fierce wearers who can't dress the way they like because they are worried about other people's eyes gather in 'Animal Crossing'," some postings were more about wanting to be recognized by Kotono-chan than about wearing the fashion they like. It seems to be seen. The article also suggests that the site was created as a place where people can dress the way they like, without worrying about what others think.

8 Discussion

As we have seen, Animal Crossing was meant to: 1. follow the norm through various avatars, 2. experiment with luxury brands, 3. propose Japanese fashion as a fashion media, and 4. create a community where people can wear their favorite fashions freely. However, no matter what gender or hair color they chose, the avatars all had a childish image that was distant from the user and to some extent uniform.

In this section, we have reflected on how people have obtained and practiced fashion information from various media, attempted to transcend the normative boundaries of self and society to realize their ideal image, and processed and given up their own bodies. Here, using Animal Crossing as a case study, I have examined the practice of obtaining and fashioning a new "body" different from one's own in the virtual space of the 21st-century game space.

Jeremy Beilenson, a leading Virtual Reality(VR) researcher specializing in psychology and communication studies, has two points in his theory of VR: (1) VR as "the medium with the most powerful psychological effects in history, producing physiological responses in the brain similar to those of real life experiences" and (2) the "potential of avatars." By examining the "potential of avatars," I want to consider the practice of fashion in relation to Animal Crossing.

With regard to the first point, Beilenson states that VR is "a medium that has the strongest psychological effect on the brain and produces physiological responses similar to those of real-life experiences [14]." The immersion of the experience alters the brain and the risks that VR poses to the brain include escaping reality, confusion between reality and VR and reduced attention span (ibid.). Even considering the practice of fashion in VR game spaces, "Animal Crossing, reality or escape?" for the whole idea of Animal Crossing, including the practice of fashion, Hiroshi Yoshida's thoughts are quoted below:

In Animal Crossing, players are forced to endlessly repeat tedious "physical labor" to pay off the mortgage. The world we face in Animal Crossing is nothing but a microcosm of the corrupt real world of the late capitalist and information age. What is so sad about having to check your smartphone, accumulate miles, and operate ATMs every day, even in the game world? There is not a shred of fantasy or fictional imagination of "another world." There is no escape from reality. If this is the case, why are we so fascinated by this game? One possible reason is that fiction substitutes for reality in a strange way. This substitution is happening all over the world right now. As the real-world freezes and comes to a standstill due to the blockade of COVID-19, we see, rather unrealistically, that the joys of physical labor, modest rewards, and consumer activities are gone. A locked life where everything is remotely controlled through monitors feels more like a fiction. In my hypothesis, Animal Crossing presents an alternative world to the real world, or an alternative object to the real world, rather than a caricature of the real world (Bijutsu Techo, August 2020, Yoshida: 119–120).

Thus, Yoshida states that "while the real world has frozen and stopped functioning due to the blockade of COVID-19, fiction (such as Animal Crossing) is substituting for reality in a strange way." This substitution is happening all over the world now. It is neither an alternative world against the real world nor a caricature of the real world (Bijutsu Techo, Aug 2020, Yoshida: 119–120). However, as we have already seen, the activity of practicing fashion in Animal Crossing can be described as an extremely "normative and realistic" practice, except for the option of how to freely choose an avatar, but the players who practice it do not necessarily reflect the same life-size self as their original selves. Rather than attempting to transcend frameworks and become their ideal selves by wearing fashion, players who play anonymously choose characters (avatars), which transcend frameworks (gender, species, body part, race, etc.), as Kaiser and Green described earlier in their discussion of intersectionality and race. However, they can also choose a character that is different from the real-life, self-handed to them in physical space, such as a man in his 50s who can choose the character of a cute girl, create a passport to match, choose a reaction, and wear foreign fashion brand clothing. The fact that other people's real-world practices that their real-life selves cannot do are possible at Animal Crossing suggests that the fashion practices at Animal Crossing are also an escape from a reality that is usually insurmountable. I have examined this further from the second point of Beilenson's argument: the possibility of avatars.

The other point concerns the acquisition of "new bodies," or "avatars." For example, he argues that young people acquire humanity's first new bodies in these VR spaces, and that the movement of the body leads to an "exchange of perspectives," increasing empathy for others (Beilenson, above: 109–145) [15]. Beilenson also discusses avatars in VR spaces. Some, such as FaceShift, a motion-capture technology acquired by Apple, instantly reproduce "realistic" facial expressions captured by a PC's built-in camera, making them avatars of CG characters. This is more data-efficient than videos showing actual people, leading to a solution to the problem of line capacity and making VR chat mainstream (ibid.: 255–262) [15]. The main reason why VR chat with avatars is favored is that it solves the "burden of exposing one's (real) appearance to the public," even if the avatar is "real" (ibid.: 263) [15]. However, although we can communicate using another "body," our real avatar, which is not anonymous and more sensitive, the disconnect between our real body and that avatar is undeniable. What does communication without an entity produce? Thus, in a VR chat, you are an extension of your real self, but in Animal Crossing, as mentioned earlier, the self you create by selecting fairly simplified body parts does not have to be an extension of your real self. This simplification of the self-image is also illustrated by Elliot, cited earlier, who, using selfie culture as a case study, says: "It can be said that selfie culture oversimplifies the self. It involves intimidation and addiction, where people cannot stand or tolerate their own complexity and take revenge on their identity [4]".

To begin with, as Narumi [16] mentions, "fashion is an act of giving meaning to the body through the two categories of "time" and "space," clothing is a technology of making the body visible, the body becomes fashion and fashion becomes the body [16]". However, in Animal Crossing, the characters leave behind their real-life bodies, or choose entirely different characters, to practice the everyday experiences of others that they cannot practice themselves in the real world.

Park [17] presents the concept of the virtual personality as a current multi-identity, transforming self-representation as ego, especially in the post-Internet (a world where the distinction between online and offline is not established). Unlike the social face used in physical space, which is considered real, the virtual space shows what is considered to be the "real self" for the user, hiding the user's actual physical, mental, and social appearance and revealing what seems to be the real self. In forming a newly created virtual and essential personality, or "virtual personality," one has to rely on the visual, and fashion plays a major role in its presentation. In other words, in VR game spaces such as Animal Crossing, it is not that "the body becomes fashion and fashion becomes the body," but that "fashion becomes the body, and a virtual identity is formed."

When considering what the avatars in Animal Crossing attempted to transcend by wearing fashion, it is necessary to consider what they were trying to transcend by abandoning their own bodies and acquiring the body of a cute 3-headed high character. What does it mean for the "body" of an avatar character with a virtual identity to wear fashion?

Joanne Entwistle, who published The Fashioned Body (2005), writes about fashion and the body.

Two key ideas are at the heart of her book: Maurice Merleau-Ponty's assertion that "to dress is to adapt oneself to a particular situation, to work with one's body in a particular way," and Irving Goffman's understanding of space, or "stage." It is based on the idea that appearance and dress matter if one is not appropriately dressed for a particular situation [12] and that "the seemingly personal and individual act of dressing is in fact rather socially oriented, as it allows people to adapt their bodies to appropriate social standards, and furthermore it can be argued that it is a preparatory process for people to remake their bodies into something that conforms to appropriate social standards, and even into a decent and attractive body [18]".

Based on this Entwistle idea, let us consider fashion and the body in SNS and VR spaces. Compared to the time when only the social figure could be practiced, the development of SNS and the formation of online microcommunities where many minorities gather has created many opportunities for me, the 'invader' of social norms, to interact with and be recognized by others. As a result, it became possible for me, as an invader of social norms, to present myself to society as a microcommunity in addition to my social normative figure, and to wear fashion to transform myself into an attractive body within it. In a VR space such as Animal Crossing, the person's body does not exist, and by wearing fashion, one can transcend various frameworks such as gender, race, age, and ethnicity, and more freely construct the attractive body one desires, appropriate to that virtual space. It is important to emphasize that in such virtual spaces, fashion can be said to be the only source of identity.

9 Conclusion

In 2005, Zygmunt Bauman contrasts the solid modernity of the early modern phase with the solid modernity of late modern society, in a book on what identity is in liquid modernity. Although what is written here was written before the widespread use of smartphones, it is highly illuminating in considering identity in the current VR era. Bauman writes: "Identity is, in the first place, an expression of various institutions such as the family, the state and the church, which are, from a Kantian perspective, the a priori of social life." In this case, the locus of such identities has been largely dismantled by contemporary mass society [2].

Bauman cites the example of Pigò della Mirandola (1463–94) (who became like the legendary Proteus, changing form after form, and whose paintings were free to range from unrepeatable possibilities of cheerfulness to those in which even the tail floats).

"It is no longer a question of choosing the means necessary to choose an alternative identity for oneself." You can go to a shop and immediately find a product that transforms you into the character you want to see, to be seen and to be recognized. [...] The real problem, and the most common concern today, is the opposite. Namely, which alternative identity to choose and, once chosen, how long to cling to it. If in the past the "art of living" was primarily concerned with finding the right means to reach a defined goal, today it is a question of tackling all the (infinite) goals that are already in our hands or can be obtained by means that are within our reach. (op.cit.,131 [2]).

As to which alternative identities in the VR space of this game are liquid modern identities, Bauman quotes management theorist Charles Handy: "These virtual communities may be fun, but they only create the illusion and pretense of intimacy" [omitted]. They are not a suitable alternative to "kneeling under the table, looking each other in the face, and having real conversations". Such "virtual communities" do not materialize the individual identity that is the primary reason for seeking them (54), but as Yoshida also pointed out earlier, on the occasion of COVID-19, in reality, neither is it possible as an alternative venue for communication, nor in reality as a place to wear the gaze of others. It would be possible to form communities that share values with oneself. Today, with the proliferation of virtual spaces, will our alternative and fluid fashion identities, which are largely shaped by the practice of fashion in these spaces, be recognized as our identity every time?

In addition, now that the possibility is opening up for people to become who they want to be by wearing fashion with avatars in various virtual and digital spaces, how will the online experience (fantasy identities) back in the online world pointed out by Solomon [3] change?

Moreover, now that the possibility has emerged for people to become who they want to be through avatar fashion in various virtual and digital spaces, how will the online experience (fantasy identity) that Solomon [2] points to influence the offline self?

As early as 2005, Bauman warned of the dangers of acquiring an identity of easy choice, quoting Stewart Hall:

The greatest danger now is that cultural diversity has gradually become the fate of modernity (omission), with (new and old) forms of nationality trying to secure identity by choosing a closed culture or community, or by refusing to engage with the difficult problems that arise from trying to live with difference arising from identity and cultural identity. The greatest danger now arises from (old and new) forms of national and cultural identity that seek to secure their identity by refusing to engage with the difficulties arising from living with difference [2].

Hall, from the British colonies, who considered "culture" as the site of a constant struggle of various relations of power, reads the multilayered situation of history in the site of cultural struggles and points out that it is critical to stay in the community without fighting because of the resistance to the dominant culture. Further, as Bauman says, does the free formation of digital identities leave us no place to return to physical space?

More than two decades later, and with the accelerating formation of digital identities, where are we heading?

How is the real world from online to offline changing and affecting us? For example, Meta's Zuckerberg, who wore a grey T-shirt for many years, says that "the advantage of dressing up in the Metaverse is that you can wear any formal wear while (among other things) physically wearing a pile shirt" (Fashionsnap.com 2022.6.22 [19]). Thus, whether it provides a self that enjoys a different style from the physical space, or—and this is also different from the present issue and will be described in another discussion—in the case of other avatars and digital identities that cannot change in Animal Crossing but are close to you, in order to like yourself more, you can. Do they further diet, have cosmetic surgery, etc., to form a self-identity that incorporates more of the other? Or will physical identity also simplify as a simpler self that relies solely on fashion to form itself? How will physical and digital identities be shared and what interactions will occur between them?

References

1. Kaiser SB, Green DN (2022) Fashion and cultural studies, Bloomsbury, New York
2. Bauman Z (2005) translated in. Identity (Zygmunt Bauman, Identity: Conversations with Vecchi B., Polity Presss, 2004), Nihon kenzai Hyoron sha, Tokyo
3. Solomon MR (2010) Digital identity management: old wine in new bottles? Crit Stud Fashion Beauty 1:173–180. https://doi.org/10.1386/csfb.1.2.173_1
4. Webb S (2001) Avatarculture, narrative, power and identity in virtual world environments. Inf Commun Soc 4:560–594. https://doi.org/10.1080/13691180110097012
5. Elliot A (2019) Digital kakumei no syakaigaku (The culture of AI: everyday life and the digital revolution. Routeledge, 2019) (Akashi Shoten, Tokyo, 20222019) in Japanese by Endo, H., et al.
6. Giet S (2005) Soyez libres ! C'est un ordre- le corps dans la presse féminine et masculine. Edition Autrement, Paris
7. Koma K (2021) Fashion as transboundary culture. in Invitation to Information and Communication Studies, Minerubashobo, Tokyo
8. Barthes R (1967) Système de la mode. Seuil

9. Fukai A ed. (1998) Sekai Fukusyoku Shi (World History of Fashion), Bijutsu Shuppansya, Tokyo
10. Bendoni WK (2017) In social media for fashion marketing: storytelling in a digital world, Pub Sa (Japanese ed. Ito, S.)
11. Matsuda Y (2009) Syufu ni natta Paris no bourgeois josei tachi, hyakunen mae no shinbun zasshi kara yomitoku (Parisian Bourgeois Women Who Became Housewives: Reading from Newspapers and Magazines 100 Years Ago). Osaka University Press, Osaka
12. Koma K (2011) Acculturation of French fashion in Japan after WWII:Fashion as a device constructing identity. Acta Orientalia Vilnensis 2011 12–1:Modern Japan and Korea seen through various media, 63–78
13. Koma K (2019) Modo wo kochiku/dentatsu suru discourse: gatekeeper to dokusha no shinkyu nokoutiku (Discourses that construct and communicate Mode: gatekeeper and the construction of a class of readers.) in Semiotopos 14, Shinyosha, Tokyo
14. Alix SJ (2021) LGBTQ youth and technology: finding their way through online communities. In: Intersect, Vol 14, No 2. https://ojs.stanford.edu/ojs/index.php/intersect/article/view/1847
15. Bailenson J (2018) translated in (Japanese ed. Kurata, U.). VR Ha Nou Wo Dou Kaeruka, Kaso Genjitsu No Shinri gaku(Experience on Demand, What Virtual Reality Is, How It Works, and What It Can Do, WW Norton & Co Inc; Reprint 2018), Bungei Shunju, Tokyo
16. Narumi H (2003) Mode to Shintai, Fashion Bunka No Rekishi to Genzai(Mode and Body-History and Present of Fashion Culture) (Kadokawa Syuppan, Tokyo)
17. Entwistle J (2005) translated in. Fashion to Shintai. in Nihon Keizai Hyoron sha, Tokyo (Japanese ed. Suzuki, N.) (The fashioned body. Polity)
18. Park N (2017) Virtual Personas wo tuujita Henshin Ganbo no Hyogen no kenkyu (Research on the Expression of Transformation Desires through Virtual Personas), Doctoral mdissertation, Tokyo University of the Arts
19. Meta avatars Store. https://www.fashionsnap.com/article/2022-06-21/meta-avatars-store/checked on June 28th 2022

Interventions in Traditional Clothing Systems Through Anthropological Perspective

V. Nithyaprakash, S. Niveathitha, and A. Thambidurai

Abstract This chapter discusses the characteristics of fashion, its Eurocentric ideals and its overbearing reference to modernity. The institutionalized meaning of fashion and its values in the history of fashion studies is summarized. The research challenges the Western perspective of traditional clothing systems. In this background, an inclusive definition of fashion from the perspective of fashion anthropology is used as an anchor to analyze the contribution of traditional or indigenous clothing systems toward the changing needs of the world. The representation of the traditional clothing styles and its textile art forms is discussed. Analysis of traditional clothing styles in the framework of art design compositions provides an opportunity to re-imagine their significance. The source of these information is collected from the different forms and domains of secondary information available in published space. The result of the analysis exemplifies the design propositions of the African traditional clothing styles: African wax prints/Ankara prints, kente cloth, Bogolanfini, traditional batik prints, Adinkra and Adire—tie and dye fabrics as an emerging aspect for promoting cultural diversity, decolonization and sustainable designs. Further the case study also explores the interactions with the fashion leaders, designers, artists, exposure to newer ways of life and relations with foreign nations.

Keywords Fashion · Traditional clothing styles · Culture · Art · Anthropology · Decolonization

V. Nithyaprakash (✉) · A. Thambidurai
Department of Fashion Technology, Kumaraguru College of Technology, Coimbatore, India
e-mail: nithyaprakash.v.ft@kct.ac.in

A. Thambidurai
e-mail: thambidurai.a.txt@kct.ac.in

S. Niveathitha
Department of Fashion Design, National Institute of Fashion Technology, Chennai, India
e-mail: niveathitha.niveathitha@nift.ac.in

M. Á. Gardetti and R. P. Larios-Francia (eds.), *Sustainability Challenges in the Fashion Industry*, Sustainable Textiles: Production, Processing, Manufacturing & Chemistry,
https://doi.org/10.1007/978-981-99-0349-8_13

1 Introduction

The concept of fashion draws majorly from the Eurocentric idea of aristocratic, lifestyle values in the context of modern industrialized landscape. However, European fashion history reveals the adoption and appropriation of Indian, other Oriental and African Ethnic designs in large scale around the eighteenth and nineteenth centuries when they exercised colonial powers. Albeit fashion exists in every country with the transcending characteristics of its own ethos and the behavior determined by its international relations and exposure to the Global market space. The growing multi-polar world order, universal need to adopt sustainable fashion designs and country-specific fashion determined by its international relations and decolonial ambitions draws upon the necessity to reimagine and redefine the fashion concept in an alternative perspective along the lines of fashion anthropology. Examining the ecumenical aspects of fashion concepts and discussion of traditional textile craft practices, art forms in perpetuating its cultural identity shall set forth a positive premise to integrate sustainable values, relativistic cultural values as fashion's entities in decolonizing fashion.

2 Characteristics of Fashion

The concept of fashion constitutes two major characteristics, namely change and novelty. Novelty is the highly valued feature of fashion and Koenig refers the fashion followers as "neophilia" attributing human's receptiveness for new as the quintessential aspect for sustaining fashion behavior [1]. Barthes regards fashion phenomenon as neomania which emerged with the advent of capitalism in an instituted manner, with novelty as its purchased value [2]. Fashion characteristics also comprise three more properties; i. symbol of wealth, ii. Measure of up datedness and iii. leisure according to Veblan [3]. Fashion is conceived as a non-occupational identity according to crane especially for women [4]. In the 21stcentury, a man's identity is linked to his occupation as well as his influential social traits such as idea of lifestyle and appearing home to multicultural aspects [5]. Crane concluded fashion is decentralized where a sole standard style of fashion belonging a specific time frame might not appeal to everyone in the context of contemporary postmodern society [4]. From sociologists' perspective, fashion is referred to concept of imitation. Imitation is a relational concept that accords diffusion of certain fashion style from the upper sections of the society to the lower sections of the society as a sign of livelihood betterment for the lower sections of the society though there is no change in their material wealth and their social status [6]. According to Simmel, fashion promotes imitation and social equalization. Rather fashion is perceived to exist amidst the incessant process of change that discriminates one pattern of lifestyle from the other pattern [7]. The outcome of this incessant process of change creates different groups of people on basis of their choice of a particular fashion style. Sumner regards fashion as "mores" a certain behavior characterizing the socializing activities likening it to a custom and traditional clothing as a "folkway" with no agenda for characterizing a certain lifestyle criterion [8]. The traditional clothing is defined as a habitual way of dressing associated with local cultural events, socially constituted practice of occasion specific etiquette and emotional disposition. The purview of traditional clothing system explains the dressing behavior in specific social and cultural contexts as each context constitutes different social meanings. The dress function of traditional clothing is to serve physiological protection and modesty.

Toennies, ascribes the dominating feature of fashion to the desire for distinction prevalent among the urban elites where the fashion outfit is different from the traditional attires and costumes of the place [9]. Fashion outfits like flapper dress, beach pajamas and sporty girl sported by the affordable upper sections of the society implies leisure and lifestyle that are markedly different from the traditional clothing functions. Bourdieu concluded that the desire for distinction is more prevalent among the bourgeoisie rather than it among the commoners [10]. The Bourgeoisie, patronize what is good style and good class for the example the Parisian upper class members of the society regarded a certain way of dressing and accessorizing as good style and good class otherwise known as Bon chic and Bon genre (BCBG). In fashion, group of people patronizing certain style are referred to a subculture. Analogous to BCBG concept, Preppy style did exist in USA, and in UK, it is referred as Sloane rangers. A subcultural group or member of style tribe does not follow any of the conventional dressing rules, in fact they are overwhelmed and opt for a dressing style that connotes their feelings and lifestyle preferences.

Contrary to the distinctive and discriminating virtues of French and British fashion, Italy's fashion is shaped by its history, artistic virtues, culture and its relationships with French and the USA. Blumer attributed the fashion quotient of the dress or outfit to be the key signifier of fashion and its acceptance among the fashion consuming public [11]. Blumer further lists consumer taste as the prime factor for defining the fashion outfit functions.

3 Institutionalization of Fashion

McCracken (1988) observed that the imitation of fashion is facilitated and fostered by a fashion system [12]. He believes fashion system is the framework through which fashion values and style are constituted and promoted. White and White, Becker concluded fashion system is constructed by several institutions [13, 14]. Earlier Blumer had stressed the role of fashion designer in interpreting the consumer tastes and constituting the fashion values in the fashion outfit where the fashion designer works along with a network of closely connected professionals in the process of design creation [11]. The fashion system constructs the image of fashion and infuses its culture among the top five fashion cities: New York, London, Paris, Rome and Tokyo. Kawamura observes that fashion as a system existed in the form of Haute couture at Paris as early as 1868 [6]. Fashion system's existence lies in the network of people connected with creating fashion, proposing changes to fashion and those who adopt it. The various agencies associated in the process of constructing the fashion image are creative artists in the cultural industry, photographers, journalists, fashion forecasters, models, etc. The French fashion system consists of different organizations with a hierarchy among those who design clothes: Haute Couture, Prêt-à-Porter for women and Prêt-à-Porter for men. It is always a group of people who are involved in the process of co-creating and legitimizing the fashion values. Fashion carries a culturally constructed meaning, created and disseminated by the likes of journalists, models, designers, fashion photographers and the creative industry professionals. The interpretation of fashion is exclusive to the traditional clothing and its association with the symbolic components of the culture. The French fashion trade organization that organizes fashion events, fairs and other fashion rendezvous activities also plays a pro-active role in choosing designers for creating fashion, engaging the fashion industry professionals for producing them and mobilizing the fashion agents: photographers, models, journalists, etc., for propagating fashion. This institutionalized framework sustains itself in the fashion business by perpetuating what is fashion and lifestyle for every season through its network of professionals. Bourdieu associates' fashion with three types of capital as social capital, cultural capital and economic capital [10].

3.1 Spread of Fashion—Influencing Factors

Europe and USA experienced rapid changes in fashion as a result of nineteenth-century industrialization. Industrialization of fashion manufacturing produced new fashion quickly and inexpensively. The social structure of the Western world underwent a great change in the eighteenth and nineteenth centuries: the population increased, productivity soared, money economy developed due to the growing division of labor, technology improved, commerce expanded and social mobility became possible [6]. Since fashion acts as a symbol of social relevance in a particular context [15], the drive for owning fashion clothing grew higher and higher than ever before. The classical perspective of fashion is the premise that fashion spreads from higher classes to lower classes. However, the contemporary view of fashion holds a different perspective that ascribes Individual's wish to update the tastes in the emerging scenario as fashion [11]. The individual's rate of change is characterized by the pattern of changes in politic, art and technology that constitute the contemporary society [16]. Roach higgins observes that the awareness for a fashion change has to be felt and nurtured by a group of people not an individual [17]. The convergence of webspace technology has brought fashion news to every individual's doorstep. Thus, in the interconnected Global scenario, fashion activities and the space occupied by them draws no parallel activity equal to its status symbol in the society. Bourdieu also observes taste as key signifiers of social identity.

3.2 *Fashion as a Concept and Its Exclusivity in Fashion Studies*

Prior to 1980s, the fashion studies were confined to what happened in and around the European continent with the emphasis on industrial development, historic social changes in North America and Europe [18]. The influence of other culture's textile art forms on European fashion was hardly noted [18]. Rather knowledge of fashion and its possession was used as a criterion to rank the progressive level of a country and its civilization. Thus, fashion was used by the Europeans and the Americans to discriminate the indigenous cultures and their textile art forms [19, 20]. Wilson ascribed capitalism and growth of urbanization in late Medieval Europe for the emergence of fashionable dress that symbolized leisure and lifestyle. Hence, the idea of fashion is markedly different from the traditional clothing and its social value. The fashionable dress is created with the purpose of embodying the bourgeois values, to express the aristocratic lifestyle tastes which were anointed by the fashion institutions and the associated agencies. These fashion values were markedly different from the cultural expressions, emotional dispositions and social values of the traditional clothing [21].

Fashion history accounts state fashion as place and time specific since it existed in several forms such as court fashion, Bourgeoise fashion and consumer fashion. So, contextualizing fashion with modern capitalism alone does not encompass the other associated values of fashion. Moreover, the extended associations of fashion with photography, creative arts and other cultural systems raised hopes about more inclusive definitions of fashion which resulted in much broader contextualization [22]. Twenty-first century witnessed more extended definitions of fashion such as Oriental fashion, Islamic fashion, African fashion, Latin American fashion through the Anthropological perspectives of fashion theory journal. Thus, fashion influences as a two-way process either from top to bottom or bottom to top came to be acknowledged [22]. However, empirical research did hold on to the fact that fashion changes were more frequently noticed in European and North American cities than in the other cultures and less industrialized countries. Based on the accounts of the findings stated by Jennifer craik, fashion is also associated with cultural identity of style [23]. This phenomenon was possibly more relevant among the aspiring societies of developing countries. Classical example is the changing semiotic values in Indian society. Post 2000's the new Indian woman adopted fusion styles; a blend of Indian and Western silhouettes which also furthered the contemporalities of ethnic silhouettes [24]. Charlotte Jirousek in her studies ascribes the emergence of mass fashion in Ottomon Empire to the changing local socio-economic conditions and a kind of lenience toward bourgeoise values [25]. Leslie rabine also highlighted the relevance of trans-national values among the clothing styles of African fashion and its diasporas [26].

Finnane's findings about the Chinese dressing style changes between sixteenth and nineteenth century also records the evolving socio-cultural patterns and transnational values among the urban people [27]. Penelope Franck's study based on the evolution of Japanese kimono styles of eighteenth century Japanese society owing to the industrialization also cites the shift in socio-cultural values [28]. The Japanese fashion was deciphered as Avantegarde and the reason behind its cultural identity by greenwood with reference to the creations of Japanese designers Rei Kawakubo, Isse Miyake and Yohji yamamoto. These designers incorporated the Japanese elements such as the folds, pleats and wraps with appealing material fantasy that Rivalled the Eurocentric origins of fashion.

In a multiracial society like Jamaica, the contemporary clothing style is an evolution of the dominant afro-European clothing styles an outcome of the continuous interaction between European and African cultural elements along the process of development movements, changing patterns of political and social life [29]. The different genres of Kanga textiles which reflect the socio-cultural values of East African people evolved from late nineteenth century onward with influx of Indonesian batik motifs, Swahili script and Indian printing techniques [30]. Indonesian batik-printed national dress is a confluence of Islamic culture, Western culture and the local dress. The Filipino national dress evolved from sarong and sari style wrap skirt to fuller skirt embodying the cultural values of modesty brought forth by the Spaniards.

Mexican fashion that reflects broader version of Latin American culture is attributed to co-existence of multiple cultures like native culture, European culture and African culture. The Mexican clothing co-opted the Western designs within the existing native elements allowing itself to adapt to newer ways of life and culture. The shift in socio-cultural clothing values and the resultant semiotic evolution of clothing entity of Indians, Chinese, Africans and Japanese connote a different identity and property to fashion apart from its Eurocentric ideology. Thus, the fashion identity appears to accommodate the pluralistic values of culture.

3.3 Fashion and Subculture

According to Deschamp's fashion devices: Society and norm go hand in hand; in other words, if society changes, norms changes and if the norms changes, fashion changes or vice versa [31]. Though style in clothing is a combination of personal expression and social rules, neither a member of subculture group nor a movement follow any of these social rules [32]. The societal changes, social movements and subculture are regarded as cultural phenomena [33]. Along the discourse of fashion history, there has been numerous accounts of subcultural movements and social movements inspiring fashion or being central to fashion concepts. These subculture movements especially Japanese style tribes has contributed a lot to the development of unique Japanese fashion characteristics in its revival during the 1990s.

3.4 Unconventional Aspects of Fashion

Frida Kahlo the Mexican artist chose to express disability, overwhelming passion for life and fashion through the traditional Tehuana dresses and contemporary jewelry. The choice of traditional dress to hide her disability, provide movement comfort and accommodate support braces expresses the convenience of traditional dresses in manipulating them according to the needs, deep affection for own culture and the choice of contemporary accessories brought forth her complex style statements in unapologetic way. It showcases her cross-cultural aesthetics which influenced Givenchy, Dolce and Gabbana. Alexander Mcqueen added a dimension of raw emotions in his designs including sufferings, darker side of romantic life and unconventional aspects of beauty. Thus, fashion is not complete without these extended expressions of beauty, passion, imperfections of life. And it is true that fashion is perceived as culturally constructed and embodied identity.

4 Fashion Anthropology

Fashion anthropology in the context of cross-cultural, holistic and cultural relativistic perspectives promotes inclusive definitions for fashion. It considers fashion as a universal system with a wide range of local variations in the same way that political or economic systems are universal with local variations. Eller recommends classifying the human behavior on the foundations of his/her diverse ways of being and behaving in life that is otherwise referred as cross-cultural perspective [34]. A cross-cultural study by Elodie Gentila et al. suggests that youth of France and US preference for luxury fashion brands share commonalities irrespective of their different origin [35]. Another cross-cultural study across two countries of vastly different cultures like Tehran and Berlin indicated that the preference for fashion seems to be inspired by their own traits and local culture [36]. The multidimensional attributes of the luxury fashion clothing is equally preferred by Generation Y cohorts of Germany and China, irrespective of the value differences perceived by them according to Maximillan Fraschan et al. [37]. The luxury fashion attributes, its global identity, perceived social value influences the purchase intention of Americans and Thais provided thais religious commitments is low according to cross-cultural study by Jieqiong Ma et al. [38]. And the attempts to explore cross-cultural communication abilities of fashion design concepts showed it is possible to interpret the concepts from a cross-cultural background [38]. Further studies also hold the fact that a fashion clothing irrespective of its Narrative content and perceived original value might appeal to other cultures with varied connotations [39]. The interpretation of fashion values differs from one region to another region as the way the information is structured is not always understood in the same context and fashion hardly follows any hierarchical concept in its deployment of icons and symbols [32]. The Western fashion dress idea constitutes intangible values visualized through pictogram, icons or mood-based imagery is open to the consumer perspective and his/her knowledge while deciphering the different levels of fashion subject [32]. The visualization of fashion design and its narrative connoting the intangible values is anchored by the subject matter contextualized in the Aesthetic views of Art. The discourse about fashion and its narrative is disseminated extensively the social media sources, fashion magazines, fashion shows, music, cultural events, festive occasions and soap operas.

Cross-cultural comparisons by way of cross-temporal methods and cross-generational methods are used for investigating the cultural patterns and changes [40]. The devices of fashion such as society, gender and norm are also common among the objects studied in cross-cultural perspective. So cross-cultural perspective might be useful in ascertaining the assimilated human behavior patterns and interpreting the generic commonalities of universal value. The generic commonality of human behavior is also analogous to "Zeitgeist" phenomenon, that is central to different areas of social life across a specific period. Zeitgeist themes is characteristic to overlap and escalate in complex ways across different cultures. [33] Zeitgeist is also represented through the dominant theme, meanings and objects seen in fashion trends.

The holistic perspective refers to the process of studying a particular culture as a wholesome existence instead of ascribing to a narrow list of traits. The concept considers all aspects of culture such as politics, economics, religion are interrelated and interdependent on each other [41]. Given the state of fashion's desirability defined on a set of wide-ranging values [42] such as social, politics, sports, inclusivity, nationalism, youth, etc. nevertheless further strengthens the interdependencies between the different parts of the culture.

Cultural relativism points out that the standards of one culture cannot be applied for evaluating another culture rather permits relative assessment of a well understood phenomenon of one culture with another culture. If social psychology over individual psychology is ascribed by Yuniya Kawamura in the institutionalization of fashion values, likewise interpreting fashion's ideological aspects: gender, human body in the perspective of cultural relativism shall add new dimensions to fashion. Fashion by itself has no universal values and varies greatly across disciplines [32]. The stereotypical masculine or feminine qualifications are socially constructed representations of gender, based on what society expects of each sex [32]. Breaking away from socially constructed representations of values and qualifications is referred as perversion. So when Hippies or beatniks symbolized freedom and liberation from stereotyped gender characteristics by sporting long hair, people belonging to the conventional social culture interpret it as perversion of gender. Likewise different values are associated with the dresses in different cultures.

Our living environment irrespective of its culture origins has evolved continuously owing to an extent through industrialization and post World War effects in the twentieth century and certain extent through acculturation resulting out of globalization. And twenty-first century witnesses the promotion of multicultural practices for inclusive human resource development of different ethnicities, decolonizing fashion values and gender equality that is analogous to Postmodern fashion where all and sundry ways of making of fashion is appreciated rather than aligning with the idealistic perspective of making art. Due to the influences, People's living environment changes and so does the attitude and outlook of life which culminates in the fashion choices an individual makes [43]. Yet traditional clothing and textile art forms is not naïve to these changing perspectives about life, lifestyle and attitude. Today's traditional clothing styles and art forms not only constitute the social values of the local culture and its entity but also subtly reflect the assimilated values of Global events perceived to affirm a sense of appropriateness.

Hence from the fashion anthropology perspective, the intent is to examine the African traditional clothing styles in the framework of art design compositions to posit a Universal narrative with its sustainable characteristics essential for its progress but also to preserve its indigenous representation, its cultural identity and not to decide its superiority or inferiority. In this case study, the essence of traditional African clothing styles: African wax prints, kente cloth, Bogolanfini, Adinkra, Adire and African traditional batik is investigated through the secondary published sources. Their cultural content, knowledge systems and the scope for design value propositions in the context of sustainable textile art forms is elaborated. Understanding the nuances of these African traditional textile clothing styles could serve as the microcosm of the much larger "sustainable fashion phenomenon." Oriental concepts of clothing style, African patterns and it myriad of colors, traditional fabric designs down the timeline since ages, the cult status of Ponchos, Bohemian style, etc. have influenced fashion clothing more than once. Representing the traditional clothing cultural values is also crucial to decolonization of fashion values.

5 African Traditional Clothing Styles

According to Webster the word tradition is an inherited, cultural feature handed over from the past [44]. Traditional clothing acts as medium to render the cultural expressions and fulfills dress functions of the ethnic representatives of the society. According to Yoder, in folk culture or traditional society, the clothing reflects the age, sex, relationship status of the individual rather than his or her personal style statement [45]. WIPO acknowledges the original owners of the traditional cultural clothing; knowledge systems associated with the traditional clothing manufacturing practices and recommends rightful interpretations of the traditional clothing elements [46].

5.1 African Wax Prints/Ankara Wax Prints

The traditional wax-printed fabrics of Africa is inspired and adopted from Indonesian batik-printed fabrics. West African soldiers posted in Indonesia choose this batik-printed fabrics for Africa. Batik is an ancient resist wax dyeing/printing technique, where the molten wax is applied on the fabric surface in specific places that constitutes a non-colored distinctive part of the design landscape, and later, the other colors are added in the remaining spaces to complete the design. The African wax prints are known by other names like Java, Ankara and Hollandais. The African wax prints are indeed based on Javanese mechanical resist printing techniques with its artistic style borrowed from several countries. The intricacy of the batik print design stems from the tools and technique used for making the pattern, especially the traditional pen like tool "tjanting" holding liquid wax used by the Indonesian artisans is said to produce the most intricate design patterns for the wax print designs. Since the design pattern creation using the traditional hand held pen is slow and tedious task, the Dutch converted the process into roller inscribed design patterns using chemical resists to enhance the production. The traditional wax print fabrics are produced through mechanical resist process, where the molten wax is applied in the non-colored distinctive part or the lighter part of the design prior to dyeing process. The dyeing process is repeated several times by hand according to the print intensity and depth required for the producing the multicolored print design story. In the later stages of dyeing, the fabric is boiled to get rid of the wax applied on the cloth and hand dyed again for imparting color in the wax applied areas. The multiple stages of dyeing deposits one color layer after another color layer to achieve deep saturated hues and multiple colors. Indigo pigment is the dominant color source where each color is ascribed to a particular dye source; indigo or mixture of two or more sources. These wax prints are machine made adaptations of original hand-printed batik fabrics of Indonesia. In the adapted machine-made fabrics, the wax was replaced by resins which was applied on both the sides of the fabric prior to printing [47] (Fig. 1).

Contrary to the industry-scaled wax prints, the handmade wax prints color fastness quality was more superior [48]. These wax-printed fabrics introduced by the colonial rulers contrast the more indigenous Yoruba adhire/Adhinkra cloth/Asooke, yet has become part of African cultural legacy. Irrespective of the origin, wax-printed fabrics not only reflect the African ideas, symbols and local art subjects, but also canonized African beliefs and culture [49]. The local African woman ascribed the cultural themes to these Dutch-produced fabrics. The African wax prints are kitschy composition of multicolored patterns juxtaposed adjacent to one another. Each wax print design composition connotes cultural expressions, emotional dispositions, memorable moments and interactive experiences. Alphabet design wax print symbolized colonial education and mathematical literacy, fly swatter design wax print represented power and prestige, the iconic wax print "Darling, don't turn your back on me" indicate the values of the local cultural milieu [49]. The traditional wax-printed fabric designs reflect the African proverbs and social values with each design marking a special purpose and occasion of their usage. "Leaf trail design' wax-printed fabric symbolizes the African cultural saying "empty barrels make much noise", whereas a blue motif wax print represents pregnancy phase of women in Ghana, the family of hen and its chicks fabric design refers to Happy African family [50]. The African wax-printed fabrics counts to around 54 types with each type connoting a social value and purpose.

The wax print fabric irrespective of its foreign origins permeated the African ethos with relishing interactions of its cultural milieu and acquainted itself the Ethnic

Fig. 1 Ankara wax print. *Source* (https://commons.wikimedia.org/wiki/File:Mills,_Obama,_GHANA-USA_wax_print_GTP_cloth.jpg)

African lifestyle. The Africa's emotional disposition about the theft of Asante King's Sword which lies now in British museum was expressed in a classical wax print design named "Staff of Kingship" [51]. The subject of the wax print design could be commemorative function with cultural signifiers or more abstract with social signifiers [52]. Thus, the wax print fabrics got absorbed in the African culture by continuously incorporating the local cultural elements as the subject of the print design [53]. Analogous to how Choco Chanel gave the New modern look, leading African ladies infused the cultural subject to the Dutch wax-printed fabric.

Anthropologist Nina Sylvanus perceived that Africans relish a colorful design palette with a shimmering surface and a crispy fabric that contrasts the tactile comfort and pliable preference of Europeans [54]. Among the machine produced wax print fabrics meant for Africans, the intrinsic organic motif arrangement was replaced by the geometrical motifs with defined margins that acted as color blocking lines. And the evolution of African wax print designs further incorporated commemorative designs, current events, public art subjects, social art subjects and emotional dispositions. Therefore, every African wax print design artwork evolved as a consequence of interaction between the artist, local fashion influencers, manufacturers and the traders. The artistic style of the wax print design and its meaning is thus influenced by the market forces [55, 56]. The market forces in turn operate within the framework of the PESTEL factors, namely Political, Economic, Societal, Technology, Environment and Legal. In the earlier days, the African wax print followed the Vasari's model of art development [50]. Vasari proposes that every artwork has a lifecycle with concept of inception, maturity and decay [57]. In the twentieth century, Wolff contemporized the lifecycle concepts as art concept creation by the artist, concept maturity associated with the consumer [55]. According to modern lifecycle concept, the developmental stage of artwork is guided by the continuous interaction between the artist and the consumer similar to Codesign concept. Consequently, the artwork embeds a functional value in the product value stream creation but also in the mindset of the consumer. However, the prevailing market systems do not guarantee the existence of such artwork models instead prone to the market disruptions such as the Cheaper Chinese wax prints.

At every stage in the life process, an artwork assumes a functional value which is aesthetic, social or economic in nature. The development in stages and changes in value are facilitated by the interaction of the artwork with the creator and the consumer. The African wax prints originally inspired by Javanese batik work, shared an Indigo base was strikingly similar to African tie and dye techniques [58]. Today these wax-printed designs are applied on silhouettes like jumpers, T-shirts along with other Western coordinates to produce everyday casual look [59].

5.2 *Mud Cloth or Bogolanfini*

Bogolanfini, otherwise known as mud cloth, derives its name from the dyeing process using fermented mud. Bogolanfini derives its name from the literary translation of mud-cloth in Mali. The fabric is assemblage of several woven strips of around 10.5–14.5 cm width and around 21–24 m length upon which the decorative motifs are arranged [60]. The Bogolanfini designs are basically graphic patterns in rustic red, beige and earthen colors. The woven fabric is made of spun cotton and woven in narrow strips which are later converted into a wider width by joining together the strips using handmade stitches [61]. The assembled fabric is dyed with organic extracts sourced from tree branches. The fabric base is smeared with mud to impart the characteristic color at the desired location according to the arrangement plan.

The whole design pattern is developed one after another across several part stages. Generally, the mud painted/dyed parts constitute the background. Initially, the fabric is soaked in a liquid which is prepared from leaves of N'gallama tree. N'gallama extracts absorbed by the fabric acts as a fixative for further color dyeing process. Black color on the fabric is produced by a combination of mud and tannic acid. Different tones of colors are prepared by varying the mud layer compositions. The graphical patterns are hand illustrated prior to dyeing. These graphical patterns comprise abstract leaf motifs, grid lines, geometrical shapes and point clouds. The Bogolanfini is referred to mark birth, marriage and other occasions along with leather accessories in the traditional folk lore (Fig. 2).

The design pattern arrangement is not confined to any style, all forms starting with graphics to symbols, different compositions using various resist painting techniques are encouraged. The Bogolanfini characters comprise traced outlines and footprint like patterns of household objects, local flora and fauna. These characters do not serve any as any signage nor any art world interpretations, rather their sole purpose is to convey the message about life; a characteristic set of beliefs that have a unanimous appeal among the cultural milieu. The local Mali people revere the Bogolanfini in their beliefs and consider it to be representative of earth, river and forests [62]. The tools used in producing Bogolanfini are either made of wood or other naturally available materials of animal source. A study by Mukta V et al. confirmed Bogolanfini method and materials as a potential for futuristic dyeing and processing techniques [63].

5.3 *Kente Cloth*

Kente cloth is a handwoven four-inch-wide narrow fabric made by the Ashanthi tribes. These narrow width fabrics are joined together according to the required design by hand stitches [64]. The kente fabric connotes the history, etiquette and cultural expressions of the Ashanthis, Ewe and other Northern parts of Africa [65]. However, the Kente design patterns vary from one region to another. It refers to kente fabric among the Ashanthi, Fugu among the Northern parts of Africa and kente among the Volte region. Though the same weaving and assembly techniques are used, the motif shapes, the pattern of arrangement against the respective color background are different from one another. In traditional manufacturing setup, the men take care of the weaving process and women are engaged for hand spinning the yarns that make up for the loom weaving [66]. The kente fabric derives its color from the organic dyed yarns used in weaving the patterns. Along the weave, iconic shapes of symbolic animals are also interwoven in the fabric using figuring techniques. The narrow strips of woven fabric are assembled by lapping the fabric strip edges equal to 1/8th of an inch and joining them by zig zag stitches. The edges protruding the seam width are folded and hemmed. The kente fabrics for men is an assemblage of 24–25 strips and 11–12 feet in length with gender distinctive draping and wrapping style [67]. For women, the fabric width and length are smaller compared to the men's

Fig. 2 Bogolanfini, the mud cloth. *Source* (https://commons.wikimedia.org/wiki/File:S%C3%A9gou_(11).JPG)

fabric that is wrapped around the body along with a blouse. The wrapping style of the kente fabric resembles the Roman Toga wrapping style [68] (Fig. 3).

In the contemporary scenario, the kente fabric is converted into tailored Western silhouettes for wedding occasions and African party themes. Nowadays, the kente design is incorporated in African wild silk fabric base to enhance the Aesthetic value of the product [69]. The traditional vegetable dyed cotton yarns are replaced by silk and spun rayon yarns to add more prestigious value to the kente cloth [70]. The kente fabric design is identified in diverse products such as scarves, hats, stoles, bags and footwear in the contemporary scenario [71]. Traditional kente fabric design pattern

Fig. 3 Kente cloth. *Source* (https://commons.wikimedia.org/wiki/File:Kentecloth.jpg)

is a block arrangement of several pattern repeats tending to represent a moire of grids with embedded motifs. In the modern-day usage across artefacts and products, the block arrangement of pattern repeats, their size and shape amidst the background landscape is manipulated to connote the desired aesthetic function. The kente design and its pattern arrangement draws semblance to modern Pop art concepts with switching colors in blocks. The kente design is also combined with Adinkra print fabrics with the purport of adding more visual depth [72].

5.4 Adinkra Fabrics

Adinkra fabrics comprises the Adinkra symbols which are either woven or hand stamped on the fabric. The Adinkra symbols are seen in other products like Jewelry, wood works and artworks [73]. Adinkra fabrics are produced either by hand block prints or screen prints. Unlike other print designs, adinkra design features an array of symbols and icons arranged on a red colored base or black colored bases that acts as a visual communication language. The dark colored pigment used in the print paste is obtained from the badie tree bark and the roots of kuntunkuni trees [74]. Adinkra symbols are systematically designed and organized in five different disciplines: Education, Politics, Spiritual, Funeral and Social [75]. The Adinkra symbols are abstract 2d shapes of 3D objects composed from lines, points and shapes exploring the figure ground principle in its rendered existence. Though Adinkra owes its origin to the Asante tribe, it is equally preferred by other tribes of Ghana [76]. Adinkra symbols in its visual communication context express the attitude, mood and social values of the cultural milieu. The arrangement of the symbols and the visual communication is designed with reference to the nature of the event and the etiquette. Other than the traditionally prepared red or black color base fabric, imported fabrics in different color bases are also identified in the modern-day usage. The choice of symbol color and the base color draws from the color scheme where the intent is to provide apparent contrast that enables the observers to gaze and feel the theme [76]. The Adinkra symbols were also found in the 19th-century British clothes design as motifs [77]. However today, the adinkra symbols take several variations while incorporating the influences of foreign cultures. In the current scenario, the adinkra symbols finds itself in logos of corporate institutions, Conglomerates, National agencies. Hill [78] considers symbols as very old forms of visual communication to convey a meaning or the mood. Popular paintings like the starry night by Vincent van Gogh [79] and Guernica by Picasso [80] represented the larger life themes symbolically through the contextual elements of the scenario enacted. Wiz Kudowor [81] one of the celebrated artists of Ghana uses Adinkra symbols along with the subject matter to convey the themes in his paintings. While the Eurocentric fashion reflects on the lifestyle and evolving tastes, the Adinkra primarily conveys the emotional disposition and the behavior patterns. Adinkra symbols and their meanings serve as inspiration in shaping the cultural identity of Ghanian youth [82]. Analogous to the Bauhaus concept of designing economically feasible functional products where aesthetics compliments the function, Adinkra prints using the abstract symbols to convey the message is synthesized by the craftsmanship and technology of the Asante tribe. The Adinkra symbols non only connotes the cultural heritage of Ghana people but also has the potential to convey larger themes besides fulfilling the aesthetic functions [83]. The meanings of these Adinkra symbols are constituted from the social cultural knowledge systems of the Ghana Culture [84]. And the symbols share an undisputed mutual understanding in its meaningful existence among the Ghanians [85]. The vast repertoire of the Adinkra symbols and its abstract expressions lays out a premise for complex nonverbal designs in the contemporary existence [86] (Fig. 4).

5.5 *Traditional Batik of Africa*

Traditional batik of Africa is hand block-aided resist wax print style where the wax is applied by hand. The block patterns of African batik are replete with line grids, ornamental designs, abstract figurines [87]. Among the African batik prints, the wax is melted and applied on the distinctive non-color areas of the print either using a wooden block or a paint brush and dyed several times. The wax is made from the starch of cassava plant or rice paste and the design is illustrated on the fabric by means of bird feather, stick, bone, or comb-like structure [88]. In the traditional method, dye is extracted from the indigo plant. Cotton fabric is the base for batik printing albeit occasionally wild silk is also used.

African batik tradition is passed from the Mother to daughter along the family lineage [89]. The African batik print pattern is inscribed on a foam block as well as

Fig. 4 Adinkra cloth. *Source* (https://commons.wikimedia.org/wiki/File:NtonsoAdinkra.jpg)

wooden blocks. The flexibility of the African batik method lies in the application of dye using the block or paint brush that opens up the design space for both elaborate and intricate arrangement of the design elements in desired colors and style.

The art production style is more spontaneous and draws semblance to artist style expressed in romantism art movements. Artists Dr. Abraham ekow Asmah and Sylvester Lumor add a contemporary Western art touch to batik style by painting the entire Human art subjects on the fabric using batik resist techniques. Such art works portray the technical capabilities and the finesse of batik style to render complex art subjects in dynamic postures [90]. Resist and stamp method, hand painting and stencil dyeing methods fit the process of producing the batik prints.

5.6 African Tie and Dye—Adire

The African tie and die is referred as Adire Yoruba woman of Africa, is an indigo dyed fabric comprising of tie and dye patterns. The tie and die technique embody the local practice of producing the design pattern deploying thread to tie the fabric at design specific non-colored ground areas to resist dye indigo intake. The tying techniques are called by the names of Gara and Adire [91]. The Adire fabric unlike other textile art forms exists as a cultural production and heritage across several African nations. The multiple hues of the Adire pattern is developed over multiple layered stages where the fabric is dyed in indigo color initially, later the resist areas are marked and

Fig. 5 Adire. *Source* (https://commons.wikimedia.org/wiki/File:Aso_Adire.jpg)

tied prior to dyeing the fabric in subsequent colors. Thus, there is no single ground color exposing the pattern rather a multitude of overlapping grounds created through dyeing in several stages. As larger patterns require more depth of color across a vast area, the fabric apparently heavy [92]. The traditional Adire attire of the royal king features blue and white colors, where blue symbolizes rain and white stands for peace [92]. The tying work at the resist locations on the fabric deploys many techniques such as 1. binding: where the portions of fabric are held firmly in round or square shapes using thread, 2. knotting: the fabric portion is twisted and self-knotted and 3. sewing: a heavy thread stitch is made reminiscing a line or curve shape holding the neighboring fabric parts in gathers. Over the years, the African textile artists have incorporated the Japanese Shibori techniques of stitching and clamping to produce new age expressions and complexity of tie and dye designs [93] (Fig. 5).

5.7 Significance of African Traditional Clothing Styles

The African traditional clothing styles reveal the emotional dispositions and social meanings of the design patterns and symbols deployed in communicating a message or two. The cultural significance, etiquette of the occasion and its environmental setup acts as the norm for the choice of the design composition in an African traditional clothing. The interaction between the Native African artists, fashion leaders/influencers and textile craftsmen provides the platform to choose the desired design patterns and structure the socio-cultural meaning of the Ankara prints. In other words, this process is sort of co-design activity set out to put forth the design scheme appropriate for time and occasion. Therefore, Ankara prints is classical example of cross-cultural aesthetics. Such interactive process approach sustains the cultural production of African traditional clothing styles and acts as a source of equitable income generation for ethnic Africans. Therefore, sustaining the African traditional clothing styles production facilitates smooth transition of their manufacturing knowledge systems from one generation to other generation.

The African traditional clothing styles: kente cloth, Adire, Adinkra and African traditional batik's Structure and semiotics has the potential to draw parallels along with contemporary art styles. Kente cloth assemblage with Adinkra symbols is analogous to pop art assemblages. The interventions in Adire have enabled it to express more complex visual renditions on the lines of Japanese style. The Spontaneity of African traditional batik clothes has inspired the contemporary artists to deploy its techniques for producing Artworks and Tapestries out of them. Therefore, these traditional clothing styles possess the capability to co-build and branch out beyond their traditional forms with market energy design propositions and sustainable material sources. Irrespective of these interventions, the African textile art forms retain their uniqueness, in the framework of Art-based compositions.

6 Scope for Decolonialization of African Fashion Values

It used to be the practice of the Afro-Americans those who represented HBCE culture prevailing in the historically black colleges and universities to blend the African colors and other signifying elements in a respectable fitting manner with the Eurocentric fashion silhouettes as a means to be at par with the White man's sartorial standards [94]. In other words, only certain aspects of African signifiers were fitted in the fashion entity that served the Eurocentric interests. The Anthropologist Sandra Niessen believes that the existence of binary concept: Eurocentric fashion and the other fashion in fashion parlance will only enhance the opportunities for defining the distinct identity of other fashion [95].

Decolonization refers to the process and practice of restoring the indigenous peoples rights and livelihood through cultural, psychological and economic freedom [96]. Dominique drakeford reckons reclaiming the indigenous cultural identity through inclusive representation of traditional African cultural values for decolonizing fashion [97]. The African traditional clothing: Adinkra, Bogolanfini, kente, Adire and traditional batik are indigenous textile art forms that has sustained through continued craft practices since a long time even through the hard times of colonization. And the discovery of handwoven cloth fragments in the Tellem caves of Mali, only reproves the historical precedence of African clothing [98]. The cultural continuity of craft techniques, symbols and motifs that defines the indigenous cultural identities and ideologies of the people serves as a measure of anthropological self referentiality [99]. Such continued practice of producing African traditional textile clothing brings forth not only the knowledge of the craftsmen, but also highlights the preference for visual aesthetics of the textile craft and the cultural context of the themes used in them among the indigenous people. The Kaba and Slit, the prominent women's traditional dress silhouette of West Africa, comprise three garment pieces: top that resembles a blouse, a long skirt and a cover cloth [100]. This iconic women's dress features vast range of print patterns and color which has evolved in line with the changing preferences of the customer is referred as classic dress/outfit for women in West Africa. One can find this silhouette used regularly for multiple occasions like social gatherings, cultural events, festival holidays, family get togethers. Such traditional silhouettes rooted in the cultural context and social milieu convey a cultural identity and has perpetuated the indigenous beliefs and way of life. Meanwhile, the kente fabric remains the signifier of Pan African identity and has also pushed forth outward influences on the global sportswear designs [101]. Hence, the rich resource of African visual aesthetics present in the African traditional clothing like traditional batik, Adire, Bogolanfini, Adinkra and kente provides vast opportunities for the budding African designers to further enrich, recraft and extend its visual aesthetics in a culturally relativistic manner. Oppel, Marie Claire South Africa's fashion director opines that the African designers have learnt the art of interpreting the traditional prints and giving them new extended forms. African fashion designer Chris seydou's design interventions in Bogolanfini, contemporarized the motifs and their stylization, thereby adding exquisite design value to an indigenous clothing [102]. Designers like Maki Osakwe combined the traditional African clothing elements and the Western construction to lend a contemporary look to her designs [103]. Bwesigye bwa Mwesigire relates the prevailing African fashion identity to its contemporaneous culture: a blend of African traditional clothing elements, its culture and the Western notions of clothing etiquettes; its silhouettes [104]. The indigenous African designs are hand-made and belong to the slow design mode. Additionally, to represent the cultural identities of African traditional clothing styles in a non-discriminating and non-exploitative manner, it is also necessary to engage and aid the anthropological self-referential craft practices, the motifs, symbols through holistic perspective. The African designer Sindiso Khumalo believes in the social empowerment of the African Artisans by employing them to create modern sustainable dresses laced with its cultural elements in a Western silhouette [105]. Lisa Folawiyo's garments are worked upon by the local craftsmen in the African visual aesthetics which are listed in Selfridges [106]. Maki Osakwe hand-dyed indigo colors, and repetitive prints are constructed with the help of traditional craftsmen [107]. These design interventions refashioning the traditional African clothing elements not only increases the cultural and social mobility of the local African culture but also acts as a means of building their identity. Further, the immense success of brands like Black Panther, the fictional characters of Wakanda and Avengers: Infinity war has established African visual aesthetics much beyond the notion of being a mere trend [108].

7 Conclusion

The extended discussion of fashion reveals the unconventional aspects amidst the culturally constructed and embodied identity. As fashion in post-modern culture encourages variation and flexibility in the cultural sphere, its disposition naturally supports cross-cultural aesthetics. And the perspective of Anthropology within the framework of art design compositions underscores the need for promoting cultural diversity, decolonizing fashion values, preserving indigenous dress functions. Hence, the discussion on the African traditional clothing style and its textile art forms: kente cloth, Adinkra, Adire, Bogolanfini, traditional batik prints and Ankara or African wax prints proves their versatility to incorporate evolving cultural values through interactions with the fashion leaders, artists, exposure to newer ways of life and relations with foreign nations. The case study findings suggest African traditional clothing and their textile art forms is routed in the cultural and symbolic interpretations beyond its dress functions of physiological protection and modesty. Further sustaining the continued craft practice of producing African traditional clothing not only acts to preserve traditional knowledge systems but also serves to improve cultural and social mobility of the local African culture. So it is inferred that sustaining the growth of traditional African clothing styles and its inclusive representation from an anthropological perspective addresses the growing universal need to adopt sustainable fashion designs and encourage non-discriminating fashion values. However, further studies are required to extrapolate this concept of traditional clothing styles, textile art forms and their applications to more wider corridors of fashion institutions and industry.

References

1. Koenig R (1973) The restless image: a sociology of fashion, trans. Bradley, London: George Allen & Unwind, Ltd
2. Roland Barthes (1984) The fashion system. Farrar Struass, Giroux, New York
3. Veblen T (1899) The theory of the leisure class. Originally published by Macmillan New York
4. Crane D (2012) Fashion and its social agendas: class, gender, and identity in clothing. University of Chicago Press. 89:294. https://doi.org/10.2307/2700908
5. Adler P (1998) Beyond cultural identity: reflections on multiculturalism. Basic concepts of intercultural communication: Selected readings, 225–245
6. Kawamura Y (2018) Fashion-ology: an introduction to fashion studies. Bloomsbury Publishing
7. Simmel G (1957[1904]) Fashion. Amer J Sociol LXII, 6:541–58
8. Sumner WG, Keller AG (1906) Folkways: astudy of the sociological importance of usages, manners, customs, mores, and morals. Ginn and company
9. Toennies F (1961[1909]) Custom: an essay on social codes, translated by A. F. Borenstein, New York: Free Press
10. Bourdieu P (2018) Distinction a social critique of the judgement of taste. In: Inequality classic readings in race, class, and gender, pp 287–318. Routledge
11. Blumer H (1969) Fashion: from class differentiation to collective selection. Sociol Q 10(3):275–291

12. McCracken GD (1990) Culture and consumption: new approaches to the symbolic character of consumer goods and activities, Vol 1. Indiana University Press
13. White HC, White CA (1993) Canvases and careers: Institutional change in the French painting world. University of Chicago Press
14. Becker HS (1982) Art worlds. Univ, Berkeley
15. Barnard M (2013) Fashion as communication. Routledge
16. Baudrillard J (1976) Symbolic exchange and death. Sage
17. Roach-Higgins ME (1995) Awareness: requisite to fashion. Dress and Identity
18. Hobson JM, Hobson JM (2004) The Eastern origins of Western civilisation. Cambridge university press. ISBN: 9780521838351
19. Hollander A (1975) Seeing through clothes. New York: Viking Press. ISBN: 0670631744
20. Niessen S (2010) Interpreting 'civilization'through dress. Berg Encyclopedia of World Dress and Fashion 8:39–43. ISBN: 9781847883971
21. Wilson E (1985) Adorned in dreams: fashion and modernity virago. ISBN: 0860685527
22. König A (2006) Glossy words: an analysis of fashion writing in British Vogue. Fash Theory 10(1–2):205–224. https://doi.org/10.2752/136270497779754589
23. Craik J (1994) The face of fashion: culture history in fashion. ISBN: 0415052629
24. Nithyaprakash V, Niveathitha S, Shanmugapriya V (2021) Indian handloom design innovations and interventions through sustenance lens. In: Handloom sustainability and culture, pp 221–249. Springer, Singapore. https://doi.org/10.1007/978-981-16-5967-6_10
25. Jirousek C (2000) The transition to mass fashion system dress in the later Ottoman Empire. Consumption studies and the history of the ottoman empire, 1550–1922: An Introduction, 201–242. ISBN: 0791444317
26. Rabine LW (2002) The global circulation of African fashion. Bloomsbury USA Academic. ISBN: 9781859735985
27. Evans H (2008) Changing clothes in China: fashion, history, nation. Antonia Finnane. London: Hurst & Company, 2007. xviii 359 pp. The China Quarterly, 195, 712–713. ISBN: 978-85065-860-3. https://doi.org/10.1017/S0305741008001021
28. Francks P (2015) Was fashion a European invention?: The kimono and economic development in Japan. Fash Theory 19(3):331–361
29. McKenzie JO (2003) Jamaican ethnic dress: an evolution of cultures from post emancipation period 1938 to independence 1962
30. Ressler P (2012) The Kanga, a cloth that reveals-co-production of culture in Africa and the Indian Ocean Region
31. Akdemir N (2018) Deconstruction of gender stereotypes through fashion. Europ J Soc Sci Educ Res 5(2):259–264. ISSN: 2411-9563
32. Kaiser SB (2018) Fashioning identity: status ambivalence in contemporary fashion. Bloomsbury Publishing. ISBN: 9781350104686
33. Krause M (2019) What is Zeitgeist? Examining period-specific cultural patterns. Poetics 76:101352. https://doi.org/10.1016/j.poetic.2019.02.003
34. Eller JD (2020) Cultural anthropology: global forces, local lives. Routledge, ISBN, p 9781138775527
35. Gentina E, Shrum LJ, Lowrey TM (2016) Teen attitudes toward luxury fashion brands from a social identity perspective: a cross-cultural study of French and US teenagers. J Bus Res 69(12):5785–5792
36. Iran S, Geiger SM, Schrader U (2019) Collaborative fashion consumption–a cross-cultural study between Tehran and Berlin. J Clean Prod 212:313–323
37. Faschan M, Chailan C, Huaman-Ramirez R (2020) Emerging adults' luxury fashion brand value perceptions: a cross-cultural comparison between Germany and China. J Glob Fash Market 11(3):207–231
38. Ma J, Hong J, Yoo B, Yang J (2021) The effect of religious commitment and global identity on purchase intention of luxury fashion products: a cross-cultural study. J Bus Res 137:244–254
39. Dai Y, Zhou T (2022) The variation of clothing culture and the transformation of fashion brands in cross-cultural communication. In: 2021 international conference on social development and media communication (SDMC 2021), pp 1457–1461. Atlantis Press

40. Kashima Y (2014) How can you capture cultural dynamics? Front Psychol 5:995
41. Esaine NO (2021) Anthropology of consumption, fashion and identity in emerging class. Revista Geintec-Gestao Inovacao E Tecnologias 11(4):3120–3133
42. Maria A (2015) Fashion anthropology: challenging Eurocentricity in fashion studies, fashion tales 2015 conference, 18–20 June Milan, Italy
43. Bhumika S (2020) Cultural sustainability: the future of fashion forecasting 6(3),IJARIIE-ISSN(O)-2395–4396
44. Webster dictionary (1993), pg. 2422
45. Dorson RM (Ed) (1972) Folklore and folklife: an introduction. University of Chicago Press
46. https://www.wipo.int/wipo_magazine/en/2019/04/article0002.html. Accessed 12 June 2022
47. Julia F (2012) The curious history of tribal prints. http://www.slate.com/articles/arts/design/2012/03/african_fabric_where_do_tribal_prints_really_come_from_.html. Accessed 12 June 2022
48. Clara spencer (2020). https://www.bbc.co.uk. Accessed 12 June 2022
49. Tordzro ND (2021) Decolonising African costume and textiles: naming, symbols and meaning in the Ghanaian context. MPhil(R) thesis, University of Glasgow. Available at: https://theses.gla.ac.uk/82319/. Accessed 12 June 2022
50. https://www.adinkrabrand.com/blogs/posts/african-wax-prints-and-the-story-behind-their-names. Accessed 10 May 2022
51. Willem A (2010) The Wax print its origin and its introduction on the Gold coast. Master thesis, University of Leiden Faculty of Humanities Institute for History. Available at https://www.academia.edu/31417064/the_waxprint_its_origin_and_its_introduction_on_the_gold_coast. Accessed 12 June 2022.
52. https://www.vam.ac.uk/blog/museum. Accessed 12 June 2022
53. Made in Holland: the Chanel of Africa, INGE OOSTERHOFF, (October 30, 2015). https://www.messynessychic.com. Accessed 12 June 2022
54. Sylvanus N (2007) The fabric of Africanity: tracing the global threads of authenticity. Anthropological Theory—ANTHROPOL THEORY 7:201–216. https://doi.org/10.1177/1463499607077298
55. Perani J, Wolff NH (1999) Cloth, dress, and art patronage in Africa. Oxford; New York, Berg
56. Trager L (2004) Money struggles and city life: devaluation in ibadan and other urban centers in Southern Nigeria, 1986–1996. Int J African Historical Stud 37(3):564
57. Vasari G (2007) The lives of the most excellent painters, sculptors, and architects. Modern Library
58. Wax Prints are based on Javanese Batiks (2017). Available at: https://www.journeytobatik.org/2017/09/wax-prints-are-based-on-javanese-batiks. Accessed 12 June 2022
59. Ezinne Chukwu Anyasodo, https://nkitidesigns.com/blogs/stories/complete-beginners-guide-to-african-wax-prints. Accessed 12 June 2022
60. The Michael C. Rockefeller Memorial Collection, Bequest of Nelson A. Rockefeller, (1979). https://www.metmuseum.org/art/collection/search/312395, Wrapper (bogolanfini) Accessed 12 June 2022
61. Natalie Kimani, (November 12, 2018). https://tdsblog.com/bogolanfinimalimudcloth. Accessed 12 June 2022
62. https://www.theethnichome.com/the-bogolan-mudcloth/. Accessed 24 May 2022
63. Limaye MV, Bacsik Z, Schütz C, Dembelé A, Plea M, Andersson L, ... Bergström L (2012) On the role of tannins and iron in the Bogolan or mud cloth dyeing process. Textile Res J 82(18):1888–1896
64. https://www.khanacademy.org/humanities/art-africa/west-africa/ghana/a/kente-cloth, Dr. CourtnayMicots, Beth harris and Steven Zucker. Accessed 12 June 2022
65. Lartey RL (2017) Cross cultural weaves potential for Ghanaian cultural identity. ADRRI J Arts Soc Sci 15(4):1–22
66. Adegbite SA, Ilori MO, Aderemi HO (2011) Innovations in the indigenous textile weaving firms in Southwestern Nigeria. Int J Bus Manag 6(12):243

67. https://www.midwestglobalgroup.com/pages/background_inprogress, Kwaku Ofori-Ansa, 2003, History of AshanthiKente cloth/kente symbolism/meanings, more than a piece of fabric: a culture. Accessed 12 June 2022
68. Kwakye-Opong R (2014) Beyond ethnic traditions: philosophies and socio-cultural relevance of the Ashanti and the ewe kente cloths. Res Humanit Soc Sci 4(26):150–165
69. Fening KO (2006) History of kente cloth and its value addition through design integration with African wild silk for export market in Ghana. In: Trainers course and 4th international workshop on the conservation and utilisation of commercial Insects Duduville, Nairobi (Vol 1, pp 62–66)
70. Dickson A (2009) The influence of European elements on Asante textiles (Doctoral dissertation, Department of General Art Studies, Kwame Nkrumah University of Science and Technology)
71. Kwekudee (2019) Kente cloth Ghana's cultural contribution to World's trends. https://www.safari254.com/Kente cloth—Ghana's cultural contribution to World's trends, Accessed 26 June 2022
72. El Anatsui. https://www.contemporary African art.com/kente fabric. Accessed 26 June 2022
73. https://www.shopafrothreads.com/blog/four-types-of-african-fabrics-to-know. Accessed 26 June 2022
74. Kissi SB, Fening PA, Asante EA (2019) The philisophy of Adinkra symbols in Asante textiles, jewellery and other art forms. J Asian Scientific Res 9(4):29–39
75. Valentine Tette, Adinkra Symbols the cultural symbols of Asante People, http://www.stlawu.edu/gallery/education/f/09textiles/adinkra_symbols.pdf. Accessed 26 June 2022
76. Owusu-Sampah C (2014) RIT scholar works (Doctoral dissertation, Rochester Institute of Technology)
77. Martino AJ (2020) From Kings to Kids: Refashioning Akan Adinkra symbols as African motifs in a Nineteenth-century British Cloth design 51(1), Textile history
78. Hill S (2006) Understanding the context of a painting brings appreciation of art (online) Available from https://suite.io/suzanne-hill/2wp211. Accessed 28th May 2022
79. Vincent Van Gogh—Biography and paintings, https://www.arthistoryarchive.com/arthistory/expressionism/Vincent van Gogh. Accessed 26 June 2022
80. Guernica, (1937) by pablo Picasso (online) Available from https://www.pablopicasso.org/Guernica.jsp. Accessed 26 June 2022
81. Wiz kudowor biography (online) Available from: https://www.wizkudowor.com/Wiz/Bio.html. Accessed 26 June 2022
82. Blount TN, Brookins CC (2021) Adinkra symbolism, printmaking, and the cultural identity of Ghanaian emerging young adults. J Creativity Mental Health, 1–20
83. Adom D (2016) Adinkra: an epitome of Asante philosophy and history. Res Humanit Soc Sci 6(14):42–53
84. Marfo C, Opoku-Agyeman K, Nsiah J (2011) Symbols of communication: The case of Àdìǹkrá and other symbols of Akan. Language Soc Culture 32:63–71
85. Ruby J (2022) Mastering Adinkra symbols: an establishment of positive Ghanian personality in the next generation. Global J Human-Soc Sci G linguistics Educ 22(2)
86. Owusu P (2019) Adinkra symbols as "multivocal" pedagogical/socialization tool. Contemporary J African Stud 6(1):46–58
87. https://globalmamas.org/handcrafting/hand-batiked-textiles/history-african-batik/. Accessed 26 June 2022
88. Batik transitions, batik guild (2006). https://www.batikguild.org/uk/batik/Africa. Accessed 26 June 2022
89. https://artsandculture.google.com/story/thecreationofhandmadebatikinGhana. Accessed 21 May 2022
90. A glimpse into batik art from other continent, (April 4, 2021). https://www.abirpothi.com/Aglimpseintobatikartfromothercontinent. Accessed 26 June 2022
91. https://elodietravels.wordpress.com/2018/02/03/traditional-textile-art-tie-dye-in-west-africa. Accessed 21 May 2022

92. Gausa S, Abubakar E (2015) Tie-dye (Adire) among the Jukun people. Mgbakoigba J African Studies 4:1–13
93. East African smes learn shibori tie and dye technique, fiber to fashion (2017). https://in.fashionnetwork.com/news/East-african-smes-learn-shibori-tie-and-dye-technique,829965.html. Accessed 12 May 2022
94. Darnell JA (2022) Styling the quad: fashioning the legacy of HBCU culture. Available at https://fashionandrace.org/database/styling-the-quad/. Accessed 12 August 2022
95. Niessen S (2003) Afterword: re-orienting fashion theory. In: Re-orienting fashion: the globalization of Asian dress. Berg, London
96. Mary Frances O'Dowd & Robyn Heckenber (2020). Explainer: What is Decolonization? Available at: https://theconversation.com/amp/explainer-what-is-decolonization. Accessed 14 August 2022
97. Melisa Jun Rowley (2020). What decolonizing fashion has to do with black lives matter, Available at: https://www.forbes.com/sites/melissarowley/2020/06/23/what-decolonizing-fashion-has-to-do-with-blacklivesmatter. Accessed 16 August 2022
98. Madeleine Jance. From Kente to Colonialism: African Fashion over time. Available at https://adht.parsons.edu/fashionstudies/bias-post/from-kente-to-colonialism-african-fashion-over-time/. Accessed 16 August 2022
99. Marlene Pop (2021) The cultural sustainability of the textile art object. In: Handloom sustainability and culture. Springer, Singapore. https://doi.org/10.1007/978-981-16-5967-6_10103
100. The Kaba and Slit fashion dresses in Ghana (2019). Available at https://seeafricatoday.com/culture/african-fashion/the-kaba-and-slit-fashion-dresses-in-Ghana. Accessed 19 August 2022
101. Sporting Kente cloth, by Fashioning the self. Available at https://fashionandrace.org/database/sporting-kente-cloth/. Accessed 16 August 2022
102. Chidera (2021) Designer Spotlight: This Is Why Chris Seydou Is Regarded As The Father Of African Fashion. Available at: https://AfricanVibes.com/Designer-spotlight-This-is-why-Chris-Seydou-is-regarded-as-the-Father-of-African-fashion. Accessed 17 August 2022.
103. Maki Osakwe. Available at https://www.Makioh.com. Accesses 16 August 2022
104. Decolonising dress and fashion: African wear in Ghana, Nigeria, Kenya. Available at https://www.thisisAfrica.com. Accessed 17 August 2022
105. Bromley WJ, Brynteson P, Magumbe A, Hawkes PJ, Parmanand S (2021) Prints with a purpose: how Sindiso Khumalo is reshaping sustainable fashion. Available at https://www.net-a-porter.com/en-sg/porter/article-1972d90ffdda39ba/fashion/art-of-style/sindiso-khumalo. Accessed 17 August 2022
106. Available at https://hautefashionafrica.com/designer/lisa-folawiyo. Accessed 18 August 2022
107. Esther (2021) Designer Spotlight: Maki Oh exposes the secret language of Prints. Available at: https://AfricaVibes.com/Designer-spotlight-Maki-Oh-exposes-the-secretlanguage-of-Prints. Accessed 18 August 2022
108. Alexis Chaffe Mey (2018) Looking at African fashion beyond Wakanda. Available at https://www.thehindu.com. Accessed 18 August 2022

Foregrounding the Value of Traditional Indian Crafts: Voices from the Fringe

Shalini Gupta and Ishi Srivastava

Abstract Craft in India is as old as the civilization itself and contributes to preserving art, culture, identity and storytelling (Dayinee and Priya 2020). The rich textiles tradition of India, with an abundance of natural resources, diverse climatic conditions and indigenous communities, have contributed to its ubiquitous and unique craft heritage. They are laden with perspectives of wellbeing and co-existence, a living heritage, expressing beauty and culture, community happiness and sustainability [1]. Indigenous communities create shared experiences within, socially transiting traditional values and identity associated with craft practices, from one generation to the next [2]. Madhavi Gandhi in her talk states, 'All objects have a story to tell, and with objects that are handmade, the stories are personal, sometimes emotional and largely cultural' [3]. Different crafts in India communicate unique varied embodied experiences and express emotional values. Additionally, Indian craft has emerged as a major export in recent years, proving its potential to meet global consumer demands [4]. However, India faces a paradoxical situation with reference to its crafts [2]. Craft here is often pushed outside everyday fashion practice, making going to market a challenge for artisans. Although crafts do not exist in isolation to mainstream fashion, they are often linked to other parts of the creative economy and are accompanied by a sense of pessimism, perceived as representative of a 'sun set sector' [2] and sometimes equated with manual labour. Contrarily, the United Nations Educational, Scientific and Cultural Organization (UNESCO) posits that this industry has massive potential for income generation by fostering its cultural assets (Dayinee and Priya 2020). However, as documented in the Sustainability of Rural Artisans journal, India continues to see large scale migration of many craft communities to urban centres in search of unrelated but more profitable employment opportunities [5, 6].

S. Gupta (✉)
Styling and Beauty, School of Fashion, Pearl Academy, New Delhi, India
e-mail: shalini.gupta@pearlacademy.com

I. Srivastava
Fashion Styling and Image Design, School of Fashion, Pearl Academy, New Delhi, India
e-mail: ishi.srivastava@pearlacademy.com

M. Á. Gardetti and R. P. Larios-Francia (eds.), *Sustainability Challenges in the Fashion Industry*, Sustainable Textiles: Production, Processing, Manufacturing & Chemistry,
https://doi.org/10.1007/978-981-99-0349-8_14

Keywords Fashion · Unheard voices · Cultural appropriation · Cultural legacy · Sustainability · Tradition · Craft · Decolonisation

As an attempt to foreground the voices of the indigenous craft practitioners in India and the value of their products, in the last quarter of 2020, the Crafting Futures India Scheme Funded Project, Raising Awareness of Value (RAV): Women and Crafts in India, planned and delivered a series of workshops with local textiles artisans from Bhuj, India. The project aimed at co-producing visual content, still and moving image along with social media posting plans and in some cases brand names and logos for seven artisans from the region. The project was run online using multiple web conferencing and knowledge sharing collaboration platforms. The participating artisans were interviewed at length, to understand where they saw themselves within the larger global and local fashion ecosystem. Through the experience of the RAV project and the subsequent Craftisan Project (an online collaborative project between students and staff of Pearl Academy, India and Manchester Fashion Institute, Manchester Metropolitan University, UK and Indian textile artisans) this chapter is an attempt at emphasising the importance of those on the fringes of the fashion system and why their unheard voices matter. It throws light on how indigenous practices that support social enterprise and ecologically sensitive production can contribute to the ever-increasing need for a more sustainable and responsible fashion industry.

1 Introduction

1.1 Craft in India

Considering the rise in consumer demand, Indian handicrafts have unquestionably illustrated their potential for growth [4]. Craft is a matter of appropriate expression through skills [7] and has been ubiquitous in India. In India craft is not limited to hand skills but is a powerful instrument of storytelling and cultural preservation. Additionally, handicrafts are the second largest source of employment locally, employing 23 million people [8] and are deeply connected to social transformation and empowerment. The chronicles of several indigenous crafts and practices have traversed generations in India. The climatic conditions, resources, creativity and inherent or inherited skills have helped nourish different crafts across this vast region. The practice of craft especially here can thus be appreciated, as a reflection of diverse perspectives, personal stories, self-expression and a living, evolving heritage [1].

Furthermore, the growth of craftsmanship steers the way to including more sustainable practices within the local fashion industry. This has been discussed in the article 'Fashioning Wellbeing Through Craft: A Case Study of Aneeth Arora's Strategies for Sustainable Fashion and Decolonizing Design' by Arti Sandhu [9].

Craft today, is being looked at, as a force of change towards a more sustainable society [10]. It looks at positive activism underpinned by the narration of strong stories with high emotive values. Communities own traditional knowledge, as practical and spiritual dimensions of life overlap within indigenous groups. Knowledge is created through efforts by people in a community and sharing practices [11]. Craft practices are brimming with shared knowledge, storytelling and draw deeply from social identity. They make up the socio-cultural fabric of society. They represent heritage that has roots in religious and folk narratives of indigenous communities. Through its model of social enterprise, craft has been responsible for community unity, empowerment and financial self-reliance.

1.2 Craft in Relation to Fashion and Design

Indian crafts have, in the recent past, seen somewhat of a revival with numerous young brands leveraging the techniques, value chains and processes inherent to the craft. These new age fashion brands predominantly work on a lean production model with limited stocks, production on demand and smaller teams [12]. However, even with all the strides that such brands, governments and local organisations are making, craft continues to frequently be placed outside of, or on the fringes of everyday design and fashion practice. This makes it complicated for independent craft makers and designers to demand a fair price for their products or be recognised as the brands they rightfully are. The newer brands leveraging crafts are often very well versed with the use of image making and storytelling to draw the consumer into the brand's eco system, be it through compelling visual narratives or content focusing on process, people and raw material [13]. The same, however, cannot be said about most independent craft producers and artisans, who, up until the COVID-19 pandemic, relied almost solely on selling to retailers and through physical exhibitions. Online sales if any were dependent solely on online aggregators who also cut into the meagre profit margins. Through aggregators and middlemen, the artisans procure lower wages, anonymity and insubstantial returns on the investment they put into their products [14]. Globalisation and changing domestic preferences have brought the crafts to face many challenges [4]. The rise of industrialisation and the stigma of inferiority associated with the crafts sector has led to a decline in the relevance of handicrafts in India. The grim reality is that many artisans today are struggling to sustain themselves and to preserve their craft against machine made replication available in local and international markets. The need of the hour is to create employment, preserve craft and to hear their stories. Opportunities need to be created without uprooting the youth from their environments [15].

1.3 Craft: Value and Social Image

At the UNESCO general conference, approximately 120 states voted for an international convention to safeguard Intangible Cultural Heritage [2]. Intangible heritage also includes craft as it is an integral part of daily practice for many communities. Beyond production and consumption, craft adds meaning to society and holds high its cultural value. Crafts deserve more recognition and the recultivation of a crafting culture. In many countries like Germany, Italy and Japan craft is no longer perceived as only tangible products acknowledging skill but is additionally recognised for the stories it is capable of unpacking. This change of narrative is essential if craft practice and products are to be drawn out of the shadows and become central to the creative economy [2]. Stronger policies and programmes are required in India to raise awareness and to strengthen the craft ecosystem [16].

India's heritage draws inspiration from the environment, respects local ecology and its traditional crafts have deep roots in techniques like up-cycling, reusing, recycling, waste minimisation and material optimisation. They involve using natural materials and processes in crafting that are low on carbon emission, making craft more favourable to preserving natural resources [17].

2 Artisans: The Unheard Voices

Given the ecological crises we are faced with presently, local and ethical products and practices are critically important to sustain. It is now more crucial than ever for the artisans, who have carried the torch for these traditions for centuries, to be able to share their stories. Furthermore, craft practitioners and producers have been pushed to the background while professionally trained designers have come to the fore due to decades of exploitation and appropriation. This gap needs to be addressed and opportunities for transparent co-creation and collaboration between the traditionally trained and the formally taught need to be encouraged if traditional artisan designers are to gain prominence and a voice.

2.1 Cultural Appropriation and Migration

Borrowing cultural elements from a civilisation and recognising it for its value and meaning is important towards preserving the story and legacy of its crafts. However, borrowing an aspect by being uninformed about its significance and value may lead to misrepresentation a complete lack of assignment of its value [18]. Post colonisation, in India there has been a dominance of western culture which has taken away from its craft heritage. Another important aspect is the social image and value projection of craft to the world through the media. Intangible aspects of crafts like its story, heritage, emotion and value need to be communicated appropriately with the intent of preserving them [11]. It is a well-researched and documented fact that Indian handicrafts play an important role in the global market with a $7 billion USD industry of exports [19]. Artisans must deal with many brokers and dealers which has lowered their income and can be attributed to why many artisans are migrating to cities for different forms of employment [19]. Post mass industrialisation, in the last three decades alone, Indian artisans have decreased by 30% [20].

2.2 Design Practices Within Indigenous Craft Communities, Design Schools and Fashion Brands

Local and Global fashion brands are selling tangible products enveloped in intangible values that the products hold. They are also engaging with contemporary issues that plague society. Fashion in some pockets, is focusing its energies on ethical and social practices that are relevant and important for society today. Consumers are not only looking at tangible products today but also the intangible experience the brand offers associated with its identity. Individual identities that brands are building today reflect their values. Fashion is influenced by its interaction with culture, and craft is a sizeable element of culture [21]. It has the capability to shape the socio-cultural context [4]. Design schools in India have craft cluster expeditions and intervention programmes where students study the craft for a few weeks and document or look at applying different design ideas to develop innovative or new products. Artisans are perceived as skilled craftsmen by brands, designers and design school students alike, however artisans are looking at equality and inclusivity within the design communities. In an interview with the RAV Project team, artisans alluded to the same challenge, where they are approached by design schools and fashion brands for skill-based work. However, they assert that they have much more to contribute to design than only their skills. They share their stories, their art and culture through craft, and are constantly working towards innovation and creativity through their practice [22].

3 The Craftisan Project

In October of 2020, the British Council, Crafting Futures India Funded Project, Raising Awareness of Value (RAV): Women and Crafts in India, launched its first stage of workshops with local textiles artisans from Bhuj, India. The group of artisans are highly skilled weavers, dyers and embroiders who produce textile-based garments such as stoles, scarves and saris. During the COVID-19 lockdown these artisans were able to develop their textiles but their primary reliance on face-to-face channels for sales, such as exhibitions and craft fairs, meant that opportunities to sell their goods, instantly diminished. During the primary research stage, one of the artisans, Adil

Khatri, shared his experience of interactions with established fashion designers. He disclosed that many designers saw him as a skilled artisan but not as a creative contributor and a designer or innovator within the textile space. 'I don't replicate designs,' he categorically stated and added that he had his own unique design process and was looking to refashion the term artisan to the more befitting 'artisan designer' in its stead.

In its first stage, RAV focused on the promotion of these artisans' current products via photography, digital marketing, and brand development workshops. The workshops were aimed at empowering the artisans themselves to create compelling social media content to be shared on Instagram and Facebook. The artisans were also provided with 'Home Studio Kits' comprising of a sewing machine, a backdrop stand, a dress form, a tripod and two studio lights. This was done with the aim of helping them become self-sufficient in being able to promote their own products digitally and virtually. More specifically for the female artisan designers, considering the local socio-cultural norms, the home studio kits proved to be a boon as they were traditionally, not encouraged to travel to external studios, unaccompanied.

A long-term approach and consistency are paramount to being able to leverage social media as a channel for promotion and sales. What the content should be, how often it should be posted, how one draws a consumer into the craft brand's ecosystem, are all important to know for novice brands. This establishes a strong need for sustained empowerment of social storytelling. Social storytelling takes information that may not be exciting and transforms it to make it important, impassioned and relevant. In the case of hand-crafted products, it has the power to move individuals at a personal level. It removes anonymity and focuses on stories of the people behind the products. Lived experiences are carefully crafted into compelling narratives that leverage history, timelessness and value. This also gives the craft producers control over the narrative and builds awareness, leading to the assignment of fair value to their products. The 2nd stage, The Craftisan Project (February 2021 to June 2021), which is reported here, was a live global co-creation project developed by staff from Pearl Academy (PA) in India and Manchester Fashion Institute (MFI), Manchester Metropolitan University (MMU) in Manchester United Kingdom, in collaboration with their students and the Indian artisan designers. The Craftisan Project takes ahead the RAV project that focused on training artisan designers from Bhuj in image making, photography and storytelling for social media. The intention for the Craftisan Project was for students at both universities to collaborate with the artisans and produce social media campaigns and posting strategies that could provide a more structured approach for the artisans and their brands when engaging online, specifically looking at sustained social storytelling. With students and teachers from two very diverse cultures, social and economic backgrounds, these varied perspectives were fundamental to creating globally apposite outputs. To this end, the intention with the development of this project was to create a community of practice and unique methods of visual expression via storytelling. The students, with the help of the artisan designers, defined the target audience, the aspirations of the producers and how they may resonate with the group of people being communicated to.

3.1 *Structure and Learning Outcomes*

The study of art and design as an academic and intellectual pursuit, develops a range of cognitive abilities related to the aesthetic, ethical and social contexts of the human experience. Engagement in the study of art and design is therefore a commitment to improving the quality of one's own and others' experiences, and this statement succinctly acknowledges a responsibility towards sustainable development and towards equipping students to work in a way that contributes to society, the economy and the environment, both in the present and for the future [23].'

The primary project planning team comprising of Shalini Gupta (PA), and Dr. Elizabeth Kealy Morris (MFI), started by examining the need of the artisan designers and how they may benefit from a project with students. This led to the formulation of a very specific brief, learning outcomes and deliverables expected from the students. They investigated the outcomes produced by the artisan designers in the previous RAV project and built up taking ahead the groundwork laid there.

The students were to work alongside the artisans to support the storytelling of their brands, communicate their goals, aspirations and aims and work together to develop a rationale for the brands and products seeking answers to the following and other questions:

- What drives, motivates, and inspires the artisan?
- How can we work together with the artisans to tell the unique stories about their brands and products?

The staff, students and Indian craft producers were encouraged to work co-creatively using digital platforms and virtual learning environments. Delivery was planned through masterclasses, collaborative production meetings, demonstrations and virtual visits. The students were supplied with statements from the artisan and images of their product ranges.

The students were tasked with developing promotional campaigns for the artisans' brands which were to include:

- A strategic plan of Social Media posts on relevant platforms.
- An art-directed, styled and storyboarded pitch of fashion film/video of the artisans at work to highlight the skill of the makers and the heritage of the products produced.
- Edited test films.

3.2 *Findings/Development/Analysis/Evidence*

A survey was conducted among the 20 students selected for the project to establish a baseline and have a sense of what they felt they were going to gain during the project and what they wanted to gain coming out of it. When asked what they understood of these terms, the participants believed that collaboration and co-production were a coming together of diverse perspectives towards a common goal wherein the outcome was enriched on account of the differences in perspective. The term 'artisan' to most meant an individual who was skilled especially in the handmade.

> Sustainability means meeting our own needs without compromising the ability of future generations to meet their personal needs. In addition to natural resources, we also need social and economic resources. Sustainability is not just environmental. Embedded in most definitions of sustainability we also find concerns for social equity and economic development [24].

The above statement published in a report by the University of Alberta, Office of Sustainability, sums up the responses by the participants who referenced it in part or whole.

The participants expected to learn more about craft, the process, the people, and what drives and motivates the artisans to continue this journey despite the hardships they have to face. They also hoped that in some small way they may be able to have a positive impact on the life of another person. Understanding sustainable practices and learning new skills were the highest motivators for the group.

To see the application of technology in pushing crafts ahead was a welcome and much needed antithesis. It was planned as a four-week project within which the first few weeks followed a mixed pedagogical approach of information transfer through masterclasses, followed by studio-based learning through group work and discussions and reflective practice through group tutorials with an assigned mentor. Collaboration and co-production with the artisans, mentors and peers were the intended learning outcomes and the backbone of the project. The projects also gave insight into the need for innovation, sensitivity, empathy, problem solving, collaboration and inclusivity within the wider fashion industry.

3.3 Input and Outcomes

The first session in week one was an introductory session where the students were briefed about the project, made familiar with the virtual platform and the resources. These included the project handbook that housed the links to all the masterclasses as well as group work and tutorial rooms. The folders where they could find weekly and general reading and viewing resources and where they could upload their weekly and final tasks.

In the same session Ms. Harroop Grang (PA), and Dr. Elizabeth Kealy Morris (MFI) were invited to talk about the current state of craft in India and the UK. They discussed the evolution, transformation and relevance of craft today. This helped the students understand the deeper meaning and significance of craft in their lives. It laid a good foundation for students who didn't have a background in and knowledge of craft.

Session two by Mr. Paddy Lonergan (MFI) discussed various promotional theories and market segmentation. This was very helpful for students in understanding their artisans, the artisans' clients, and target audience as per demographic, geographic, behavioural and psychographic parameters. It helped them in formulating an in-depth analysis of the client as a base to develop the final promotion plan. They studied the artisans' market, worked on the market segmentation, and prepared a promotion plan for social media.

In week two, an art direction and story boarding masterclass was conducted by Ms. Poppy Cartwright (MFI). Students had to plan a film for the artisans, documenting their story and the story of their craft using technology tools. This masterclass helped them understand the process of creating a story board and the importance of creating one for their film. They worked with their groups and created story boards (Fig. 1) in the given format for the film they were planning to create.

The second masterclass in this week focussed on brand styling with Ms. Sonia Genders (MFI). The class threw light on how every brand has a unique style that conveyed the brand's identity. The masterclass also focused on identity and storytelling as important aspects of image making. Students looked at examples from varied brands to study, analyse and interpret styling within different contexts (Fig. 2).

The masterclasses in week three by Mr. Amitesh Singhal and Mr. Aditya Mittal (PA) focussed on film planning and editing techniques. Students learnt framing, colour correction and editing from a creative and storytelling perspective. They compiled short clips shot and shared by the artisans into a short film for their brand.

Fig. 1 Storyboard prepared by students to communicate the story and process of an artisan's (Adil Khatri) craft

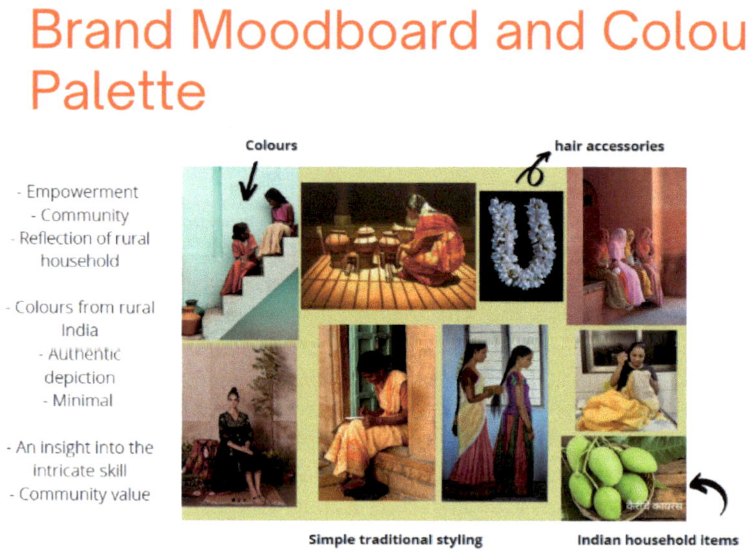

Fig. 2 Style Board created by students for artisans to communicate their visual language

The videos were shot by the artisans using storyboards (Fig. 3) created by the student groups.

The pitch was the culmination of this project. Every group created a pitch presentation for their artisans. Artisans shared their feedback and ideas on the pitch. The nuances of making an effective pitch presentation were taught in a masterclass by Dr. Elizabeth Kealy Morris of MFI.

Fig. 3 Screenshot from a film created by Group 1 for the artisan Krishna Vankar. The film captures the story and process behind her craft of weaving. Link to the film: TAAGA by Krishna- The Film.mp4

For the pitch, in addition to the film, the student groups created and presented a social media promotion plan (Fig. 4) for every artisan based on their primary and secondary research and artisans' requirements. Students presented eco-friendly, cost effective, zero waste packaging ideas for products created by artisan designers and in some cases presented their clients with new brand names, logo ideas (Fig. 5) and website design ideas (Figs. 6 and 7).

Additionally, they suggested relevant hashtags, captions, posting schedules, interactive elements on Instagram like reels, stories and posts, for maximum engagement. As part of the final pitch, they emphasised on the importance of creatively expressing the process and story behind the craft as it connected the target audience to the brand, product and people. The films the student groups created for each artisan focused on this very idea of presenting process, heritage, culture and people.

A mid project survey was conducted with the student group to see how they felt about the project, collaborating with the other students and the artisans. One student said on the project experience this far '…people working globally on a project that has

Fig. 4 Screenshot of social media posting plan created by students to show details of artisan's (Zakiya Khatri) craft along with establishing their identity as a designer

Fig. 5 Proposed logo Designs

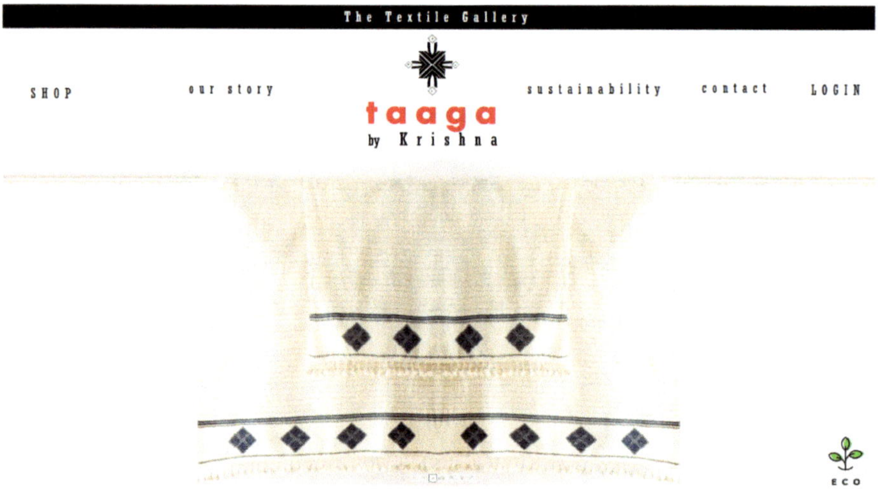

Fig. 6 Screenshot of the mock website created by Group 1 for artisan Krishna Vankar. The website covers aspects like storytelling and sustainability for the craft of weaving

a real impact over someone's life is an amazing thing to do.' Although the students believed that the process of co-production and collaboration had been an enriching experience, they would have liked more time with both the artisans and their group of peers.

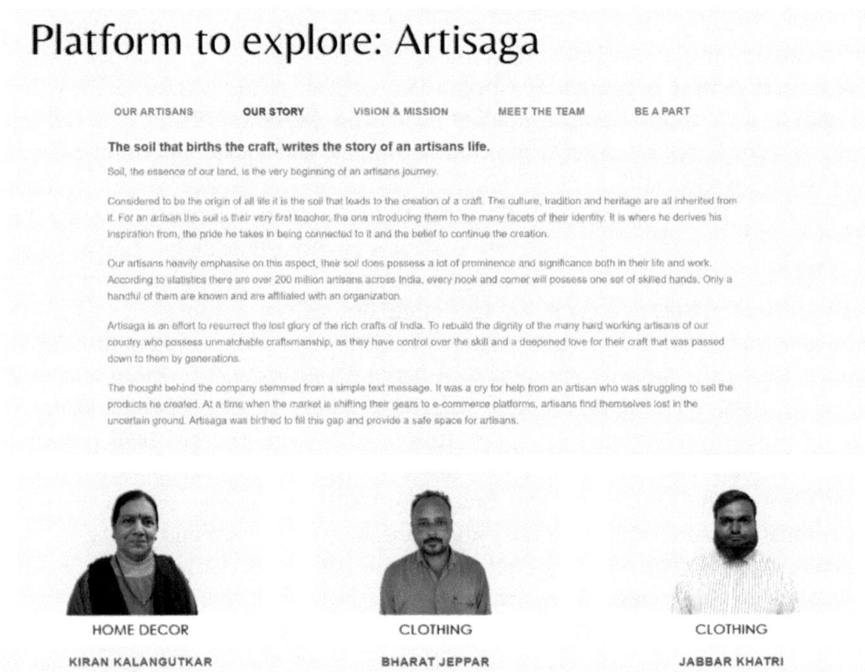

Fig. 7 A page from the mock website created by students for one of the artisans' that highlights the story of artisans and shares their images so they can be recognised for their crafts and their stories are heard

4 Conclusion

It is important to consider artisans as equal to designers as they bring skill, creativity and innovation through craft. Reclaiming traditions and skills, and respecting people and processes, is the perfect antidote to today's accelerated cycles, and many of us have our own familial or community heritage craft handprint to inspire us: it's about unravelling this thread, looking back and moving forward [9]. The 17 Sustainable Development Goals (SDGs) [25] are integrated—they recognise that action in one area will affect outcomes in others, and that development must balance social, economic and environmental sustainability. This collaboration between educational institutes and Indian artisans creates a community for upskilling, exchange of knowledge that helps in promoting lifelong learning opportunities for all (SDG Goal 4). It also aims at a more inclusive space where new ideas are welcomed, acknowledged and executed for the growth and development of all.

The Project team worked with five women artisans. Laxmi Pawar, a participant, based in Bhuj, learnt the craft of Suf embroidery from her mother and now has a community of women who practice the craft alongside her. The craft is women centric, and her purpose is to empower women of her community economically as well as socially. Historically and even today, women play an important role in craft communities across India. Even though, socially, gender inequalities are deep rooted, and change is gradual, it is important to discuss and raise awareness foregrounding their contribution. Empowerment is an important objective which also looks at helping women artisans get equal economic and social opportunities within the system (SDG Goal 5).

The book Less is more [26] discusses the idea of an economy where people produce and sell useful goods and services, an economy where people make rational, informed decisions about what to buy; an economy where people get compensated fairly for their labour; an economy that circulates money to those who need it; an economy where innovation makes better, longer-lasting products that reduce ecological pressure. This project was aimed at being a step in this direction and to see ourselves once again as part of a broader community of living beings [26]. This project primarily attempted to equip artisans with tools to create and promote a sustained, inclusive and sustainable growth environment (SDG Goal 8). Spoken language is not universal. However, images both still and moving, are understood by all. By this measure, voices are not only heard, but can be seen and touched.

Technology is often seen as the antithesis to the hand made. However, here we used technology to create cultural impact, to promote the exchange of stories, knowledge, skills and information within the project and in support of value communication and social promotion of the artisan designer's brands online. Such projects prove that it is possible to nurture a socially driven mindset among young design graduates that could lead to increased collaboration between the formally trained and the traditionally trained. These traditionally trained artisan designers inherently practice zero waste, inclusive, collaborative and ecologically friendly methods of production and the more these two groups communicate and work together the more this could send young people into the fashion industry with the drive to fix the well documented ills that plague it.

Finally, more visibility to artisans and craft products can motivate young people to stay with the traditional art forms and preserve a heritage that is fast becoming obsolete dew to a dwindling market [27]. There is a potential market to expand to, considering the growth of global and domestic tourism and spending power. Artisan will need access to more markets, buyers and technologies [4]. The project has brought to light the need for digital and visual literacy among Indian artisans if they must keep up with a globally influenced audience. Expanding the boundaries of human community, our culture and our consciousness is important from a social perspective. This project was an attempt to change the way they see the world, and this requires listening, empathy, dialogue and collaboration. [26, 28]. Craft is a powerful voice for the artisans, it carries stories, heritage, culture and promotes co-existence with environment and community [1] and this project is a step towards communicating the unheard stories of artisans through the application of technology.

Acknowledgements The authors would like to thank the British Council, Pearl Academy, India and Manchester Fashion Institute. Manchester Metropolitan University, United Kingdom for their support. This project was only possible because of their constant support. We are thankful to Shalini

Gupta from Pearl Academy (PA) India, and Dr.Elizabeth Kealy Morris of Manchester Fashion Institute (MFI), Manchester Metropolitan University (MMU), United Kingdom for planning and bringing this project together. The supporting team from both institutions comprises of Sabbah Sharma, Parag Goswami, Henna Parimoo, Amitesh Singhal, Aditya Mittal, Harroop Grang, Megha Khanna, Ishi Srivastava (PA) and Sonia Genders, Poppy Cartwright and Paddy Lonergan (MFI). The IT teams from both institutions helped us with all platforms and technical aspects. We thank the artisan designers, Krishna Velji Vankar, Laxmi Puwar, Suresh Velji Vankar, Adil Khatri and Zakiya Khatri for sharing their heritage, craft and knowledge with us. They have supported the project with immense faith in our abilities. Our students have been wonderful throughout the project, we thank them for their dedication, commitment and hard work to make the film and pitch presentation. Lastly, we thank all the authors we have cited in this paper.

References

1. Gage K, Pasricha A (2020) Editorial—fashion in India. J Design Creative Process Fashion Indus 12(2):163–171
2. Kotlipalli P (2018) The values of craft: towards an economics of the common good
3. Gandhi M (2017) Narrative design—Learnings from India
4. Liebl M, Roy T (2003) Handmade in India: preliminary analysis of crafts producers and crafts production. Econ Political Weekly 38(51)
5. Bellampalli PN, Kaushik R (2018) Rural urban migration in India: types. challenges and opportunities. Manpower J II(3):59–71
6. Solanki S (2008) Sustinability of rural artisans. Econ Pol Wkly 19(43):24–27
7. McCullough M (1998) Abstracting craft: the practiced digital hand. MIT Press
8. Globallinker (2017) The CraftMark initiative for Indian handicrafts
9. Sandhu A (2020) Fashioning wellbeing through craft: a case study. J Design Creative Process Fashion Indus 12(2):172–192
10. Ravasi D (2021) Configurations of craft: alternative models for organizing work. Acad Manag Ann 15(2)
11. Finger JM, Schuler P (2004) Poor people's knowledge: promoting intellectual property in developing Countries. World Bank & Oxford University Press, Washington
12. Khandual A, Pradhan S (2019) Fashion brands and consumers approach towards sustainable fashion. In: Muthu DS (ed), Fast fashion, fashion brands and sustainable consumption, pp 37–54. Singapore: Springer, Singapore
13. Amritha B, Suresh K (2020) Sustainability is the new black: exploring website communication practices of Indian sustianable fashion brands. Fashion Style Popular Cult, 539–558
14. Ptal (2021) Indian handicrafts diverse, complex & colourful well this one is healthy too !!
15. HeartforArt Trust (2018) How you can help make India's arts & crafts community globally competitive & improve livelihoods
16. Chamikutty P (2013) Problems facing the Indian artifacts sector
17. Kala I (2017) TBI blogs: how supporting India's Handcrafted products can also help protect the environment
18. Arewa O (2016) Cultural appropriation: when 'borrowing' becomes exploitation
19. Kaul S (2018) Indian arts and crafts—$4.5 billion domestic industry, how its dynamics are changing with 'Mobility'
20. Dasra (2013) Crafting a livelihood: building sustainability for Indian artisans
21. Khaire M (2011) The Indian fashion industry and traditional Indian crafts. Bus Hist Rev 85(2):345–366
22. Khatri Z (2020) Interview with Zakiya Khatri. (J. Morris, & S. Gupta, Interviewers) Delhi, India

23. QAA (2019) Subject benchmark statement: art & design. Retrieved from QAA: Subject Benchmark Statement: https://www.qaa.ac.uk/docs/qaa/subject-benchmark-statements/sbs-art-and-design-17.pdf?sfvrsn=71eef781_22
24. University of Alberta, Office of Sustainability (2013) What is sustainability? Edmonton, Alberta, Canada
25. UNDP (2022) UNDP
26. Hickel J (2020) Less is more. Cornerstone digital
27. Gupta S, Srivastava I (2021) Crafting narratives: co-producing content for social storytelling of indian, craft-based brands. Global fashion conference, Warsaw, Poland. Warsaw: Global Fashion Conference
28. Kolb D (2015) Experiential learning: experience as the source of learning and development. 2nd Ed. Pearson Education

Change, Imitation and Novelty. Notes on Cosmotechnics and Mass Communication in the Face of Fashion and Its Political Possibilities

Ileana Jalil Kentros, Laura Camila Ramírez Bonilla, and Edward Bermúdez Macías

Abstract In this interdisciplinary work, we explore the inherent nature of fashion as a global, cultural, symbolic, and psychological phenomenon. We delve into the conditions and interactions that allow fashion to interact with different social groups, contexts, and the environment, and start by differentiating the social phenomenon it represents from the simple practice of dressing and personal grooming. This leads us to consider fashion as a matter inherent to the human being, which regardless of time and space could be defined as global, cultural, symbolic and psychological. However, its development does not refer to a uniform transformation, since particular conditions are required to extend its capacity for interaction with different social groups, contexts and the environment. On these conditions and interactions, we follow with a set of reflections that address change and imitation in fashion as traceable conditions in time, accelerated by mass communication and technological development since the late nineteenth century.

In *After a Fashion* [1996: 6], Joanne Finkerstein comments that it is wrong to think of fashion only in relation to clothing, for example, fashion involves elements of language, nonverbal expression or body movement. Similarly, Rene König in *The Restless Image: A Sociology of Fashion* [1974: 40] says that we must destroy the prejudiced idea that fashion only concerns the external covering of people, when it is really a social institution that affects and shapes individuals and society as a whole.

In this article, we will define fashion as a social issue where change is systematic, aimed at novelty in social practices mediated by design; and design, as a cultural political project that configures the artificial world created by human beings [1].

I. J. Kentros (✉) · L. C. R. Bonilla · E. B. Macías
Universidad Iberoamericana Ciudad de México, Mexico City, Mexico
e-mail: ileana.jalil@ibero.mx

L. C. R. Bonilla
e-mail: laura.ramirez@ibero.mx

E. B. Macías
e-mail: edward.bermudez@ibero.mx

M. Á. Gardetti and R. P. Larios-Francia (eds.), *Sustainability Challenges in the Fashion Industry*, Sustainable Textiles: Production, Processing, Manufacturing & Chemistry,
https://doi.org/10.1007/978-981-99-0349-8_15

This systematic change operates under a dualistic dynamic of social imitation and individual distinction, which is enabled and accelerated by technology and communication. For this to occur, in a modern and Western sense, it requires people belonging to a social context in which they can exercise with certain freedom the manifestation of their individuality. In direct relation to this, and assuming social mobility as a possibility of the system, change and novelty become something desirable, constituting an aspirational ideal.

These conditions mean that fashion is not a homogeneous or timeless phenomenon, since, in different times and places, including pre-modern societies, there have been human groups that prohibit variation in housing practices, work or personal grooming, where change and novelty are not desirable, but on the contrary, symbols of alteration to the norms. This is the case of some Taliban or Amish communities, where modifications are determined by community rules and not by the free exercise of the individual, so there are no conditions for the phenomenon of modern fashion to operate, since, even if changes occur, they cannot be considered as systematic, nor in constant acceleration to be integrated into the global phenomenon.

At the same time, we distance ourselves from the Eurocentric view of the origin of fashion held by classical authors such as Georg Simmel [2]; from his perspective, fashion began in the middle ages due to the dynamics of change and novelty in personal grooming for the courts, where the anatomical difference of the sexes was already marked in the dress, through a system of modeling for the garments. Other studies such as that of [3] Van Dyk Lewis about the Mursi and Omo people in Ethiopia or David Inglis have shown that this dynamic of transformation and social distinction can also be traced in ancient communities outside Europe, such as the Aztecs or the Byzantine Empire [4]. His work allows us to speak of diverse manifestations of non-institutionalized fashion, also known as *pre-fashion* or *proto-fashion.*[1]

Based on these parameters, this text aims to analyze the relationship between change and novelty in modern fashion, taking as a guiding thread the intervention of technology and mass communication. In dialog with the concept of *cosmotechnics*, we seek to evidence mechanisms by which the speed of innovation has increased, becoming an imperative of voracious consumption and not of collective improvement. The analysis of mass media allows us to trace the power of imitation in fashion, conditioned by the efficiency and capacity of diffusion of the novelty. It is thus a constant over time, deepened by the technological possibilities that have made it possible to accelerate the transmission of messages, establish remote exchange in real time, broaden coverage, strategically intervene in the image (illustration, photography, moving, animated, etc.), reduce the cost of resources for interaction and, in recent years, the production of content in the hands of consumers. For its part, technology is present in an increasingly sophisticated way, both in mass production systems and in the design that conceives these products and gives them functional and aesthetic characteristics that represent novelty.

[1] Space between fashion as socioculturally universal and Western and modern fashion pluralistically in both times and places [4].

This text is divided into three parts: in the first, a succinct definition of fashion as a social phenomenon is presented; in the second, a reflection on cosmotechnics and fashion is presented, through historical processes and examples that problematize two factors that have fueled the acceleration of change: technology and the media; in the third part, the reader is invited to play an imaginative game that allows him to approach two possible future scenarios: dystopia or utopia, the gateway to a reflection on the future of fashion, which will allow him to approach two possible future scenarios: dystopia or utopia, the gateway to a reflection on the future of fashion: technology and the media; in the third part, the reader is invited to a game of imagination, allowing him/her to approach two possible future scenarios: dystopia or utopia, gateway to a critical reflection on accelerated change and imitation in the world of fashion; finally, in the last section our readers will find a segment of conclusions.

1 The Social Phenomenon of Fashion and Systematized Change

The present analysis does not seek to date the emergence of fashion in historical terms, since it is an open debate with multiple edges; as Yuniya Kawamura points out in *Fashion-ology* [5]: "fashion is a social and psychological mechanism that lacks a point of origin". On the contrary, we are interested in establishing what are the conditions that make the phenomenon of fashion possible, to analyze its relationship with the technologies that enhance it, and how this leads to a certain understanding of the relationship between human beings and their environment.

For Foley [1893], modern and pre-modern fashion are not differentiated in principle by change and social distinction, but only at the empirical level by systems of production, distribution and consumption. Considering that in both cases simple change is not synonymous with fashion, this implies an active awareness of the participants about what they take and reject in their social intention to construct identity from imitation and differentiation. Nor, is it just a human drive for symbolic and material distinction, a system of novelty supply is required. Adspers and Godart suggest that it is not just change per se, it requires continuity over time.

The key for the phenomenon of fashion to be installed with all its power in modern societies, even as a total or institutionalized fashion [5], is centered on the means of production. If fashion is a "manufactured cultural symbol in an institutionalized system" [5], the key also lies in manufacturing. From this perspective, modern fashion is characterized by massification and industrialization, in addition to the absence of sumptuary laws of restriction and prohibition in personal grooming. As a historical process, it is tied to the consolidation and expansion of capitalism [6], which reinforces the exercise of individual freedom as a right. The panorama becomes even more complex when the consumption of goods and services is constituted in modernity as a factor of individual and collective identity.

In this context, fashion manages to materialize through design, directly related to the industrial revolutions, since industrial production required specialists to configure the shape of the products that could be mass-produced. It is through these means of massification and industrialization that systems of supply of novelty are propitiated, which Kawamura speaks of as an "institutionalization of fashion". For Blumer [1969], fashion operates in an institutionalized and systematic way, within industrial society, made possible by the technology that has not ceased to advance to date. From our analysis, to this factor are added the communicative processes that, through diverse media, potentiate the speed of change and adoption of fashion. For example, the arrival of the bicycle, in the late nineteenth century, stimulated the use of pants in women who could use the new artifact as a means of transportation, thus prioritizing comfort and technology over the dominant ideal of the skirt and dress [7].

Taking this into account, the essential features that we highlight of modern fashion are:

1. The freedom of the individual to make decisions about aesthetics and social practices, within the framework of the society in which he/she lives and those aspirations of social mobility
2. The occurrence of increasingly accelerated systematic change
3. The incessant search and desire for the new, enhanced by technology and the media.

2 Fashion, Technology and Communication: A Reflection on Cosmotechnics

Derived from Heidegger's philosophical reflection on technology and the ontological turn in anthropology, Yuk Hui raises the concept of *cosmotechnics* as the unification of the moral order and the cosmic order through technical activities [8: 19], in order to question whether all peoples think of technology in the same way. Finally, these two orders are not the same in all cultures and, therefore, technology in each does not represent the same thing either. In this way, Hui [9: 11] formulates a duality:

> Thesis: technology is an anthropological universal; it can be understood as externalization of memory and liberation of organs, as anthropologists and philosophers of technology have formulated it.
>
> Antithesis: technology is not an anthropological universal; it is enabled and constrained by particular cosmologies that go beyond functionality or utility. Consequently, there is no single technology, but multiple cosmologies.

This allows him to distinguish a modern and western understanding of technology that, in agreement with Heidegger, considers nature as an accumulation of resources available to be exploited. The whole effort to impose this model as part of capitalist globalization corresponds to ontological reflections that, as in the case of Escobar [1] on design, speak to us of the *one world*. The promotion of these items in advertising, with illustrations and photographs, reflected a new lifestyle available to the majority.

A style that reflected the well-being or happiness of people thanks to the action of the factories, which displaced the objects that made up the everyday and represented the apparently outdated cosmotechnics. Hui calls this dominant model *monotechnology,* and the set of its opposites: *technodiversity.* Thus, as for Hui, behind the slight differences in form and branding there is the same type of technology, monotechnology, it is the same design approach, the same conception of our relationship with the natural and the moral, which determines the current phenomenon of fashion. In this uniformizing and globalized version, small differences are only part of the whole that leads to the destruction by depletion of resources, and of societies or cultures themselves by the reduction in their capacity for discernment and formulation of visions or futures.

For its part, communication is revealed here in two starting points: on the one hand, the natural act of transmitting signals through codes with meanings, indispensable for interaction between humans; and on the other, the means by which that transmission has the possibility of spreading to multitudes, remotely. In this intertwining, it is not only the communicative act that brings the news of novelty or change to a receiver or the technological resource that allows this transfer of information in an effective way (immediate in recent times), but also the way in which the message is received by individuals, incorporated or rejected, modified or controversial. In other words, audiences become protagonists not as receptacles of messages, without criteria, but as agents in decisive interaction with what is received.

This exercise requires us to think about the communicative process in an integral manner, assuming that socioeconomic, emotional, symbolic, moral and political factors are involved in the relationship with the contents, the media and the receivers, which do not always circulate in the same direction, predictable and uniform. Mass communication in the last century not only lives in technological convergence but also in cultural and social convergence [10], which superimposes layers of interpretation to the multiple relationships that a consumer may have with a product. When we talk about communication and mass media, we are not talking about impositions at all costs, between good guys who submit and bad guys who submit, but about complex interactions modified over time, with negotiations, conflicts, hierarchies, flows and gaps in which all parties assume a role. What is expected? That this role be one of interpellation, that action be present. That the disposition is not assumed from passivity.

With the intention of applying and exemplifying these reflections to the phenomenon of fashion and providing evidence of design and its intention to contain meanings in objects, we offer a brief review of three historical moments on the acceleration of change through technology and the media. Why does thinking of fashion in terms of change entail a historical reading? Why does thinking of it in terms of imitation and novelty involve technological and communicative processes? In what way have these two sectors influenced the acceleration of change as an imprint of fashion? Understanding fashion as a social phenomenon, the implications of change and its gradual acceleration involve alterations in socio-cultural environments, modes of production, consumption, social imaginaries, political projects, labor conditions or environmental footprints, among other aspects, which require us to think about convergences and historical tensions.

Next, we will dwell on three historical macro contexts and three specific examples, in dialogue with the Western context, allow us to analyze and allude to the complex mediations that technology and mass communication have played in the acceleration of change in the social phenomenon of fashion, an argument that we have followed throughout the text: (1) the fashion newspapers of the late nineteenth century, (2) television in the fifties and (3) the socio-digital networks of the early twenty-first century.

3 Macro Historical Contexts: Fashion, Design and Technology

- **End of the nineteenth century: Second industrial revolution and the diffusion of fashions through printed media.**

The second industrial revolution, considered between 1870 and 1914, was a period when technological change completely altered the social and economic structure of the West. The systematization of industrialized mass production, the speed of movement of humans and materials through new means of transportation (automobile, airplane, steel-hulled steamship and railroad), and the expansion of communication by printed media, are just a few examples of the fundamental transformations. The acceleration of production and communications coincided with the professionalization of the designer as the link between these new industrial products and the people who would consume them, through functional and aesthetic considerations, his activity became strategic. Because of industrialization, the new products required mass media to communicate their novelty and promote imitation. Therefore, going beyond the natural capacity of the human being through machinery and the consequent motivation to achieve more, on a larger scale and at greater speed, is a reflection of the dominant relationship of the human being with his environment; that is, the cosmotechnics of the time.

- **World War II and technology in search of application**

The technological development that boosted and speeded up World War II with the war industry was followed by an unprecedented application of consumer products accessible to mass groups of the population. The technological innovations of the war strengthened pre-existing industries on a larger scale, invigorating production capacities and innovations in materials for the expansion of the supply of everyday objects. The social need for economic reactivation in the post-war period required the expansion and diversification of mass consumption, taking advantage of the fact that the figure of the consumer had already been popular since the 1920s. The United States of America demonstrated that production capacity was greater than the desire for consumption. This meant setting in motion the machinery of seduction to achieve the almost sole objective of profit, above even others such as simply satisfying a functional need [11: 69].

These products included household appliances that not only facilitated the daily life of the rising middle class, but also diversified the income possibilities of some families as was the case with the sewing machine or the iron. It was a period of boom in synthetic and artificial materials, which determined the design of massive products and new clothing. The promotion of these items in advertising guidelines, accompanied by illustrations and photographs, reflected a new lifestyle now available to the majority; a style that reflected the well-being or happiness of people thanks to the action of the factories, which displaced the objects that made up the everyday and represented the apparently outdated cosmotechnics. The resulting cosmotechnics focused on facilitating daily life through technology applied to massive and accessible products and processes, where the user is the one who seeks access to this novel offer.

To keep the possibility of change active, planning the obsolescence of products came into play, both in the functional and psychological sense, something that Higgs [11: 75] rescues from Vance Packard, who wrote about it in 1960. By that time, advertising had established itself as a creator of needs, procuring better revenues for corporations that controlled large distribution chains and direct sales. Thus, design became the protagonist, both in the conception of products with the technical limitations that ensure their disposal, and in influencing through esthetics the formulation of characteristics that could easily be perceived as obsolete before the launch of another novelty. In that sense, a practice documented and revisited by Kerryn Higgs [11: 69] is the pursuit of opulence in ever-widening social circles, thanks not only to a greater availability of products, but "to the propulsive power of envy" promoted by the display of these in spaces such as department stores.

- **Present time and technology for the imperative of consumption by novelty**

The new change did not erase what preceded it; there was only an overwriting. The current dominant cosmotechnics consists of the administration of technology to promote novelty, with the promise of individual distinction, as an impulse for a consumption that must always keep growing. To this end, a continuous flow of innovations involving technological and aesthetic modifications is generated, which are dosed by means of planned and regular dates. Technological breakthroughs that change everything do not happen, and they already exist. They belong to the companies, and they are dosed according to consumption needs. What is new is always the best, fashion permeates all areas of life, and the flow of novelties is multiple and constant through digital media.

Currently, technological changes are managed in advance to maintain a controlled pace of changes in fashion and promote consumption through the media and social networks. These advanced and managed changes in technology can be seen in all areas of life, such as in the new cell phone models with gradual and scheduled updates, where technological innovations are managed year after year regardless of how long the technological development has been in place, which are not usually disruptive but only a continuation to encourage consumption. This managed and planned technological change is internalized as a hegemonic dynamic, where novelty is always considered desirable. It is common to find queues for the purchase of the new iPhone, to enter the new Shein store in Spain or the immediate *streaming* premiere of a movie on Netflix.

The practice of hegemonic fashion for consumption, which after pursuing novelty leaves a ballast of material or virtual waste, has generated a dominant cosmotechnics, our way of making sense of the artificial world we have created by technology and the moral order it implies, indicates that novelty is always desirable regardless of its effects. Today, it is a voracious fashion phenomenon. It is so fast-paced in a planned way that it is difficult to grasp and analyze as it happens. Added to this are phenomena such as disinformation that distort or hinder awareness of the consequences, and lead to strategies such as green washing or to the assumption that the digital world is a neutral form of consumption with respect to natural resources.

4 Cases and Examples: Fashion, Design and Communication

- **The global press of the nineteenth century**

In the last two decades of the nineteenth century, a newspaper from France arrived weekly in American ports, in perfect Spanish. "La moda de la elegancia parisiense" was a publication edited by De Mourgues of the Union de la Prensa Hispanoamericana, which operated on the Avenue de l'Opera in the French capital. With a printing of four issues per month, the newspaper boasted of bringing to its readers more than 1,000 engravings in the text, 48 illuminated figurines, and 24 patterns with the representation of the most updated fashion: "it is the only fashion newspaper made in Spanish in Paris and that is what gives it its supremacy", it pointed out about itself at the end of its pages.

The need to keep Latin America up to date with the novelties of the fashion industry had two aims: first, to ratify an imaginary of sophistication and luxury monopolized by the France of the time, and second, to confirm a spirit of renewal, creativity, and permanent advancement of the industry, as if it were an evolutionary scale, in which change was announced with exaltation. What trend would be set for the beach season? What novelties would arrive on hats? What would be the invention for wedding dresses?

From the center of Europe, the signals of transformation and the haste to bring the good news were sent. "The increase in the subscription to this fashion newspaper made in Paris, and therefore, without rival in its genre, imposes duties to the company that hastens to satisfy them (…) we do not neglect anything to make incessant improvements to a publication", it announced in 1885.[2] The reports did not only refer to textiles, clothing, apparel, and perfumery. Novelty being the hallmark of its communication, the newspaper took great pains to showcase household items that

[2] Advertisement in *La moda de la elegancia parisiense*, July 1, 1885, p. 200.

acquired the status of modern: the spoon-knife. "The unusual zeal for novelty, which seems the latest expression of English inventors, has produced a new object in which two necessities of the table are satisfied. It consists simply of an ordinary knife with its handle, and at the end of its blade it carries an appendage with the hollow of the spoon."[3]

References to other countries multiplied in the weekly, emphasizing the globality represented by fashion and the press itself. Parisian elegance for the beach season, for example, exhibited a daring model of Japanese robe for women, made with Chinese silk,[4] while the hats brighten up their look with straw from Italy.[5] The transatlantic trip that the newspaper undertook each week spoke of the magnitude of the project and its international sense (Figs. 1 and 2).

The publication, which circulated in Mexico until the early years of the twentieth century, maintained a narrative style dominated by the four-column text, balanced with detailed illustrations of models, patterns, materials and others. The quality of the latter, with the urgency of accuracy, required printing one page per issue in color, which allows us to appreciate the high level of the work, which also made editorial production more expensive. The newspaper arrived in times of cultural afrancesamiento [12, 50–52]. The Porfirian bourgeoisie, eager to know, ventured to experience one of the vital signs of fashion: imitation, propitiating the conditions to be like engravings and figurines exhibited in the European press. This was evidenced by the arrival of department stores in Mexico City, inspired by the commercial model of Paris. El Palacio de Hierro, the first of its kind, was a standard bearer of new forms of consumption in the modern city and understanding of fashion in Mexico [13, 32–35].[6]

- **See you on TV**

Since the late 1940s, the broadcast of fashion segments was common on American television. Faye Emerson did runway show narrations for NBC in 1948 as a public interest event, while Adelaide Hawley presented the 30-min broadcast of "Fashions on Parade" between 1948 and 1949 for DuMont and ABC ([14], 166). The screen slots were gradually expanded. In 1951, CBS presented the show "Fashion Magic" hosted by Arlene Francis and Robin Chandler. The show was advertised in the fashion section of *Life* magazine,[7] as synonymous with entertainment and current events.

The TVDAYS platform, on the other hand, preserves fragments of "lost fashion shows TV" from the 1950s, lost programs whose exact date and origin we cannot date, but which allow us to recognize practices, styles and visual and discursive trends of

---

[3] *The fashion of Parisian elegance,* "Knife and spoon in one piece", September 4, 1985, p. 199.

[4] *The fashion of Parisian elegance*, August 1, 1885.

[5] *The fashion of Parisian elegance*, September 4, 1885.

[6] Its production and sales promotion strategy came to include the hiring of French personnel in the last decade of the nineteenth century (Sánchez 2017). Similar segments are available on YouTube, from British programs in the same era. One of them shows Dior's London catwalks in 1950. https://www.youtube.com/watch?v=YOAGTKgA7-4.

[7] Advertisement in Life, August 27, 1951. p. 55.

LA MODA DE LA ELEGANCIA PARISIENSE DEL CORREO DE ULTRAMAR

1885 — 1er Nº DE JULIO

ANUNCIOS : la linea. . . . . francos 0 75
RECLAMOS : la linea. . . . . idem 2 50

PARIS
DE MOURGUES
ADMINISTRADOR-DIRECTOR-GERENTE DE LA SOCIEDAD
DE LA UNION DE LA PRENSA HISPANO-AMERICANA
Avenue de l'Opéra, 14, Paris

TOMO XVII — Nº 25
48 números 24 Patrones 48 Figurines al año
PARA LAS SUSCRICIONES DIRIGIRSE Á LOS AGENTES DEL « CORREO DE ULTRAMAR »

Crónica de la moda

SUMARIO. — Las novedades para las fiestas del bosque. — El modelo « Bello Silvano ». — Un vestido de encaje de Génova tejido de oro. — Una manteleta de extraordinaria gracia. — Sombreros elegantes. — La lencería de lujo. — La moda en las habitaciones parisienses.

La elegancia parisiense ha reservado este año, como de costumbre, todas sus novedades para la fiesta del « gran premio ». Entre tanto modelo como hemos visto, señalaremos algunas muestras. Hé aquí en esta primera página un modelo llamado « Bello Silvano » hecho de paño gris plomo y encaje de lana. La pechera es de lana y una pequeña drapería de siciliana plegada guarnece la pechera de encaje y acaba en punta en el borde de la faldeta. Dos vueltas en forma de oreja de perro adornan los lados del pecho sujetas con botones La faldeta viene en punta y sube abriéndose hácia las caderas. Un volante de encaje redondea el contorno de la faldeta. El costado delantero termina con una vuelta cubierta de siciliana que cae sobre la faldeta de espalda. El costado de espalda y la espalda son cortos, al mismo nivel que la faldeta del delantero, y para completarlos se añade una faldeta de siciliana plegada en el contorno de la prenda. Cuello derecho de terciopelo y manga de codo con pequeña bocamanga de siciliana. Sombrero redondo de paja satinada con plumas y galon de oro formando muchas lazadas.

Son tan numerosos los modelos dignos de citarse que verdaderamente la eleccion es difícil. Entre cien creaciones à cual mas seductoras nos fijaremos en estas dos que merecen seguramente los honores de la crónica. Primeramente un vestido de encaje de Génova tejido de oro, y siciliana azul de Francia. Falda de encaje sobre un viso de seda blanca, y á guisa de túnica un verdadero delantal de siciliana. Este delantal forma una pechera prendida con dos alfileres de azabache y abullonada al talle. A la izquierda cae derecha, un poco corta, y luego se prolonga à la derecha y viene á formar draperías sobre la cadera hasta el recogido que adorna la parte de detrás de la falda. El cuerpo de siciliana está guarnecido por delante con un gran fichu de encaje sobre el que se prende la pechera, y en los pliegues del fichu de encaje aparece una pequeña camiseta de crespon azul. Una cinta de moaré azul ata el delantal á la izquierda y da una gran lazada que cae sobre la falda. Manga de siciliana recortada en dos ondas bajo las cuales asoma un encaje, El sombrero propio para este vestido es un paillasson azul con guirnalda de pétalos de rosas y grueso lazo mariposa de cinta de moaré azul de Francia.

El otro modelo es de gasa heliotropo y faya verde. Falda de faya. Sobre el delantero cae un paño derecho de gasa heliotropo. Toda la falda de faya queda así velada de gasa; pero á cada lado se muestra un adorno de faya bordado de perlas verdes. Un cinturon de perlas da vuelta al talle y cae en ancho motivo á la Valois, hasta el bajo de la falda. El cuerpo es de faya velado de gasa, se cruza y se abotona à la izquierda, y le adornan por delante y por detrás dos tirantes de perlas, anchos en los hombros y que acaban en punta en el talle. Cuello de terciopelo y manga de gasa con jockey de perlas.

Con este vestido de alta distincion se lleva la sencilla y preciosa manteleta que figura el doble grabado número 2. Es una prenda hecha con nada, un hermoso volante de encaje rizado y un cinturon de azaba-

Nº 1.

Fig. 1 Front page of La moda de la elegancia parisiense, September 4, 1885

Fig. 2 Color plate La moda de la elegancia parisiense, September 4, 1885

fashion in the new media. Among the videos stands out a Christmas exhibition, which replaced the long catwalks with a studio decorated with a sumptuous Christmas tree, from which hung pearl necklaces. While a *voiceover* narrated the outfits and offered a price list, models showed off their "elegant and sophisticated" fur coats and jackets for the season, matching refined jewelry and beaded dresses.[8]

Televised fashion was a novelty within a novelty. While cinema had already brought moving images to exhibitions, as well as imposing trends through its artists, TV offered unique opportunities for the industry: domesticity [15, 18–19] and advertising. From its beginnings, the technical and business development of television aimed to ensure the cheapness of receivers, a luxury item in the forties and fifties, to stimulate its incursion into homes and, with it, the sale of advertising space that would make the industry profitable. This conquest took decades, but it allowed the new technology to acquire notoriety in the family environment, multiply messages and create its own language of communication. Television offered a narrative based on experience and closeness to the public. Bringing fashion to the home through screens was a commercial and creative desire for designers and entrepreneurs. Seeing the most daring outfits live and direct, without leaving home, on women and men models redefined the relationship of the common individual with fashion (Fig. 3).

The expansion of television sets in Europe and America in the sixties and seventies, combined with the consolidation of a middle class with greater purchasing power and a growing mass culture. The media convergence [10] accentuated by television represented for fashion the possibility of reinforcing two of its main characteristics: repetition and massification. With time and technical advances, the ability to reproduce images with sound constantly to a growing audience strengthened the brotherhood between the world of fashion and the small screen, consumption and desire. The arrival of color added realism to the broadcasting strategy. While satellite transmissions made it possible to know the novelty more quickly, from remote geographical areas. Now, if at the end of the nineteenth century the line to follow came from France, in the middle of the twentieth century, the reference was the United States. An *American way of life* was then reinforced in Latin America favoring the textile sector, already adapted to industrial mass production and imitation.

- **The digital stir**

If the world of television was strategic as a communicative scenario for fashion, the digital culture has been challenging. On March 24, 2022, Metaverso Fashion Week began, organized by the Decentraland platform, with the participation of the most prestigious clothing brands in the world. Among the activities of the day was the Fashion District, a virtual environment organized in neighborhoods where fashion shows and pop-ups—temporary stores with exclusive offers—are offered to passers-by, potential buyers. With the aim of dressing an avatar with high fashion garments, users sometimes compete for the ephemeral novelty of the season, paying large sums of money.

[8] TVDAYS videos are available on its official Youtube channel. https://www.youtube.com/watch?v=ZuoQyy_h0m0.

Fig. 3 Image captures from "Lost fashion show TV". Presumably owned by NBC, 1950s. TVDAYS.COM Youtube channel. At: https://www.youtube.com/watch?v=ZuoQyy_h0m0

The experience implies a transgressive notion of time and space compared to that perceived with television, which allowed the remote transmission of moving images, or that exposed in the interoceanic press, which risked sending printed copies from Europe to America once a week. The need for immediacy makes online resources indispensable for the fashion business, almost a guarantee of existence, while the physical space is eliminated so that the *e-commerce* can play in the most attractive possibilities of buying and selling. Paradox is part of the bet. While designers–usually younger–are opting for virtual pieces that reduce the environmental impact of textile production and exploitation in sweatshops, the big brands are promoting immersive experiences with the consumer as a potentiator of a business model disrupted by the Covid SAR-2 pandemic. In an event like Metaverso Fashion Week, the disposition for sumptuous consumption that emphasizes exclusivity and individuality, soars. At the same time it promotes the emergence of new creatives and achieves, through alternative platforms, the democratization of certain experiences, services and products. There is no single influence here, as was the case in the late 19th or mid-twentieth century. But what does the digital world imprint on the most common characteristics of fashion? The answer is: speed. The acceleration of change processes, as discussed above, was amplified by the incursion of the internet into societies. Just as the urgency for novelty grew both in those who produce it and in those who consume it, the time for that novelty to be integrated into our daily lives and, almost at the same time, become obsolete, decreased.

5 Future Scenarios of Fashion and Its Cosmotechnics

"Passions, conflicts and tensions may be immaterial, but design and art practice can manifest their presence" [16].

In the previous section, we analyzed examples and historical moments where the relationship of fashion with mass communication and the dominant cosmotechnics was expressed. In the same way, we observed how the interrelation of technology with the means of communication and production accelerated and expanded change and the imprint of novelty in the fashion sector, understanding it as a social phenomenon.

The temporality in the change of these cosmotechnics was about 70 years: the first, in the last two decades of the nineteenth century; the second, in the middle of the twentieth century and the third in the present time. Will we witness a new temporality? What tensions do future possibilities and past processes hold? Against this background, is it viable to speak of disruptive futures or of continuity?

Futures studies tell us that the future is a multiplicity of possibilities, not a determined destiny that is passively arrived at. Thus, the act of imagining different possible futures is a kind of creative and political agency, where decisions are made that lead us in one direction or another [17]. Even if one follows the passive path toward destruction, a constant nowadays, this passive attitude does not remove responsibility from the subject, since this inaction and inertia is also a decision that leads to defuturization [18].

Although the possible futures of fashion could be hundreds, as Amy Twigger Holroyd's Fashion Fictions project[2] demonstrates, in this case we propose two opposite extremes for illustrative and pedagogical purposes, a utopia and a dystopia where fashion, mass communication and cosmotécnica are articulated. It is not about reducing reality to two opposite poles, which have no intersections, we only assume a game where the contrast is deepened to think about the gray areas and contradictions. We want this section to function as a political and participatory exercise where the reader is invited to imagine and position himself within the spectrum that opens up between two antagonistic extremes.

On one shore, dystopia, an undesirable but probable future based on the prevailing conditions of the present time; and on the other, utopia, an ideal and desirable scenario, which, although improbable to achieve, presents a vision of the future that allows us to imagine that things could be different. The following is an exercise in imagination for our readers:

6 Dystopia: Linear Path of Today's Defuturizing Hegemony

In this scenario, the technologies that accelerated fashion in the past and were administered to the public to encourage consumption enter a *hyper*, much more rushed and globally unified state, driven by more immersive and present social networks and media.

This speed does not go unnoticed in the processes of social distinction, where it is increasingly evident and rejected who is not up to date with fashion. The dynamic of imitation and distinction has become a fierce dynamic of inclusion and exclusion, aggravating social antagonisms, where the dominant fashions have established an aesthetically and culturally totalitarian state: whoever is not up to date is socially marginalized.

In this spectrum, the desire for consumption and distinction is greater, so much of life is dedicated to achieving greater income or resources to achieve the desired novelty. This hyper-state acceleration has made the meanings attributed to products even more ephemeral, so that once they are detached from the trend that brought them, when they are displaced by the next one, they are immediately rejected, becoming massive waste.

Entire nations have been marginalized [19] from the hyper-fashion totalitarian state of the marketplace, migrating to constantly discarded product dumps. The old sustainable impulse has been totally excluded as a hindrance to technological and economic progress. Territories affected by lack of water, soil erosion or natural disasters have been abandoned and large global migrations are a constant that is often oppressed or treated as a humanitarian emergency.

Social atomization consolidated individualism to the extreme. What best allows an acceptable social link between individuals are the current practices of fashion, while they last in a state of novelty: the way to eat, dress, entertain oneself, exercise or love oneself is only acceptable according to the fashion of the moment, through the brands of corporations, which are already more powerful than the States.

The need to exercise individual uniqueness for social distinction is now supplied by markets. Identity is a consumable, for which there are customizable packages that can be purchased to choose a lifestyle with all the products and practices that surround it, such as the *yogi-geek-monocarnivore-mystic* or the *multi-planetary-interspecies-fluid-performer*.

Global neoliberalism has finally reached a state of purity where states serve corporations. The macroeconomic figures of the hyper states have never been better and the population is subjected to the desire for consumption. Priority social values are linked to technology and acceleration, where efficiency, precision and momentum replace the idea of conservation, slowness and reflexivity. Change is communicated at all times in public and private areas of individual and collective life, with no possibility of disconnection. Social networks and mass media are expanded out of gadgets, cell phones and computers. Now all devices, vehicles, clothing and public spaces are connected to the internet and communicating the news non-stop.

Cosmotechnics is a disregard for nature as archaic and unpredictable, with a blind trust in technology and progress. The new ethic is bio-renewing rather than bio-conservative, so transcending and mastering nature into modified products or beings is best.

7 Utopia: Other Fashions for Life

"When fashion manifests creativity, respect, allegiance, or membership, the relationships that it fosters are potentially democratic" [20].

As the states proved ineffective in addressing the environmental disaster, and after the global anti-corporatist rebellion, there was a shift to post-capitalist communitarian social structures. These smaller-scale community societies, now increasingly present around the world, have diversified their ways of living in contrast to the old hegemonic regime. Regionalisms are more important than the remaining global traits, as planetary and inter-relational sustainability have been prioritized, limiting international travel, exports and exploitation to territories. Social and environmental restoration processes are visible in all parts of the world, where forests and jungles are rehabilitated and rivers and seas are purified. For the first time in decades, global warming is being reversed and the ozone layer is being restored. In the midst of this panorama, the phenomenon of fashion has been reconfigured as a means to exercise freedom, which is no longer understood only as consumption, and to practice relational collectivity in a creative and participatory way.

In these conditions of global degrowth for sustainability, the speed of production, marketing and consumption were dramatically slowed down, so that permanence and care are the dominant values [21]. The processes of repair and recovery are an essential part of fashion, so that inherited or shared products are constantly intervened, altered and personalized.

There are still international brands and media that contribute products and spaces, but they are not the center of consumption or fashion. Local fashions spread more slowly through multiple analog, digital and hybrid channels at the community and intercommunity level. These mass and smaller scale media, in addition to communicating novelties, provide tools to practice fashion, they are "fashion-abilities" [22]. In digital media the code is open and emancipated from formerly dominant corporations such as Meta or Google. Garment patterns, 3D printing files, fabric diagrams, software and recipes for natural cosmetics are a central part of the content of the media, as they allow appropriation, intervention and then re-communication in a continuum of collective and creative novelties.

In this deceleration that promotes permanence and continuity, the multiple fashions and the designs that accompany them promote more meaningful relationships with the products, since they cannot be quickly discarded, and the active participation of people in the process makes the link more intimate and valued.

Customization is commonplace and social practices are renewably heterogeneous, so that there is no longer one global aesthetic or culture, but many. Fashion ceased to be imperative, corporate and vertical, being now cooperative, communitarian and horizontal, a means to redistribute power and democratize social relations, [20] while continuing to mediate between imitation and social distinction, which is exercised through the personalization and alteration of practices and products, promoting through fashion "to present the potential of its best self" [16].

There are movements of cultural recovery and restoration that constitute many of the centers and trends of fashions, such as regional crafts, gastronomy and architecture, which maintain their validity through sharing and generating communities that practice and renew them.

Cosmotechnics is interrelational and ecocentric, humanity and technology are understood as part of nature and changes occur at its pace. Technology is a tool and not an end, whose main function is for repair and conservation, products and media are used and changed freely, promoting meaningful and creative links. The change that brings the novelties in fashion now is the desire for permanence, collectivity, interrelatedness and care.

Contrasting these narrative extremes of fashion, it is often difficult to project possible utopias or a radical redirection other than the obvious consequences of the present time. Much of the collective fictional imaginary is geared toward end-of-the-world narratives, so dystopia seems more destiny than choice, just a matter of choosing a possible end: epidemics, zombies, uncontrollable warming, violent mutations, among many others. In this posture, there is a cynicism that justifies the continuity of pernicious and destructive practices, which we continually reinforce. Can we generate new narratives that are sustained by other cosmotechnics and a more democratic relationship with mass communication? Is fashion necessarily a capitalist and modern issue linked to planetary deterioration? Can fashion be a means of emancipation that reproduces counter-hegemonic and futurizing practices and products, an echo of another moral and technological order? How can change, novelty and imitation in fashion be edifying? An integral part of the solution and not the problem?

8 Conclusions

> Under the reign of total fashion, the spirit is less firm, but more receptive to criticism, less stable, but more tolerant, less sure of itself, but more open to difference, to proof, to the argumentation of the other. (Gilles Lipovetsky, 1994, p. 296).

The introduction of the term cosmotechnics allows us a different analysis of the structures underlying the operation of the phenomenon of fashion. Unlike ideology, which can be understood as a system of ideas that shape the artificial world, the cosmotechnics of a context reveals a moral order that through technology transforms the environment. It is an articulating concept that by involving morality necessarily questions the validity of an action, removes the neutral interpretation of technology as just a tool and reveals what we believe about the world by how we transform it.

Modern fashion is itself an engine of transformation, contagious and expandable, and as we have described above, ever accelerating and growing. The political implication of revealing its current and other possible future cosmotechnics implies a determination to choose and direct those changes from its internal structures in which we are singularly and collectively involved. The other possible fashions that do redirect toward a futurization result from other cosmtechniques. Trying to polish the moral and technical surface of current fashion processes would only give a sheen of partial righteousness, since it operates largely coupled with the capitalist apparatus of exploitation, in the background of which there is a disregard for nature, a priority for market profitability and the large capitals that fashion pushes to consumption through novelties.

These changes can only be shared through mass media, conventional or non-conventional, analog or digital. Hence the relevance of their intervention over time, changing and strategic. But if the media belong to a few global corporations where the interest for commercial efficiency prevails over social responsibility and relevance, futurizing change is prevented. The journey through the examples and historical moments exposed in this text allowed us to identify dominant ways in which Western societies have interacted with fashion from the imprint of novelty and change, without questioning its consequences or exclusion schemes. We do not intend to demonize the action of the media, nor to ignore its versatility and complexity, but we emphasize that its technical transformation, which allowed the immediacy of remote messages and the expansion of coverage, accelerated, deepened or enhanced the imprint of novelty and change. With time and the growth of the industry, these factors became imperatives, almost blind faith for some social and economic sectors, and not an attribute for collective improvement, overshadowing activities more respectful of the environment, diversity and resistance, without losing the creative and ethical impetus or the necessary mediation of a commercial exchange.

To question current technologies, to diversify them and even to hack them is to open the way to other cosmotechnics, which are the operative basis on which other local fashions are developed, which are not of interest or convenient to the unified monotechnological globality. The complexity of the civilizational crisis in which we find ourselves cannot be simplified or resolved, considering that current fashion constitutes the accelerated civilizational order in which we live [23]. And although the impact of fashion within this crisis cannot be resolved, we can generate adaptive actions that can diminish or neutralize its negative impact. Part of the transcendence of fashion lies in the action of the person to express his or her identity, which both distinguishes him or her from others and integrates him or her with others, in a continuum of updates. In this spectrum, fashion can be an engine of agency, rather than passive consumption, where a critical determination for change is exercised.

This global monotechnology was even clearer in the pandemic period, where messaging and communications were managed by a few global companies, generating a homogenizing, passive and unimaginative effect. Fashion is not something we choose, it is something that happens to us, so it is not about erecting another "master" to dictate what to follow, but about dismissing this dominant market "master" for the sake of the market, to allow for a diverse, creative and shared ecosystem of other fashions. Autonomous from the dominant media, but framed in communities where other codes of change and novelty in social practices are generated and shared, where renewal and change are promoted, not by consumption, but by creative, singular, collective and interrelated expression.

In this context, the effect of other fashions for life can be subversive against the dominant values: where the new that the market offers pretends to be always the best, the patched, intervened or hacked can be the preferable. Where in hegemonic fashions the products and practices are provided by large corporations and there is a recognition of these brands and corporations, the singular and collective rejection of these, replaced by an affective relationship with small local productions is a fashionable practice for life. Changes, creative and not passive, can come to the rhythm of life, to the speed of prolonged and shared use of products, to the change of seasons, to the hegemonic counter trend, to the cultural and aesthetic update promoted by the shared human creative impulse. If these futurizing practices are not exercised by shared fashions, with a viral effect, they could soon be eliminated without having generated any social impact. That they happen as fashions implies a pleasure in exercising and sharing them, a desire for renewal, updating and encounter with the other. Fashion is an engine of identity change, to live it freely, communally and creatively, it can be anotXher fashion, for life.

References

1. Escobar A (2016) Autonomy and design—the realization of the communal. Editorial Universidad del Cauca, Popayán
2. Simmel G (2000) Fashion. J Artif Soc Soc Simul 10(1):130–155
3. Lewis VD (2021) Using autopoiesis to discover the birth of fashion. Fashion Style Popular Culture 8(1):97–107
4. Inglis D (2021) Is fashion universal: towards a sociology of photo-fashion. Nova Sci Pub 34:1–34
5. Kawamura Y (2005) Fashion-ology an introduction to fashion studies. Berg, Oxford, UK
6. Ekardt P (2020) Benjamin on fashion. Great Britain: Bloomsbury
7. Jung E (2019) A dialectical journey through fashion and philosophy. Springer, Singapore

8. Hui Y (2016) The question concerning technology in China: an essay in cosmotechnics. Urbanomic media ltd
9. Hui Y (2020) Fragmentar el futuro: ensayos sobre tecnodiversidad 1st ed.—Ciudad Autónoma de Buenos Aires: Caja Negra
10. Jenkins H (2006) Convergence culture. Paidos, Barcelona
11. Kerryn Higgs. January 11, 2021. A Brief History of Consumer Culture. https://thereader.mitpress.mit.edu/a-brief-history-of-consumer-culture/
12. Vigneron MC (2016) Imaginary of the everyday. Afrancesamiento y vida burguesa en México, 1880–1920. Benemérita Universidad Autónoma de Puebla, Puebla. https://repositorioinstitucional.buap.mx/handle/20.500.12371/2206?show=full
13. Sánchez Parra, Jenny Cristina. 2017. Novelty and tradition: department stores in Mexico City and their influence on consumption, 1891–1915, El Colegio de México, Mexico City. https://repositorio.colmex.mx/concern/theses/j67313970?locale=es. https://repositorio.colmex.mx/concern/theses/j67313970?locale=es
14. Halper D (2014) Invisible stars: a social history of women in American broadcasting. Routledge
15. Silverstone R (2007) La moral de los medios de comunicación. Amorrortu editores, Buenos Aires
16. Von Busch O (2020) The Psychopolitics of fashion. Great Britain: Bloomsbury
17. Forum for the Future. July 04, 2022. https://www.fashionfutures2030.com/ff2030/scenarios/scenario-1/10415/ff2030_full_scenarios.pdf
18. Fry T (2020) Defuturing: a new design philosophy. Great Britain: Bloomsbury
19. Niessen S (2020) Fashion, its sacrifice zone, and sustainability. Fash Theory 24(6):859–877
20. Miller J (2005) Fashion and democratic relationships. Univ Chicago Press Behalf Northeastern Polit Sci Assoc 37(1):3–23
21. Fletcher K, Tham M (2021) Earth logic research-action plan for fashion. The JJ Charitable Trust, London
22. Von Busch, Otto. (2019) A crisis of power in fashion: is there agency for change? Fash Pract 11(3):302–308
23. Brand S (1999) The Clok of the long now. Time and responsibility. New York, USA: Basic Books

Telling the Indigenous Ghanaian Fashion Cosmovision: The Case of Royal Ahenema Sandals

Osuanyi Quaicoo Essel

Abstract Fashion is practiced by all cultures. The people of Ghana practiced bespoke fashion culture that has evolved overtime but still remains relevant in their lives in contemporary times. In the precolonial times, the bespoke fashion was practiced on ethnic fashion basis which begot the ethnic fashion art tradition. Ethnic groups created and exhibited unique and interesting repertoire of fashion art for everyday and special occasion usage. This gave rise to unique Ghanaian fashion classics. In effect, Ghana has a robust indigenous fashion system influenced by her fashion cosmovision different from what pertains in Western fashion. But, does not sharing the same fashion cosmovision with other nations make Ghana's fashion inferior or peripheral in global fashion? Through careful observation of available data on Ghanaian fashion practice in precolonial, colonial, postcolonial and contemporary times, this chapter presents Indigenous Fashion System Model (IFSM) and Bespoke Fashion Diffusion and Consumption (BeFDaC) model that describe how different fashions are co-created, diffused, consumed, rejected and changed in a bespoke fashion milieu using the indigenous and contemporary knowledge systems. How the Indigenous Fashion System Model (IFSM) operates is discussed using the case of royal ahenema sandals, an indigenous Ghanaian fashion accessory and yet international. The chapter also established the evolutions that have characterized the indigenous fashion system in terms of gender issues in the production and commodification of royal ahenema sandals and challenges the labeling of otherness in reference to non-Western fashion with particular reference to Ghana. It recommended that these models are tested by academics in the African contexts to check its trustworthiness and workability.

O. Q. Essel (✉)
Department of Textiles & Fashion Education, Centre for Research in Culture & Creative Arts, University of Education, Winneba, Winneba, Ghana
e-mail: oqessel@uew.edu.gh

M. Á. Gardetti and R. P. Larios-Francia (eds.), *Sustainability Challenges in the Fashion Industry*, Sustainable Textiles: Production, Processing, Manufacturing & Chemistry,
https://doi.org/10.1007/978-981-99-0349-8_16

1 Introduction

Fashion permeates all human cultures. But conception and approach to fashion differ from culture to culture. How a particular culture conceives fashion ideas and inspiration may be rooted in their way of life, including their beliefs and practices, religion, philosophical thoughts, and the general cultural orientation and understanding. The fashion practices evolved by a particular culture shapes their physical outlook. The climatic conditions that surround or pertain in a place play a major role in fashion ideas development and choice. This implies that getting a better understanding of the fashion and culture of a particular place, requires thorough examination of their climatic conditions, philosophical thoughts, religious beliefs and practices, and visual culture as a whole. These variables make up the cosmovision of a people. Cosmovision encompasses both the collective and individual conception of indigenous people in relation to their physical and spiritual world and environment that guards their lives [1].

The cosmovision of peoples in the world, generally, are never the same. Acceptance of this natural phenomenon offers a better worldview of how to approach and study the cosmovision of a people. This helps to iron out personal biases and informs unbiased ideological and perspectival grounding to developing an understanding of fashion practice of a culture. This chapter broadly tackles the indigenous fashion cosmovision of Ghana with the view of exploring how it operates and informs the fashion practices of her people. Through a careful observation of the fashion practice of the people and extensive data collected from the field, this chapter specifically presents two separate models: indigenous fashion system, and bespoke fashion model, using Ghana as a case study. For this reason, the deductions made were borne out of the cosmovision of the Ghanaian experience of fashion. The royal Ahenema sandals were used to illustrate how the indigenous fashion system model proposed operates. The chapter also establishes the evolutions that have characterized the indigenous fashion system in terms of gender issues in the production and commodification of royal ahenema sandals and challenges the labeling of otherness in reference to non-Western fashion with particular reference to Ghana.

The development of the bespoke fashion model and the indigenous fashion system was informed by the grounded theory method of qualitative research approach, which is aimed at building a theory inductively in a systematic way based on what was observed [2, 3]. Data from tailors/dressmakers and designers, fashion consumers, dress-fashion poster producers, millinery and accessory producers were used. These categories of data were supplemented with existing studies on Ghana fashion industry. Direct observation and in-depth interviews were the instrumentations used in collecting data.

2 Fashion System Power Play

A survey of dominant scholarship on fashion systems rigidly compartmentalizes fashion into a stiff binary as West and Non-West. Fashion from Europe and North America belongs to the West, while fashion outside Europe and North America fall within the non-West ([4], p. xvi, [5], p. 3). This position boxes fashion from Asiatic, African, South American, amongst others, as other fashion. Apart from sounding self-glorifying on part of those who posit that their fashion as second to none, such binary fails to acknowledge the diverse fashion systems that 'have been, and are, located all around the world, and that these have been developing in conjunction, competition, collaboration, and independently from the European fashion system' ([6], p. 1). The other fashion is seen as recent and something orchestrated by globalization. The argument remains that the other fashion has not assumed a global scale [7, 8] and is usually referenced in Western scholarship as costume or traditional, detaching them from the realms of fashion.

Fashion inspirations are diverse and sought by designers from different parts of the world. It has a multicultural dimension. Fashion from Africa, Asia and the other so-called non-West fashion has inspired designs of the West and vice-versa. The hybridization of fashion borne out from borrowing inspiration from different cultures strengthened the position that a fashion could be local and yet global. Teunissen [9], p.11 defines the cultural borrowing of inspiration that creates glocalized designs as 'fusion fashion'. Irrespective of the terminological defect that could be ascribed to fashion from particular parts of the world to make it seem inferior, it is worthy of note that fashion systems are not universal. Different cultures have different fashion systems they operate. The belief systems, philosophy, cultural orientation and cosmic understanding influence fashion systems of different cultures.

2.1 Ghana's Indigenous Cosmic Worldview

Located on the Guinea Coast of West Africa, the republic of Ghana shares boundaries with Togo (East), Burkina Faso (North) and Côte d'Ivoire (West) and covers an area of 238,535 km2 (92,099 sq mi) with a population of 31 million people [10]. The country is composed of a mosaic of multi-ethnic groups [11, 12] some of which trace their roots to different parts of Africa. From about the eighth century [13, 14], many of the ethnic groups had settled in the country. Though Ghana adopted the common law system, its legal system is pluralistic consisting of customary and statutory systems as enshrined in Article 11 of the 1992 Constitution that talks about the embodiments of laws of Ghana. It makes reference to the constitution itself, enactments made by or under the authority of the Parliament established by the Constitution; any orders, rules and regulations made by any person or authority under a power conferred by the Constitution; the existing law; and the common law. As spelt out in the 1992 Constitution of Ghana (Articles 11, Sects. 2, 3 and 4, the common law of Ghana in this context is operationalised and comprises the rules of law generally known as the common law, the rules generally known as the doctrines of equity and the rules of customary law including those determined by the Superior Court of Judicature. It also defined customary law as the rules of law which by custom are applicable to particular communities in Ghana while existing law comprises the written and unwritten laws of Ghana as they existed immediately before the coming into force of the 1992 Constitution, and any Act, Decree, Law or statutory instrument issued or made before that date, which is to come into force on or after that date.

Despite the plurality of ethnic groups in Ghana, they share a common worldview of fashion and culture largely influenced by the climatic conditions of the country, philosophical thoughts, belief systems and practices and visual culture as a whole. As a tropical country, Ghana has two main seasons: the dry and wet seasons. Due to the sunny weather conditions and partly airy atmosphere, the selection of textiles for garments is preferably cotton-based, while the dress styles are created such that it allows more air to circulate around the wearer.

The ethnic groups also have belief in God, deities, animism, life-after-death, ancestral veneration, and engage in other cultural practices such as rites of passage (birth, puberty, marriage and death) and merrymaking events. The use of proverbs and aphorisms plays a central role in their daily lives. These beliefs and practices serve as a motivation for the artistic creation of fashion art objects and accessories. For example, due to their belief in animism, they used animal skin which they associate with power and authority to create sandals, amulets, bangles, millinery, among others, for use by society with the aim that the users would derive the same power of the animal. Animism has to do with the belief that God has bestowed special powers onto plants and animals, and that using their parts such as fur and leather has healing benefits or spiritual power and protection beside its aesthetic impetus.

As a former British colony, Ghana has a colonial history of fashion power politics. When the colonialists' arrived in the fifteenth century, they condemned the dress culture practices of the people labeling, them with negative terms such as uncouth, fetishistic, idolatrous and uncivilized [15–17]. Having painted a negative public image about the dress style of the people, the colonialist positioned her dress culture as a measure of civilization [15]. Wearing of colonialists' dress styles in sunny Ghana became a symbol of civilization and as a way of signaling one's erudite status [15, 18]. In supporting the standpoint that Africa is the hardest in terms of accepting Western norms and practices hook, line and sinker through the effect of globalization, ([19], p. 7) therefore questioned that:

We [African] should not look down on some aspects of our culture due to castigations and aspersions on us by those who are not part of the culture. To look down on one's culture is to discard what defines one and worst still discards one's capacity for development for one's development is tied to one's culture.

Obioha [19] argued that a lost culture is a lost society as well as an invaluable knowledge lost and therefore advocated for cultural renaissance. By this assertion, he advocates for the preservation of indigenous knowledge.

2.2 Developing an Indigenous Fashion System Model

Fashion in Ghana dates back to precolonial times. The people of Ghana practiced bespoke fashion culture that has evolved overtime but still remains relevant in their lives in contemporary times. In the precolonial times, the bespoke fashion was practiced on ethnic fashion basis which begot the ethnic fashion art tradition based purely on indigenous knowledge systems. Each ethnic group created and exhibited unique and interesting repertoire of fashion for everyday and special occasion usage.

Through careful scrutiny of archival documents, historical texts [20–24], field data, and observation of Ghanaian fashion practice in precolonial, colonial and post-colonial times [25], a model of indigenous fashion system is proposed. In this model, the conception and creation of ethnic fashion art are influenced by factors which could be internal and external (Fig. 1). The internal factors include all aspects of educational relevancies (cultural, social, political, economic, emotional, religious, philosophical, esthetical, etc.) that necessitate the creation of the fashion to serve its intended purpose within the purview, cosmic understanding and context of the culture that created the work. The internal factors are linked to the belief systems of the people (including belief in God, deities, animism, life-after-death, ancestral veneration) and practices which served as source of inspiration for fashion creation.

Multicultural adoption and appropriation constituted the external factors. Inter-ethnic cultural influences transpired because of trade, warfare, migration [15], intermarriages, interstate and inter-racial contacts which sparked cultural borrowing from these external sources in terms of ideas and inspirations for creative thinking and innovation. Fashion art that was produced included body arts, dress, millinery, hattery and other accessories. Operating from a bespoke perspective, the designs produced were distinctive and, in most cases, context- and concept-specific to suit particular

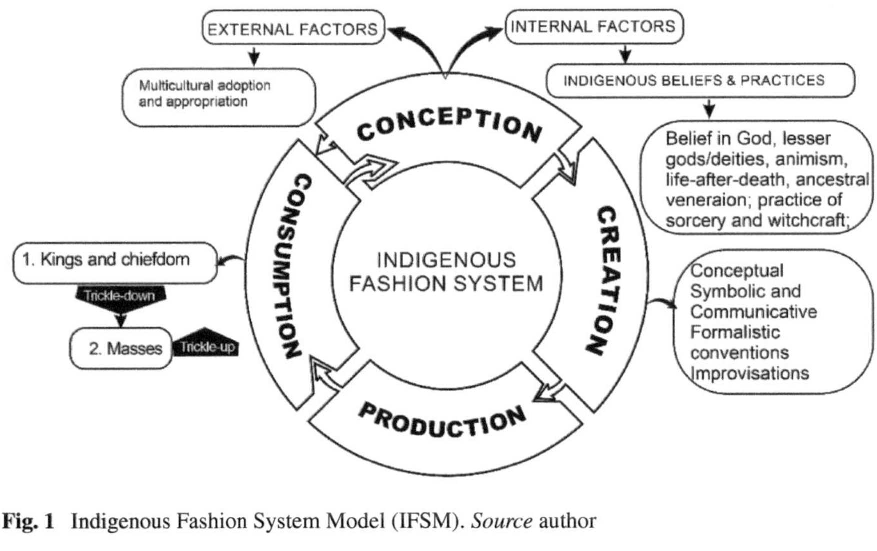

Fig. 1 Indigenous Fashion System Model (IFSM). *Source* author

occasion. No matter how simple the fashion object may look, once it is able to perform its intended function, it is considered good.

Using the case of royal ahenema sandals in the indigenous fashion system may be useful in explaining how it operates. Ahenema occupies a central place in the chieftaincy institution, customs and traditions of the chiefdom of Asante people and other Akan ethnicities of Ghana, Togo and Cote d'Ivoire [25]. It was created by the Asante people in 1700s as a utilitarian fashion object and later on ascribed symbolic and political significance based on the indigenous symbolisms used on them such as adinkra motif designs and proverbial ascriptions.

Adinkra symbol is worth a thousand words. Adinkra symbols were crafted with unique identity and Ghanaian 'philosophical thoughts and ideologies, cultural values and beliefs to the extent that they offer hints on Ghanaian worldview of cosmic understanding, religious and secular beliefs, flora and fauna plus their relationships with each other' ([26], p.31). Based on their symbolisms and language system, adinkra is grouped into symbols that depict celestial bodies, flora and fauna, human body parts and non-figurative shapes [27–29], which tells Ghanaian people's understanding and cosmic worldview. The Adinkra symbolisms used in the designing of the top part of the sandals help in determining its name. For example, the symbolism used in Fig. 2 is Gye Nyame which tells the religious attribution of the design connecting it to the supernatural power of God. Apart from the symbolism, it also has aesthetic importance. There are different designs of ahenema sandals (Fig. 3). The name of ahenema may be determined by indigenous symbolism attached to it, yet there are other fanciful ahenema designs worn for its own sake (Fig. 4). They fall within the category of fanciful designs.

When the designs are conceived, materials are sought to create the sandals. The creations may be conceptual, symbolic, communicative, communalistic, and functional. It may have the quality of formalistic conventions, and be based on improvisations in terms of materials, ideas and concepts. When a design was created and tested for its usability and comfortability, samples were produced for the chiefs/kings and queen mothers. Fashion artists lived in the courts of chiefs/kings and produced innovative designs for use by the chiefdom because they were highly revered by the society. The chiefdom used the product at durbars and other gatherings which trickled down to the masses. The fashion artist then produces for sale to the masses. With the indigenous fashion system, the trickle-down [30] theory operates widely for the products to spread to the people. In Asanteland, for example, His royal Highness Otumfour Osei Tutu II seems to have the best collection of ahenema designs. As part of the indigenous culture, it was customary and ethical to send to chiefs in the society the first portion of harvest from farm produce or artistic works as a form of reverence to the chief/king.

Though in contemporary times, such ways of showing respect to chiefs have dwindled tremendously, it has totally not wiped away. For example, the finest collection of kente fabrics in Ghana could be found in the collection of the Asante king. The king has sub-chiefs who ensure stocking of his wardrobe with the best collections. Besides, it is honorary and prestigious to fashion creators, for kings to develop interest in their fashion products. Fashion creators also create designs for the masses

Fig. 2 Gye Nyame ahenema sandal. Gye Nyame literally means 'Unless God'. It depicts the omnipotence of God. *Source* courtesy of author

Fig. 3 Ahenema designs inspired by Adinkra and proverbs. Its symbolism dictates its name. *Source* Courtesy of author

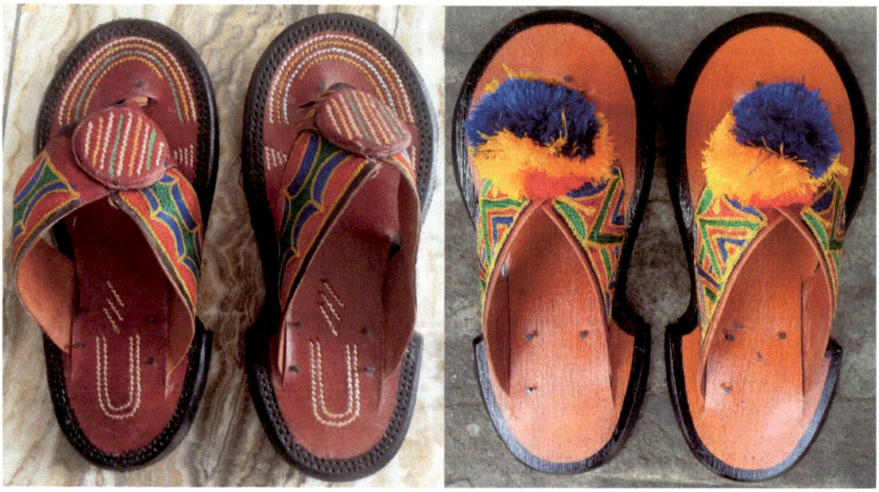

Fig. 4 Fanciful ahenema designs devoid of symbolic attributions

which gain acceptance to the chiefdom, and other people at top echelons of society. This brings into focus the trickle-up theory where fashion moves from the masses (streets) to high class in society.

3 Gender Issues in Ahenema Production

The art of ahenema sandals production took its root from the eighteenth century, during the reign of the fourth Asante king, Otumfuo Osei Kwadwo Okoawia who ruled from 1764 to 1777, and the queenship of Nana Konadu Yiadom I from 1768 to 1809 [25]. Since its inception, the design and production have been heavily dominated by males. For this reason, it became culturally accepted as a preserve for males. Ahenema designers are highly revered amongst the chiefdom. They were regarded as creative people who conceive designs for the chieftaincy institution.

In contemporary Ghana, the notion of ahenema being a preserve for males has revolutionized as some creative female designers make inroads in the design and production circles even in the heart of the Ashanti Region. One of the female creatives in ahenema shared her story:

I'm Obaapanin Adwowa Nyarkoa. I was born in Esaso (off the Barekese road) in Atwima Nwabigya District of the Asante Region of Ghana. I am Fifty-Seven years old. I have been in the Ahenema design business since 2004. My husband took me through the design rudiments and it has been very productive for me. When clients see me designing, they get excited and praise my designs. Even when the clients come to the display and sales shop, they usually prefer my designs. Some say, they have not seen a woman involved in the design and production of ahenema. For this reason, they are attracted to my designs as a female breaking barriers. They never say I have faulted Asante culture and tradition, rather, they say, what men can do women can do better! My clients are happy for me.

There have been shifting roles of participation in the production of ahenema. Once, a male-dominated occupation has opened its doors to females with interest without any discrimination, but rather admirable acceptance and patronage of their products.

4 Toward a Bespoke Fashion Diffusion and Consumption Model

The precolonial bespoke fashion culture of Ghana remains resolute due to its practice by thousands of creative bespoke designers, seamstresses and tailors. These tailors, seamstresses and designers create stunning outfits of high aesthetic quality ranging from haute couture, middle class and street fashion. With this practice, hundreds of new designs trickle onto Ghana's fashion markets on a daily basis giving consumers many options to choose from. Some of these creative bespoke fashion works such as hairstyles and garment designs are cataloged and published in fashion posters and magazines that are widely distributed and found in many tailoring shops in Ghana and its neighboring countries namely Togo, Cote d'ivoire, Burkina Faso and Nigeria [31].

The major forces that trigger bespoke are consumer preferences and quest for individualism. Intuitively, the preferences of the clientele are largely informed by the ethnic, multicultural or glocal ideals. This suggests that ethnic fashion, cross-cultural borrowing and glocal fashion displays are immediate reservoirs from which tailors/designers select, modify, recreate or introduce new designs for use. The next level which serves as a source of inspiration to look individualistic as well as medium for fashion dissemination is the print and electronic media, social gatherings and clients' own imaginations. The print media include newspapers, magazines, flyers, billboards and posters/calendars. Posters/calendars are found in almost every tailoring/dressmaking shop in Ghana even in the remotest village. The advent of television and the internet as part of the electronic media also contributes in this regard. The dissemination process is facilitated by availability of and accessibility to Image-based Communication Technologies (IbCTs) and electric power supply as pointed out by [31]. These are directly affected by consumers' socio-economic status and age characteristics. This is because consumers with little or no purchasing power could not afford IbCTs such as smart phones, laptops among others. Among persons 12 years and older, smart mobile phone (73.1%) is the most widely owned functional ICT device, and it is 10 times the proportion of persons possessing laptops in Ghana [10]. Smart mobile phone in the context of the 2021 Population and Housing Census was operationalized to mean a 'mobile phone device that performs many of the functions of a computer, typically having a touchscreen interface, Internet access, and an operating system capable of running apps such as Facebook, WhatsApp or YouTube' [10]. Kumi [32] reported that as of the end of 2016 about 82.5% of the population of Ghana had access to electricity, adding that the pace of electrification cannot guarantee the country universal access to electricity by the year 2020 unless there is an increase in the annual average electrification from 2.6 to 4.4%. Kumi's assertion is confirmed by the [10] report that the proportion of households that use electricity as the main source of lighting doubled in the last two decades from 43.8% in 2000 to 86.3% in 2021. However, access to electric power supply does not guarantee affordability as the tariff keeps soaring.

Besides, the adult (older) population seemed slow to adapt to the Image-based Communication Technologies (IbCTs) and therefore depend on what they considered to be less sophisticated avenues of image handling—posters/magazines. The multiples of image dissemination mediums used make fashion diffusion uniform, and increase the consumption rate once accepted. When a fashion becomes obsolete in the eyes of the consumers, it faces rejection which takes them back to the major forces of bespoke (consumer preferences and individualism. This sums up the Bespoke Fashion Creation, Diffusion and Consumption model (BeFDaC) (Fig. 5), which forms part of contemporary Ghana's fashion system.

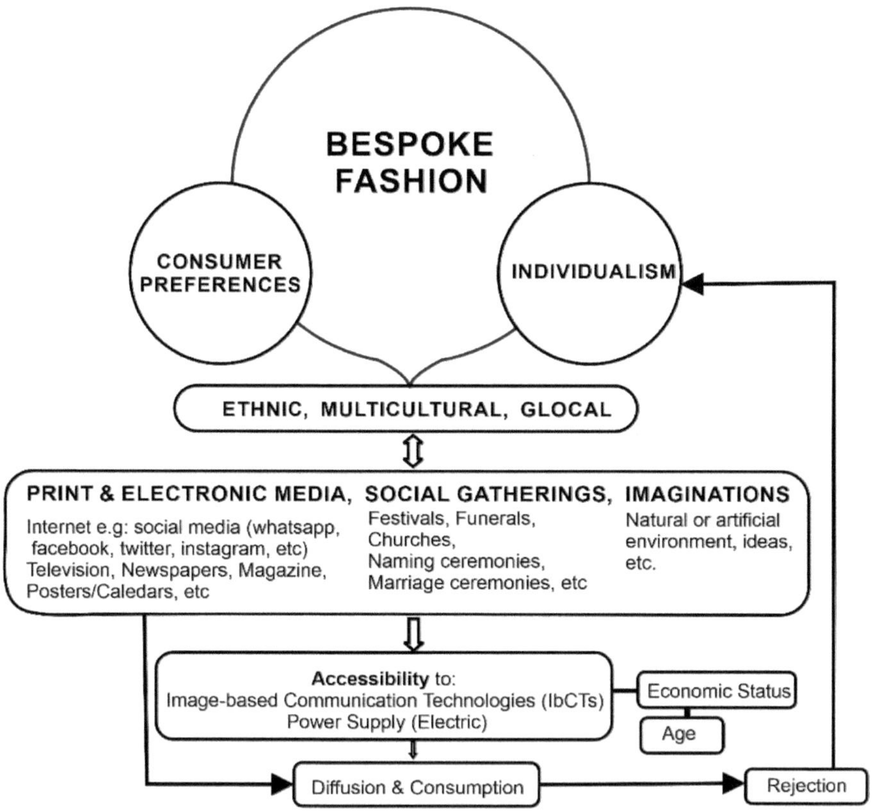

Fig. 5 Bespoke fashion creation, diffusion and consumption (BeFDaC) model. *Source* Author

5 Conclusions

The chapter proposed two models: Indigenous Fashion System Model (IFSM) which illustrates how indigenous people created and practised fashion within the primarily bespoke fashion environment using the case of Ghana. The Indigenous Fashion System Model (IFSM) has four major steps or stages which inform each other. The stages are conception, creation, production and consumption. Ethnic fashion art is influenced by factors which could be internal and external. The internal factors border on all aspects of educational relevancies (cultural, social, political, economic, emotional, religious, philosophical, aesthetical, etc.) and utilitarian functions which are mostly hinged on beliefs and practices valued by the people. The external factors touch on the multicultural adoption and appropriation caused by inter-ethnic, inter-state and inter-racial contacts and interactions which inspire fashion ideas and concepts borrowing. The creation stage factors indigenous knowledge systems which could be conceptual, symbolic, communicative, formalistic conventions, improvisation, communal or functional. Samples are created in this stage for testing after which production follows. Consumption is hooked to the trickle-down theory where the upper-class' choice of fashion influences the lower class; and the trickle up of the masses fashion choices. Using ahenema sandals, an indigenous Ghanaian fashion accessory and yet international as an example, illustrated the process through which the IFSM operates and challenges the labeling of otherness in reference to non-Western fashion. Again, the evolutions that characterized the indigenous fashion system in terms of gender issues in the production and commodification of royal ahenema sandals was established. The occupation is no longer the preserve of men. Design variations, uses and contexts of ahenema exemplified that fashion design in Ghana is by no means static.

The Bespoke Fashion Diffusion and Consumption (BeFDaC) model developed explained the socio-politico (social, economic, political and cultural) interaction that forms part of contemporary multi-ethnic Ghana's fashion system. The model describes how different fashions are diffused, consumed, rejected and changed in a bespoke fashion milieu using the indigenous and contemporary knowledge systems. It recommended that these models are tested by academics in the African contexts to check its trustworthiness and workability.

References

1. Mashingaidze S (2016) Cosmovision and African conservation philosophy: indigenous knowledge system perspective. Environ Econ 7(4):25–33. https://doi.org/10.21511/ee.07(4).2016.03
2. Schutt RK (2009) Investigating the social world: the process and practice of research, 6th edn. Pine Forge Press, United Kingdom
3. Strauss AL, Corbin J (1990) The basics of qualitative research: grounded theory procedures and techniques. Sage, Newbury Park, CA

4. Jansen MA (2021) Foreward. In: Cheang S, De Greef E, Takagi Y (eds), Rethinking fashion globalization. Great Britain: Bloomsbury Visual Arts, pp xvi–xvii
5. Welters L, Lillethun A (2018) Fashion history: a global view. Bloomsbury Academic, London
6. Jansen MA, Craik J (eds) (2016). Introduction. In: modern fashion traditions: negotiating tradition and modernity through fashion. Bloomsbury Academic, London, pp 1–24
7. Hansen KT (2004) The World in dress: anthropological perspectives on clothing, fashion and culture. Annu Rev Anthropol 33:369–392
8. Lillethun A, Welters L, Eicher JB (2012) (Re)Defining fashion. Dress 38:75–97
9. Teunissen J (2005) Global fashion, local tradition: over de globalizering van mode. In: Teunissen J, Brand J (eds) Global fashion, local tradition: over de globalizering van mode. Terra, Utrecht, pp 8–23
10. Ghana Statistical Service (2022) Ghana 2021 population and housing census, vol 3. General Report Highligts
11. Amoah E (2003) Traditional political system of Ghana: yesterday, today and the future. In: Awoyemi MO, Flolu EJ, Apreko K, Amenowode JZ (eds) Readings in arts, culture and social science education. Black Mask Ltd., Accra
12. McCaskie TC (2009) Asante origins, Egypt, and the near east: an idea and its history. In: Peterson DR, Macola G (eds) Recasting the past: history writing and political work in modern Africa. Ohio University Press, Athens, pp 125–148
13. Levtzion N (1980) Ancient Ghana and Mali. A Division of Holmes and Meier Publishers, Inc.
14. Salm SJ, Falola T (2002) Culture and customs of Ghana. Greenwood Press
15. Allman J (ed) (2004) Let your fashion be in line with our Ghanaian costume: nation, gender, and the politics of clothing in Nkrumah's Ghana. In: Fashioning Africa: power and the politics of dress. Indiana University Press, Bloomington
16. Akou MH (ed) (2004) Nationalism without a nation: understanding the dress of Somalia women in Minnesota. In: Fashioning Africa: power and the politics of dress. Indiana University Press, Bloomington
17. Nkrumah K (1963) The African genius. Speech delivered by Osagyefo Dr. Kwame Nkrumah, President of the Republic of Ghana, at the Opening of the Institute of African Studies on 25th October, 1963
18. Essel OQ (2017) Searchlight on Ghanaian iconic hands in the world of dress fashion design culture. (Unpublished PhD. Thesis), University of Education, Winneba, Ghana
19. Obioha UP (2010) Globalization and the future of African culture. Philos Papers Rev 2(1):1–8
20. Bowdich TE (1819) Mission from cape coast castle to Ashantee. Cass, London. https://doi.org/10.1017/CBO9781107444621
21. Rattray RS (1927) Religion and art in Ashanti. Oxford University Press
22. Antubam K (1963) Ghana's heritage of culture. Koehler & Amelang, Leipzag
23. Asihene EV (1978) Understanding the traditional art of Ghana. Associated University Presses, London
24. Fosu K (1994) Traditional art of Ghana. Dela Publications and Designs
25. Essel OQ (2021) Historical and sociocultural relevance of royal 'Ahenema' sandals in Asante culture. IOSR J Human Soc Sci (IOSR-JHSS) 26(04):27–34
26. Essel OQ, Opoku-Mensah I (2014) Pan-African artistic reflection in Kwame Nkrumah memorial park. Int J Afric Soc Cult Tradit 1(2):30–41
27. Arthur GFK (2001) Cloth as a metaphor: (Re)reading the adinkra cloth symbols of the Akan of Ghana. Centre for Indigenous Knowledge Systems, Legon
28. Danzy J (2009) Adinkra symbols: an ideographic writing system. (Published thesis). Graduate School. Stony Brook University. https://dspace.sunyconnect.suny.edu/bitstream/handle/1951/48176/000000570.sbu.pdf?sequence=2&isAllowed=y Accessed 10 June 2014
29. Ofori-Ansa K (1999) Meanings of symbols in adinkra cloth. Sankofa Educational Publications, USA, Maryland
30. King CW (1963) Fashion adoption: a rebuttal to the 'trickle down theory. In: Sproles GB (ed) Perspectives of fashion. Burgess Publishing Company, Minneapolis, MN, pp 31–39

31. Kraamer M, Essel OQ (2021) Creative collaborations between consumers and fashion designers: the role of fashion posters in urban and rural Ghana. In: Cheang S, de Greef E, Takagi Y (eds) Rethinking fashion globalisation. Bloomsbury, pp 191–207
32. Kumi EN (2017) The electricity situation in Ghana: challenges and opportunities. CGD Policy Paper. Center for Global Development, Washington, DC. https://www.cgdev.org/publication/electricity-situation-ghana-challenges-and-opportunities
33. Constitution of the Republic of Ghana [Ghana] (1993) https://www.refworld.org/docid/3ae6b5850.html Accessed 1 July 2022
34. Essel OQ (2013) Evolution of Ghana's dress culture (Unpublished MPhil Thesis). University of Education, Winneba, Ghana
35. Rabine L (2002) The global circulation of African fashion. Berg, UK

Artisans, Creativity, and Ethics: "Skill Regimes" in a Mumbai Fashion Export House

Matthew Raj Webb

Abstract In recent years, the position of craft-marked work within the field of fashion has become of interest to scholars, industry regulators, governments, and consumers, foregrounded in debates about production "ethics" and concerns over "sustainability," "transparency," and "decoloniality." Taking up these themes, this chapter explores an alternate sphere of ethical discourse among fashion producers in Mumbai centered on the regimes and hierarchies of "skill value" that underpin their transnational work. It asks: How do situated imaginaries about skill (quantitative) and skills (qualitative) in fashion production activities—what I call "skill regimes"—structure relations among participants in transnational design projects?

Sameer is arched over a low wooden workbench clasping a silver embroidery needle and thread. In a workshop in South Mumbai, he is putting the final touches to a line of dress designs that will be sold worldwide and attributed to the American-global fashion company Ralph Lauren. On the Ralph Lauren website, Sameer's work is represented to consumers as that of a "craftsman" and further qualified as "responsibly" and "sustainably" sourced. On the website of his employer in Mumbai, Sameer's skills are praised as those of a "master artisan," stemming from an "ancestral tradition of exquisite Indian textile art." Sameer is, however, only a second-generation textile and clothing practitioner. His father learned how to sew via a government training workshop in the 1970s. Moreover, he describes himself not as an artisan but as a "designer," proud of his creative contributions to this and many other items for globally-renown businesses.

In recent years, the position of craft-marked work within the field of fashion has become of interest to scholars, industry regulators, governments, and consumers, foregrounded in debates about production "ethics" [1, 54] and concerns over "sustainability," "transparency," and "decoloniality" [9, 28, 58, 67]. Taking up these themes, this chapter explores an alternate sphere of ethical discourse among fashion

M. R. Webb (✉)
New York University, New York, NY, USA
e-mail: matthew.webb@nyu.edu

M. Á. Gardetti and R. P. Larios-Francia (eds.), *Sustainability Challenges in the Fashion Industry*, Sustainable Textiles: Production, Processing, Manufacturing & Chemistry,
https://doi.org/10.1007/978-981-99-0349-8_17

producers in Mumbai centered on the regimes and hierarchies of "skill value" that underpin their transnational work. In asks: How do situated imaginaries about skill (quantitative) and skills (qualitative) in fashion production activities—what I call "skill regimes"—structure relations among participants in transnational design projects?[1] The data is drawn from fieldwork in a fashion export house in Mumbai in 2019, during which I observed the design and production of clothing identified primarily with American and European companies. I also draw on interviews with artisans, designers, and managers working at this site and two similar ones in Mumbai, conducted in-person and remotely over the following year.[2]

A focus on this Mumbai export house affords interrogation of several specific common assumptions about the relationship between "craft" and "design" as skilled practices and occupational identities (cf., [24, 26, 61]), as well as about the boundaries between Western and Indian fashion "worlds" (cf., [11, 47]).[3] These pairs of concepts are often presented as relational, within a schema of value defining craft as a "manual" skill and subsequently naturalized as "authentic" Indian cultural practice [13]. Correspondingly, design is presented as a "mental" skill and associated with "modern" and "creative" culture in the West. This ideology has been circulated through histories of colonial and orientalist representation, such as through literature and museums [32, 40]. It is sometimes bolstered by auto-essentializing representations produced within India, associating contemporary artisanal work generically with "tradition" and "village" life, often in patrimony to the Independence struggle [21, 22] or subsequent elite utopian nationalist projects [70].[4] But these invocations of craft value represent only one level of a relatively "high" cultural discourse through

[1] My approach deviates from scholarship engaging the concept of "skill regimes" in an economically instrumentalist and managerial sense (e.g., [68]). Rather than taking for granted the cross-cultural salience of skill categories (labels) and their associated qualities, I propose that we should pay attention to people's localized ideas and beliefs about skills, what I refer to as "skill values" [4, 50]. Methodologically, this requires attention to (i) how people label skilled practices in contexts of use and the meanings they associate with them, (ii) personal histories of skilled practice acquisition and innovation, and (iii) the structures of localized political economies through which skills are made available or unavailable. This framework for the study of "skill regimes"—that is, contexts with patterns of shared skill values—offers a bridge between objectivist and subjectivist theories of labor/work/action value (cf., [7, 10, 19, 39], inserting skill as a point of convergence between overlapping and interconnected contexts and processes.

[2] The timeframe of this research spanned the onset of the of the global coronavirus pandemic, an event that radically impacted workers along global fashion commodity chains. Throughout this time, Western companies' treatment of "informally" sub-contracted employees in India garnered global media attention (e.g., [55]).

[3] Ethnocentric distinctions between "Western" and "Indian" fashion worlds based on visual aesthetics are typically employed by global industry institutions. For example, "Indian wear" and "ethnic wear" are used by business consulting firms to analyze economic trends concerning clothing production and consumption in India, but similarly ethnicizing categories are not used in analyses of Western sites, such as Europe and America (e.g., [64]).

[4] This entanglement of meanings partially explains why contemporary fashion-critical ethical reformers hail craft production in India as having the potential to counter or slow modern (coded as Western) cultural manifestations of excessive desire, consumption, and waste production [60, 63 cf. 52 in Morocco].

which contemporary artisans' creative acts are interpolated within the field of fashion. By highlighting export house workers' everyday identifications and disidentifications with craft-marked practices and identities, this chapter advances a critical approach to the study of craft (cf., [15, 72]). I show how occupationally-designated[5] artisans critique craft both as a stigmatized occupational position and as a rubric of value that minimizes their skills. Moreover, I demonstrate that artisans from diverse Indian regions seck out work within Mumbai's fashion export houses in contrast to other craft-marked work possibilities, viewed as a more "independent" but "shared" creative platform, one that is circumscribed by a differently organized professional hierarchy.

1 The Export House as a Global Cultural Institution

In fashion industry discourse, Indian export houses are often represented as destinations for sourcing "raw" materials, identified with brokerage, low-skill labor, and manufacturing—the so-called back end of Western fashion [69]:75.[6] This mirrors the presumption that "creative" activity within the global field (e.g., "design," "planning," and "innovation") remains concentrated in Europe and America, where paradigmatic features of the industry first began [17, 56], particularly the cult of the "hero" designer [6]. Based on my fieldwork in Mumbai, I suggest a different understanding. Export houses are institutions that facilitate fashion design and production between practitioners located in different cultural contexts. That is, production processes within export houses are contact zones of "design interaction" [48] between culturally, and often socially and nationally, distanced participants.

The first export houses in Mumbai were established in the 1960s, but the large majority were created in the 1980s and 1990s during a moment of transition in national political leadership, marked by the instantiation of liberal economic policies and the deregulation of textile imports and exports [59]. Business owners often already had family connections in the USA or Europe and capitalized on joint citizenship or residence rights, reflecting their position as socioeconomic elites. Today there are approximately 25 to 30 fashion export houses in Mumbai, and several others in "Tier 1" Indian cities like New Delhi, Chennai, and Kolkata.

Work within export houses is similar to that in "garments factories" (see [45]) in that it is premised on the notion that contributions will not be attributed (or

[5] Where I refer to "artisans," "designers," and "managers" in this paper, I am referencing the occupational category to which participants' work is assigned in everyday activities.

[6] While India is often represented publicly as a site of low-cost manufacturing advantage, there have been recent attempts to market Indian fashion institutions to Western industry actors as preferential for "product development," which encompasses the phases of design and production discussed. As a recent editorial aimed at Western investors asserts: "India shows high product differentiation and development skills that make it a logical place for buyers to invest time and energy. The frenetic rate of new sampling that goes on every season in factories around India demonstrates this orientation to new product development" [27].

will be minimally attributed) to finished products as they circulate. However, like designer-named "Indian Fashion" firms [33, 34] such as Ritu Kumar or Sabyasachi, export houses do not represent artisan workers as "laborers" [*mazdoor*], instead they promote services through associations with "traditional workmanship" [*traditional kaam/kaarigari*]. Moreover, artisans are presented by export houses as a source of visual "inspiration" and often valorized as creative co-participants in "design projects." These practices sometimes garner disdain from Indian craft activists, who see the firms as working against their agendas to preserve heritage and more generally as "selling out" to the West. Artisans themselves contribute to export house projects in full recognition that their work and ideas will be "alienated" [66] as products are dissociated from production contexts and placed within Western-identified regimes of fashion value. For example, at the point of sale, techniques and even entire products may be attributed to a Western business identity (e.g., Chanel, Dior, Gucci) or even a country (e.g., French *couture*, Italian *alta moda*).

Due to the loose associations between notions of artisanal work (i.e., "hand" or "manual" labor) and "sustainability" in the fashion industry (see [1, 23]), Mumbai export houses have increasingly self-presented as socially and environmentally conscious enterprises. Some draw on accreditations from international organizations to advance this image, such as by Fair Trade International or the Social Accountability 8000 Standard (see [12]). Independent from government policies, these accreditations have "global" currency because they are endorsed by Western fashion multinationals, like Moët Hennessy Louis Vuitton (LVMH) and Kering. In practice however, because of highly complex subcontracting and outsourcing terms, as well as sourcing and assembly of apparel components across multiple contexts, exemplified below, it is very difficult for export houses to track the overall social and environmental impact of their projects.

Export houses also sometimes market themselves as niche "luxury" enterprises to marshal cultural capital, particularly embroidery focused ateliers. However, in practice most supply clients operating across a broad spectrum of markets. During fieldwork, I learned about the production of high cultural capital identified *haute couture* products for the Spanish label Balenciaga and British label Alexander McQueen, some of which are currently housed in European art museums. These were co-designed and co-produced in the very same workshops by the very same people as low cultural capital identified "fast fashion" items for American- and Chinese-identified labels like Forever 21, Urban Outfitters, and even Walmart. None of these five fashion labels publicly acknowledge the contributions of Indian participants to their clothing lines.

2 Occupational Categories and Skill Acquisition Narratives

> ... the continuity of tradition in skilled practice is a function not of the transmission of rules and representations, but of the coordination of perception and action ([26]:20).

> ... the study of artisanship and design must push back against traditionalizing and marginalizing discourses ([15]:2).

Rather than taking for granted the categories of "craft" and "artisan," we can probe the slippages surrounding how these skill- and occupation-based categories are being deployed and hail participants in export house work. My study is based on participant observation at one "full service" export house in Mumbai, meaning that it had facilities to handle all stages of design and production in house.[7] The premises consisted of three main spaces, a large, open-plan workshop—with benches, dress forms, pattern cutting tables, and sewing machines—and design and accounting offices—both with desks and computers. Other export houses may have weaving looms, knitting machines, and dyeing vats; however, at this firm these forms of labor are outsourced. Staffing the facility were approximately 120 people. 74 were identified as artisans, referred to in everyday talk through the Hindi/Urdu language term, *kaarigar*. Importantly, in this export house no skill distinction was made between artisans and tailors [*darjee*] (cf., [37]). Along with the 74 artisans and the single business owner, the firm included personnel with three other skill-identified occupational roles that are referred to in English: designers (8), managers (5), and accountants (2). Other staff were described as "unskilled" [*chotu*, lit. boy] and included transporters, loaders, security, and cleaners.

While designers and managers were female and male identifying, all artisans identified as male. This is common across the Mumbai export house landscape but certainly not the norm at all firms.[8] Few came from families who had worked in textiles and clothing for more than one or two generations, indicating shallow genealogies of skill transmission ([25]:6). Some identified with very distant ancestral backgrounds in crafts such as leatherwork [*mochi-ka-kaam*] and metal work [*lohar-ka-kaam*], which are typically identified with low-caste status in India. Some identified more recent family backgrounds working in processing of cotton, steel, or other factories. The most commonly shared identification among the men was that they were migrants to Mumbai, predominantly from rural and peri-urban locations in the "Hindi belt" states of Uttar Pradesh, Madhya Pradesh, and Bihar, with a few exceptions from Gujarat and West Bengal. A frequent rationale for migrating to take up work in Mumbai's export houses was that it was "more comfortable" [*zyaada aram*] than work in other factories, and significantly more lucrative than work in their home locations. For example, Mahir, a 35-year-old artisan from Sitapur, UP, told me he was recruited for work a decade earlier by his uncle's friend. He had passed 12th standard (upper high school) and studied tailoring at a government college. After working in a local garments factory, his uncle encouraged him to shift to Mumbai and work with the uncle's friend at the export house. His uncle funded his travel and initial accommodation in a two room *chawl* [tenement building] in the Parel district,

[7] I do not name the export house and have anonymized personal data where necessary to preserve confidentiality. I have endeavored to correctly attribute brand names. However, I have omitted some brand names per participants' requests.

[8] While there were no female-identifying artisans working at the export house in which I conducted participant observation, I interviewed 12 women working at other export houses. These women primarily accessed the work via NGO training programs—programs that recruit women based on experiences of poverty, social disadvantage, and experiences of domestic violence—or in association with their marital partners' work.

where Mahir still resides along with five other friends working in textile industries. I found Mahir's migration story to be typical among the most recent generation of men that had come to work in the export house.

By contrast, men who had been employed at the firm for a longer time, and were as such older, tended to have had previous experience working in regional "craft clusters."[9] In such sites, work is typically organized around a specific technique or style, often name-recognized and identified with regional or national "heritage"—either tacitly or through formal government certifications or NGO development projects (cf., [14, 31]). These men often narrated their decision to come to Mumbai as based on becoming more "independent" [*apne pairo pe khade hone ke liye*, lit. to stand on my own feet]. Some associated their decisions specifically with freeing themselves from the control of predatory social institutions, such as *seth* [e.g., Master Jis, cluster leaders, money lenders, see [37]:8] and had initial aspirations to start their own businesses.[10] Others commented on the general creative constraints of cluster work, framing it as both highly repetitive and overly regulated by workshop leaders and the government. For example, Nihal, an elder Hindu artisan, told me that work in the export house freed him up to work alongside and experiment with techniques and styles typically associated with people from other castes [*jaatis*], places, and with Muslims, without fear of attribution and disparagement. He gestured to the increasing presence of bureaucrats within craft clusters as the reason for constrained creativity and social division.[11] Contra this, export houses were identified as spaces in which work is experimental, exciting, and shared (e.g., "here there is no limit for new ideas" [*yahaan nya vichar ki sima nahin hoti hai*]).[12] This is an important principle, as it evidences how commonly perceived ethical imperatives of attributing

[9] The term "craft cluster" has two predominant usages in India. It is firstly used as a gloss for "sites identified with forms of work and production," tacitly suggesting the identification of particular types of persons with work activities (e.g., "Rajasthani *bandhani*" [resist dyeing] connotes an association to "traditional" work). Secondly, at a more formal level, the term references places identified for intervention via craft-focused government development projects, connoting economic disadvantage and sometimes social "backwardness" [sic].

[10] *"Seth"* is a pan-Indian caste/social identification. Within the textile industry, the term is commonly used to reference persons that control a production unit, who often have leverage over workers because of their economic and social position, typically acquired through multi-generational landholding. I encountered narratives about *seth* lending money to workers to pay for family celebrations or medical expenses, viewed as creating bonds of indebtedness. More generally, *seth*-worker relations are typically marked by tensions.

[11] Craft bureaucracy was talked about in generic terms of "government" or "NGO" presence. I asked artisan participants about projects by notable Indian craft activists, like Laila Tyabji (Dastkar), Judy Frater (Kala Raksha), Sally Holker (Rehwa), as well as historical figures commonly mentioned in academic genealogies of the field of Indian textile crafts, such as Kamaladevi Chattopadhyay and Pupul Jayaker (cf., [43]). Most did not know of these individuals or knew very little about them. In this sense, artisans often seemed distanced from the Indian craft world [70], often with less knowledge about it than the American and European clients with whom I met.

[12] Another important distinguishing factor among a few men was that they had been denied "Artisan Cards"—a government of India registration and welfare scheme, coordinated by the Commissioner for Handicrafts (see [5]). Others stated that they had not applied for Artisan Cards based on presumptions they would be denied. These participants pointed to local politics and established families as

techniques and knowledge as "heritage," a recurrent trope in recent critical ethical discourse about the fashion industry, may work against Indian artisans' interests in shared creativity as well as their concerns for freedom (see for e.g., the Cultural Intellectual Property Rights Initiative).

While some artisans had been working at the firm for over twenty years, many aspired to more lucrative employment elsewhere. This was almost always envisaged via a transition in occupation and movement away from stigmatized craft work. The average base monthly salary for an artisan was between INR 15,000 to 20,000 [USD 180 to 250], covering a six-day work week with one's day rest. On top of the base salary, artisans could earn extra income by extracting a commission from the piece rate paid to family and *biradari* [extended community, see [46]] members that they recruited for work during busy periods.[13] Beyond such forms of income supplementation, participants perceived a glass ceiling to earnings in their profession as artisans. As Sameer's statement in the introduction indicates, some younger artisans have started aspirationally identifying themselves as "designers." Some even express aspirations to create their own fashion design firms—a practice that one designer told me mirrors the growing diffusion of social media access and the popularization of fashion design as an occupational identity. To this end, a few artisans showed me Instagram accounts where they had begun amassing a portfolio of pictures of their work outside of the export house. Others suggested their future opportunities to work as textile traders [*vyapaar*] and vendors [*dukandaar*] in Mumbai or other cities. The men did however perceive significant barriers to achieving these goals: namely, financial capital, social connections, and most of all, English language skills. In this sense, rather than skill being an essential feature of identification, emplacement, and personhood ([20]:10), artisans had a "flexible" [53] stance toward their capacities and capital in the context of given social and professional opportunities.

Before turning to a discussion of export house work processes, it is helpful to introduce some very brief comparative context about designers and managers working at the export house, who tended to share what might be called a "middle class" social background (cf., [18, 42] on the problems with using this category in India). Designers and managers were roughly evenly split between male- and female-identifying, including two openly LGBTQ-identifying people.[14] All were younger than 35, and most were from Mumbai. Those who were migrants came from other big cities, like New Delhi and Bangalore. Along with urban origins, the most significant difference between designers and artisans was that designers exclusively entered export house work after formal training at a fashion design or management school (e.g., the National Institute of Fashion Technology, Pearl Academy). After receiving their degrees, they either reached out to export houses directly or their professors had

key factors gatekeeping access. The lack of access to Artisan Cards reportedly limited their ability to raise capital in their home locations to start independent enterprises.

[13] Additionally, when asked about finances, the business owner mentioned the many times artisans had asked her for loans, advances, or one-off cash sums for medical expenses or family emergencies, which she reportedly always provided.

[14] Here, "LGBTQ" is as stated by participants.

recommended them for work or an internship, thus mobilizing social and symbolic capital acquired through their education. Further, designers and managers tended to come from families with at least one parent with a white-collar work background such as law, medicine, engineering, or business, either in the state or non-state sector, indicating possession of cultural and economic capital that enabled them to access education. Some noted significant resistance from their parents after they chose to pursue design as a profession, indicating their families' perception of fashion as a low-status field to work in. While the managers tended to have longer-term career visions in export house work, the designers typically imagined their positions as transient, as an opportunity to "gain experience" and "learn how the industry works overseas," which they viewed as a necessary stepping stone toward eventually establishing their own businesses. Despite these differentiating factors, the average salary for a designer at the export house was between INR 15,000 to 25,000 per month [USD 180 to 320], only marginally higher than that of an artisan, reflecting a relative parity in value within the export house context. However, a manager may be paid slightly more, up to about INR 30,000 [USD 400], although this rate tended to be justified less by training than by entry from a prestigious "business" family background.

Finally, it is crucial to note the variation in language competencies among participants. All, including the business owner, artisans, designers, and managers, spoke fluently in Hindi/Urdu, and many also spoke a second, third, or fourth South Asian language. The business owner and managers spoke and wrote fluently in English, sometimes with an American English accent. A subtle distinction between managers and designers was that designers tended to speak English with an identifiably Indian English dialect. A few artisans spoke more limited English (Indian English dialect) and none could write fluently in English. These observations will prove important as, despite a relative social "flatness" (e.g., egalitarian ethos) achieved during design work activities, language competencies remained a defining principle of professional hierarchy through which economic and symbolic rewards were obtained.

3 The Flow of Expertise: Or, How Your Gucci Handbag is Made

Within export houses, fashion design is a highly circuitous process that takes place with regular interactions, feedback, and exchange between export house staff and subcontracting clients. There are seven broadly sequential phases: (i) swatch production, (ii) swatch marketing, (iii) purchase orders, (iv) mood boarding, (v) sampling, (vi) prototyping, and (vii) planning for production. I observed participants bring several projects through these phases, a process that could take anywhere from 2 days to 4 weeks. Through interviews and photo sharing in the months following my fieldwork, I tracked the manufacturing and completion of these projects, and in several instances, observed the clothing items being positioned in retail outlets in New York and online. When taking the export house design and production process fully into account, it becomes evident that discrete occupation-based categories, such as "artisan" and "designer," misrepresent what are in fact blurred co-contributions to creative practices. Correspondingly, skill values emerge from the mix of creative possibilities in the context, often activated by shared work across differences.

(i) **Swatch Production**

While dominant ideologies of authorship in the field of fashion commonly position the designer as the sole creative "source" [6], from the vantage point of the export house, design projects appear highly distributed. Work generally occurs in response to "swatches" dispatched by the export house owner for marketing purposes. These swatches are textile and garment samples, or sketches of patterns and silhouettes. They may pertain to part of or a whole clothing item and can take anywhere from a day to a month for staff at the export house to assemble.

Subgroups of artisans and designers are assigned to produce swatches on a rolling basis, with group members rotating regularly. This work typically takes place to the side of the workshop, a large open plan room with low work benches (for sewing and embroidering), paper and pencil supplies, sewing resources, and mannequins. As it is common for artisans to work alongside family members in the workshop, swatch production usually takes place in groups comprised of one designer and three to seven artisans, working almost exclusively in Hindi/Urdu language. The groups are surrounded by reference materials, such as art books and magazines, sourced by the business owner from around the world, as well as recycled swatches. Though mobile phones are prohibited within the workshop,[15] both artisans and designers regularly send photographed visual references for printing in the design office (via WhatsApp), a room adjacent to the workshop, to which artisans do not have access. Designers print these references and bring them into the workshop. Both designers and artisans also regularly bring samples of garments, textiles, and embellishment resources into the workshop, such as buttons and beads, collected from vendors in Mumbai or beyond—expectations are that, if selected, work or products will be sourced from the original vendor. Evident across the mix of references are similarities between designers' and artisans' sources. However, some swatches are produced with a specific client in mind. This involves online research into the client's previous fashion collections or visiting trend forecasting sites, like WGSN or Patternbank. Because computers can only be accessed from the design and accounting offices, this work is carried out exclusively by designers and managers.

Some swatches are first "laid out," which involves arranging material items, draping on mannequins, and using color chalk to mark lines for cutting fabric and stitch placements. Other swatches are first sketched and then materialized. Both designers and artisans handle textile materials as they produce swatches, although

[15] More specifically, photography is prohibited in the workshop area of the export house. Security staff perform bag checks on arrival and require workers to place their mobile phones in locked baskets during working hours. Because artisans rarely access other areas of the premises, such as the design or accounting offices, the mobile phone prohibition primarily affects them.

artisans are invariably recognized as more skilled at cutting, stitching, embroidering, dyeing (rarely required), draping, printing, and weaving or knitting (both very rarely required). Artisans are also viewed as more skilled with sketching, except when swatches are designed on the computer using Photoshop. Most ideas emerge through experiments with the reference materials and by combining visual sources. However, some artisans describe their techniques as related to family traditions or work histories. Such identifications are most common when participants disagree on the "correctness" of a technique, moments that offer a window into participants' own ideologies about skills. For example, I observed one interaction about the placement of an inner thigh seam, in which an artisan critiqued the correctness of the stitch plan by referencing, nostalgically, how his father (who was associated with the textile production hub Chanderi in Madhya Pradesh) was known for sewing loose and hanging stitches to allow the garment qualities of stretch and flex. The designer in the group, however, suggested use of this technique was impractical as it would take too long to train other artisans in the workshop; professional superiority was as such justified based on a managerial assessment of collective time use and efforts. While the technique was not adopted for this particular swatch, the artisan went on to create a swatch featuring the technique on the pocket of a pair of jeans, a design later adopted by a Swedish fashion retailer. For this, the export house owner paid the artisan a cash bonus. His claim to inherited expertise was as such transposed as a claim to original contribution in a fashion product, rewarded individually with money.

The above example demonstrates a claim to mastery of a specific skill. However, it is important to note that artisans rarely see themselves as masters of only one skill, as they demonstrate expertise across a variety of skilled practices on a daily basis. Particularly among younger artisans, this capacity for flexibility and adaptability—e.g., the working out of creative rules and codes [26] in contexts of shared practice [48]—is seen as a defining feature of expertise that enables design activity. Skill recognition is in this sense unstable and uncertain, and requires peer legitimation. Designers, on the other hand, sometimes identify themselves as having a "natural" affinity for a particular skill, like *zari* [metallic thread embroidering] or drawing outline patterns for *kalamkari* [painting on textiles with a bamboo pen]. Designers almost unliterally describe these skills as acquired through previous interactions with artisans, often in rural locations.[16] When demonstrating these craft identified skills in the workshop, designers' work was commented on by artisans with statements like "good-good!" [*acha hai*!] or "very beautiful designs!" [*bohoti khubsurat design hai*!]. While indicating the porous boundaries between craft and design practices, the expressive manner in which such statements were offered gave me the impression

[16] Such narratives mirror what anthropologist Alicia DeNicola typifies in India as professional designers' idealization of playing the role of a "cosmopolitan mediator," with responsibilities to translate between "ethnic arts of the village... and the contemporary global economy" [14]:298,see also [65].

that the artisans valued the designers' experiments, but ultimately regarded them as amateurs, as unskilled.

(ii) **Swatch Marketing**

Following completion of the swatches, the export house team assembles them into "collections" (e.g., brochures, dossiers), which they dispatch by mail (or email as photographs) either to the international clients directly or to the export house's overseas marketing offices.[17] This marks the first stage in the "repackaging"—or resignification [50]—of local creativity and notions of skill. The export house that I studied has marketing offices in New York and representatives in London. Other Mumbai export houses have marketing offices or representatives in Paris, Milan, Dubai, Singapore, and Hong Kong. The foreign clients are invited by the business owner to browse the brochures. When doing so, clients often touch them and sometimes disassemble components for closer inspection. Importantly, business owners and managers tailor the "look" of each brochure strategically to the client, imbuing them with certain narrative, stylistic, and aesthetic qualities. By this time the swatches also often have cardboard tags attached to them displaying the logo of the export house and an item and collection number, labelling and stylizing that appears to take place close to the time of display in the marketing office. Some swatches may be individually labelled. For example, in the New York office I saw a brochure destined for a British fashion house. The folder was matte-black with the client company's logo stamped on the front. Inside, swatches of lace patterns and leather garment shapes were labeled as "Demure Rose" and "Wild Willow." The ability to connect swatches with the visual and narrative identities of Western brands is viewed by export house owners and managers as part of their skill set, informed by possession of cultural capital, accrued sometimes through research and other times by "getting exposure" to (i.e., socializing among) industry-specific communities of practice in the USA or Europe.

While traces of "Indianness" (cf., [41, 65]) are thus on the whole erased from the product brochures, some traces of product origins are strategically introduced during the marketing meeting with a client. For example, the business owner (or sometimes a manager) may show photographs of Indian staff and the workshop premises to explain "how sustainable the company is," or more generally, to "explain the traditional context of the craft." The images they share typically depict very small groups of artisan workers, portrayed smiling and wearing clean uniforms, against the backdrop of a very neatly organized workshop, again, presumed to connect with a Western imaginary of traditional artisanship as well as desires for Southern workers' welfare. In my experience, these representations are far removed from the visual appearance of actual workshops. Narratives offered to clients about the background of craft skills may also reference a break from generationally deep traditions and Mumbai staff's design innovation, positioning the export house through notions of artistic modernism and development contribution. In some cases, a paper handout may be provided representing accreditations from third-party organizations, often

[17] Export houses may also participate in trade fairs internationally to exhibit swatches and gain clients (e.g., [35]).

focused on demonstrating the absence of child labor or the presence of sanitary supplies and other health and safety resources in workshops, invoking the conventions of Universal Human Rights activism.

(iii) **Purchase Orders**

This brings us to the client's "purchase order," first initiated via phone call, but arriving at the export house in Mumbai as an official document via email or express postal mail. This document is written exclusively in English. It may request anywhere from 1 garment to over 10,000 garments for a clothing line. The purchase order may wholly adopt an idea presented by the export house in a brochure or it may put forward a modified idea. It is extremely rare for a client to put forward an altogether different idea, although when this happens, it usually includes reference to previous projects by the export house. Sometimes swatch ideas intended for clothing are also reframed as part of a collection of other objects, such as for bed linens or soft home furnishings. Designers and artisans are often excited to see how their ideas have been reframed or developed by personnel at the client firms, and view the clients' requests as part of the ongoing creative dialogue.

Yet, on the whole, designers and artisans are provided with very little information about the client-side actors who they are working with. Only the business owners and managers have access to client representatives' names, and these are often masked by generic job titles, like "Design Associate" or "Research and Development Officer." Moreover, the purchase order is usually labelled with the name of a shareholding firm (e.g., LVMH, Kering) or sometimes a third-party US supplier or shell firm to obfuscate the supply chain. A client that I spoke with explained these practices are intended to "prevent copying" and generally to make it harder for the export house to provide quotes based on the cultural capital of publicly-recognized brand identities. With all this in mind, my experience was that designers and artisans at the export house nearly always knew the brand identities to which the export house work will later be attributed, demonstrated in their everyday talk around the export house and when identifying the garments under production (e.g., "the Gucci handbag," "the Dior dress").

(iv) **Mood Boarding**

Purchase orders reach the export house as a bundle of paperwork, including sketches and photographs. If a physical package has been sent, the bundle may also include small strips of fabric or embellishment objects like buttons and pins, that provide visual and tactile references, along with the original swatch provided by the export house. Sometimes a "pattern" [a paper mock-up] with measurements will be included. These materials may reference specifications for a whole garment or for a component of one that will be assembled elsewhere in the city, country, or overseas. At the export house where I conducted fieldwork, upon receiving the purchase order dossier, the owner stripped it of key financial details before passing it to a manager, who subsequently passed it to a designer. However, designers I interviewed at other export houses reported they were directly responsible for opening the packages.

Designers then assemble the materials into a "mood board"—a collection of photos, drawings, materials, and texts, intended to evoke the style or concept of a design project—to be placed on the wall in the workshop, serving thereafter as a reference point. They often add more detailed sketches of particular areas of the garment plans, and sometimes add "key words" (in English and Hindi, such as "dress," "*zari*," etc.) plus book or magazine "tear sheets" [ripped out pages] as references. These are moments of attempted translation of ideas and expectations, and as one designer told me, they demarcate the boundaries of localized creativity: "Once we have the mood boards in place we can begin to design truly.... We have possibility to sneak in our ideas. Because they [the client] know so little about fabrication, we have room to play." Such statements offer clear indications of a localized skill regime shared between artisans and designers that privileges knowledge of garment structure. Importantly, this is different from generic ideologies of "workmanship" or "craftsmanship" [62], reflecting meanings closer to "engineering knowledge" and "material knowledge." Also evident in the participant's statement is how artisans and designers define their skill imaginary in opposition to the client. For example, the clients' skills are presumed centered on non-engineering and non-material knowledge, described as only accessing "surface" elements of visual style, a notion that we will return to in more detail shortly.

(v) **Sampling**

After assembly of the mood boards, a round of conversation led by the business owner focusses on what material elements can be purchased pre-made from neighboring vendors and what will need to be commissioned externally or produced in-house.[18] Very few material supplies are held in the workshop, owing to shifting requirements and the quantities of scale required by projects. Discussions about sampling take place primarily among the business owner, managers, and designers in the design office, although sometimes artisans are called upon for input. Participants select sourced or commissioned work elements from among a trusted network of local vendors of fabrics, beads, threads, etc.[19] These local vendors are often highly specialized in terms of their products or work, based mainly in Mumbai's Byculla, Parel, or Dadar districts, sometimes further afield. For example, one vendor in Parel makes only red buttons and red sequins. Another in Dadar has a team of thirty artisans who only embroider in the *zardozi* style and work only with their own textile elements. Designers and managers often comment on the creative inflexibility of these practitioners, which they often associate with a "traditional mindset." Designers regularly comment on their efforts to "open their minds" through painstaking rounds of creative direction. Artisans, too, often reflect on their fortune to not be engaged in such repetitive work, described as being "without inspiration" [*sadharan*] and also as "without aspiration" [*ichha ke bina*, lit. without desire]. In this sense, both the knowledge of

---

[18] On rare occasions, the client will specify use of imported materials, either by sending a working fabric sample or by requesting creation of a prototype on a locally available substitute fabric.

[19] Trust is primarily based on established social relationships and precedent of their timely delivery in previous work projects. These relationships are attributed to the business owner, but some were initiated by employees and former employees.

Mumbai's vernacular networks of textile element vendors and knowledge of how to "convince" these vendors to take creative directions are valorized in the local imaginary of fashion design skills. Further, both designers and artisans clearly distinguish the export house skill regime from other contexts of adjacent artisanal work in the fashion industry.

(vi) **Prototyping**

Once sample materials are procured, the export house focuses on producing prototypes. As with the swatch production phase, a subgroup of artisans and designers begin the trial-and-error process of stretching and pinning fabrics onto mannequins or the workbench frames and use chalk to mark out patterns, before cutting, stitching, embroidering, or otherwise modifying the fabric.

This is a very quiet phase of work. Artisans and designers make sparse verbal comments but provide each other visual feedback through hand gestures and physical manipulation of the prototype. The majority of the artisans' and designers' verbal directions are in Hindi/Urdu and focus on interpreting the mood boards and deciding what material elements or techniques to use. For example, I observed debates over a beaded tunic produced for a French luxury fashion company. The purchase order specified modification of a swatch prepared by the export house, and in particular the substitution of a white bead instead of a yellow one. The order did not specify the weight, density, or texture of bead elements, or the stitching technique to be used to fasten the beads. Hrithik, an artisan, suggested a "Gujrati" style bead made of a seed pod, with multiple beads fastened in a linear row—this is typical to the *toran* [wall hanging] commonly used across India. Apeksha, the only designer among the group, countered with a critique that this would be too difficult to fasten under the sleeves, and would easily break with abrasion from arm movement. She instead suggested a white painted glass bead, pre-fastened in strips of five beads and sewn at each end. Another artisan, eldest among the group, Rashad, then commented that this would create too uniform a look, and the look needed to be more "real" [*asli*, lit. original]. Taking Rashad's cues, the group eventually settled on a hollow metal bead, painted pearlescent white but "pre-aged" by the supplying vendor, to be attached to the cotton garment using a looped stitch common to *muqaish* [heavy metal bead] embroidery.

Such discussions are typical of the prototyping phase. They evidence the overall social "flatness" of interactions achieved in this moment and participants' valorization of collective "problem solving" as a design skill (cf., [38]). They also show that decision-making over aspects of visual form take place via a code-mixed [multiple South Asian-language] vocabulary of socially-and place-identified materials and techniques. *Muqaish*, for example, is associated with Muslims entry to North India in the sixteenth century, and the Gujrati bead is identified with the northwestern state of Gujrat, although practiced across India. Importantly, because objects and styles continue to invoke reference to the identities of producer communities (see [71] on "socially dense" objects), creative negotiations often conjure perceptions of social relations and social effects. This is most evident, for example, when artisans and designers limit their creative choices. As the fieldwork discussed herein extended through early January 2020, I observed, for example, decisions to limit the use of *chikankari*, a style of embroidery typically associated with Muslim producers in Lucknow, because of perceived interruptions of supply caused by Muslim workers' participation in protests against the government of India's Citizenship Amendment Act [3]. On other occasions I observed decisions to use material objects/styles to direct economic resources to particular producer communities, and hence sustain social relationships. Design negotiations may also invoke participants' personal moral views. For example, I learned about the highly fraught negotiations surrounding a one-off design for an exhibition by a well-known British fashion company that featured sexually explicit imagery. Several artisans refused to work on the garment, while others feigned ignorance after adjusting the design to be less explicit, to the disappointment of the client.[20] In this sense, we see that the creative economy and skill imaginary within the export house enables a flattening of social space while allowing participants to make decisions and contribute based on different types of value-based preferences. The assertion of preferences may take place in verbal dialogues or tacitly through haptic play with materials and symbolism (e.g., objects, visual styles) that are imbued with meaningful social capacities (cf., [2]).[21] Knowledge of how play with materials and visual styles effects communities, practically and morally, is thus a defining principle of skill among export house designers and artisans. And correspondingly, clients are presumed to be socially abstracted, as their work on so-called surface aesthetics (e.g., sketches and models) are viewed as detached from locally defined ethical (cultural) codes.[22]

Once a working prototype has been assembled, managers or the business owner are called into the workshop to approve the work or provide input. This evidences a shift in the effective hierarchy within the setting and invocation of the ultimate hierarchy of authority within the firm, proscribed along lines of professional roles/titles. Managers' critiques and directives rarely pertain to visual style and often center on financial concerns, such as reducing waste fabric or substituting for cheaper materials (e.g., lower "pic" count [number of threads]). Their directives are often resisted by artisans and designers, viewed as diminishing the quality of design projects. When a final first prototype (the so-called "dummy") has been "signed off" by the business owner, managers send photographs of it to the client. None of the projects I observed received immediate approval from clients, and some went through two or more rounds of feedback and revisions. Feedback from the client tends to come in either visual or written form, always in English. Visually annotated comments are usually hand drawn but can be supplied via photographs of a modified dummy, sometimes "faked" in Photoshop. Managers respond to this feedback, either accepting the comments in writing or providing counter-suggestions, leading to further rounds of prototyping.

---

[20] This one-off fashion piece is in the permanent collections of a British museum.

[21] This mirrors what anthropologist Don Brenneis once called "social aesthetics," that is, "intellectual sense-making activity [fused with] local aesthetic criteria for coherence and beauty" [8]:237).

[22] This is an inversion of the common design/labor value paradigm in the fashion industry, assigning low value to design work (i.e., what artisans critique as "surface" work) (see [17] on the "frontstage" vs "backstage" of French *couture* fashion; [44] on British fashion as an "image industry" vs "rag trade").

Alternatively, the client may send written comments. These tend to be focused on the color or overall "feel" of the prototype. Such comments are relayed by managers to designers, who subsequently relay the feedback (translated into Hindi/Urdu) to artisans. Urmilla, a designer, told me that these latter rounds of feedback are often the most frustrating and hard to "pin down."

Evident here is that designers rarely correspond with the client directly and artisans never do. When I asked participants about the multiple layers of mediation and translation in their day-to-day work, I received quite different evaluations. Owners and managers invariably said the hierarchy protects confidentiality and clients' proprietary knowledge—which makes little sense, given the general distribution of knowledge about brand identities and products in the workshop.[23] The main criteria policed seems to be the names and contact details of clients. On the whole, the secrecy seemed like more of a performative gesture, aimed at representing the export house as "professional" (i.e., secure, discrete). Designers, however, tended to offer a slightly different justification. They suggested that artisans' lack of English language proficiency prevented them from being able to talk with the client directly, and moreover, that their "unprofessional" self-presentation would diminish the image of the firm as a space of luxury. This mirrors a broader tendency for designers to comment on artisans' styles of interaction and speech as curt, ineloquent, and abrasive. When I asked artisans explicitly about their perspectives on the issue of secrecy, some did say that they would like to meet with the clients directly. One artisan suggested WhatsApp calls would make their job a lot easier. However, on the whole, artisans tended to not be interested in knowing more about the clients. Some even upheld the designers' stereotyped assumptions, offering statements like: "we don't know how to talk to those big people" [*vo bade log hai, hum nahi baat kar sakte hai*] or "we are not professional [like them]" [*hum itne professional nahi hain*]. It may be tempting to see such statements as internalized bias. They certainly bolster the business owner's and managers' self-images of being necessary intermediaries, e.g., framing Indian work and workers in ways that connect with Western imaginaries of professionalism and civility. Yet, in the context of artisans' other statements about the value of work in export houses, I am more inclined to take them at face value. They reflect artisans' knowledge of the racist and classist stereotypes that often underpin clients' knowledge of their socioeconomic position, and business owners' and managers' work to create commercial relations across such ideological structures.

(vii) **Planning for Production**

When managers and the client agree on a final prototype, the export house issues an invoice and work moves into the production planning phase. Importantly, despite the significant outlay of labor and resources up to this point, there is no guarantee that the client will pay for services rendered until they sign the invoice. Conversations about production take place in the design office away from artisan workers. Participants

---

[23] There have been a series of high profile "leaks" of sketches and garment designs from Mumbai export houses, occurrences that export house owners guard against, or at least claim to do so in their public messaging (e.g., [57]).

include the business owner, designers, managers, and accountants. These discussions pertain firstly to logistical arrangements for production and transport (including export/import), and secondly to quality control, policy compliance, and accounting.

For smaller clothing lines, production may take place in-house, utilizing regular artisanal staff and resources sourced from local vendors. During "rush" periods (notably in the lead up to New York, London, and Paris Fashion Weeks) artisans may be requested to source assistive labor, which takes place often through family and *biradari* networks. For larger lines, some aspects of garment stitching, dyeing, and embroidering may be outsourced to local factories, although the business owner disfavors this, as it increases the potential for leaks of the designs. Managers and designers are primarily responsible for establishing outsourcing relationships, and they take the lead in communicating work instructions, evaluating work quality, and arranging payments. As with local vendors, relations between the export house and local factories are fraught with assumptions of "inflexibility" and "traditional mindsets." Because of the secrecy surrounding *haute couture* garments—one-off pieces, usually for catwalk shows and museum displays—these are always wholly constructed in the workshop, often with intensified security and mobile-phone policing during production.

The final stages of production evidence the most rigid moments of professional hierarchy, but understood as necessary to standardize products and actuate the shared creative vision. The assumption in this moment is that creativity has ceased, and both artisans and designers can no longer modify products. The flat social space of the workshop transforms, and artisans and designers are assigned to different work activities: artisans primarily occupy the position of laborers (or "manufacturers"), and designers fulfil the role of overseers, enforcing conformity to the prototype. A common issue that arises is that artisans who regularly work side-by-side can develop similar "hands"—meaning that their work may develop an appearance of visual congruity. Designers sometimes ask artisans to shift between tables and benches to standardize the unstandardized appearance of the products—a decision rationalized as "exposing them to others' hands" (cf., [30] on "craft" vs the "factory hand"). When the products are complete, managers and designers are responsible for coordinating "finishing" (e.g., attaching tags and labels, folding, packaging) to prepare items for export. Artisans are never trusted with this task. Similarly, the export paperwork and customs duty forms are prepared by managers and accountants, designers are not given access to these financial documents. In this sense, professional hierarchy during production works in service of "fixing" collective accountability to the prototype—comprised of earlier labors, and which importantly consolidates inputs from artisans, designers, managers, and the business owner with those from the client.

4 Concluding Remarks: Recognizing Skill Regimes in the Field of Fashion

Despite deep-seated ethnocentric assumptions about creativity and skill and how these properties are socially and geographically distributed, the global field of fashion is changing. We need a theory of ethical (cultural) transformation that can encompass both older categories for understanding labor and craft, and that can be respectful of the actual categories through which contemporary workers desire to be recognized and aspire. In just these few examples of what I call a skill regime, we can sense the frictions through which Indian export house participants are negotiating, collaborating, and improvising creative possibilities that are emergent across differences. Rather than work towards a vernacular or national industry (e.g., [34]) or a parallel regime of fashion value (e.g., [36]), export house workers see their contributions as generative of the shared "global" field.

By exploring Indian export house workers' practices and perspectives, I have sought to demonstrate some of the ways their conceptions of skill in "craft" and "design" exceed common imaginaries about their work. Notions of skill in the export house, for example, prioritize value in un-attributed work, technical flexibility and adaptability (e.g., non-specialization), as well as a concept of mastery rooted less in "manual" prowess than to associations of "engineering" and "material" knowledge (counterposed to the "surface" knowledge of Western designers). Further, they allow for the combination of ephemeral visual references and references from verbal peer feedbacks with inherited knowledge and social knowledge; and may further be shaped by perceptions of accountability to social and political groups or personally held moral views. In this sense, my discussion demonstrates how Indian export house workers invoke the labels of craft and design—artifacts of a hegemonic global discourse about creativity—while pursuing everyday local possibilities that cut against the grain of these categories as imagined both within the Western and Indian contexts.

We also see how the export house, as a hierarchical institution with multiple "faces," enables the movement of localized knowledge and objects through different cultural fields, such as craft, fashion, and even, in the case of one-off catwalk and museum objects, art. Shifting, context-specific hierarchical dynamics and strategic practices of mediation and translation—or at least the appearance of translation—are essential to achieving these goals. They enable local participants with different forms of capital (economic, social, cultural, and linguistic) to collaborate, co-constituting a proxy platform from which to strategize and engage Western industry participants in terms of their, often ethnocentric, skill imaginaries. Similarly, these distinctive workplace structures facilitate movement between localized cultural fields. Artisans perceive freer, more exciting, and socially "flattened" creative co-participation in export house work, and in turn disidentify from other craft-marked work possibilities, such as in government-identified regional clusters or for urban vendors and factories. Rather than negatively-valued practices of divergence from purist, state-authorized visions of cultural and historical representation (cf., [29]), the possibilities of shared work and collective experimentation across (or in spite of) differences within the export house are viewed positively.

For those wishing to advance a more ethical future for the field of fashion and its stratified transnational production arrangements, I suggest we must therefore be careful not to uncritically embrace categories—or "match" categories—across differences. Closely contextualized ethnographic attention to localized skill regimes as well as how skill regimes are positioned in relation to each other, reveals the slippages surrounding how everyday work categories (e.g., practices, occupational roles, identities) are being deployed and hail participants across social and national contexts. The skill regimes framework is above all a critique of abstract and universalist theorizing of practices, such as craft, design, and even fashion (e.g., [16]. Just like the category of artisan [24]:5, these labels do not have "natural" content but need to be understood in terms of their local social and historical articulation. Knowledge of how local actors perceive skill and recognize themselves in a complex, culturally hierarchical web of transnational relationships will thus be key to working through the intercultural frictions and paradoxes raised by a changing and uncertain, but essentially shared, global field.

Acknowledgements I would like to thank all in Mumbai who lent their time and patience to this project. To those who recognize themselves as Mahir, Hrithik, Nihal, Rashad, and Sameer, I am particularly grateful to you for sharing your expertise. At New York University, I thank Tejaswini Ganti, Fred Myers, Faye Ginsburg, and Bambi Schieffelin for being generous readers, teachers, and mentors. This chapter also benefited from valuable critical correspondence with Aarti Kawlra and Jane Schneider and editorial feedback from Ihaab Syed and Nikita Shah. Sections were presented at the 2022 American Anthropological Association Meeting, for a panel on "The Craftiness of Craft," co-organized by Sowparnika Balaswaminathan and myself. The research was supported by a travel grant from the NYU Graduate School of Arts and Sciences in 2019 and a Society for Visual Anthropology Robert Lemelson Foundation Fellowship during 2020-2021. All errors are of course my own.

References

1. Alvarez Astacio P (2015) Moral fibers: the making of (Trans)National fashions in post-authoritarian. PhD Diss., Peru University of California Santa Cruz, Anthropology
2. Alvarez Astacio P (2021) Tactile analytics: touching as a collective act. In: Ballestero A, Winthereik BR (eds) Experimenting with ethnography. Duke University Press, Durham, pp 15–29
3. Anon (2020) Thousands join anti-CAA protest in Mumbai. Times of India, January 12
4. Appadurai A (ed) (1986) The social life of things. Cambridge University Press, Cambridge
5. Banerjee A, Buhroy S (2020) Majuli weaving cluster: a study on weaver dropout using industry competitive analysis and concepts of behavioral economics. Fash Pract 12(2):193–218
6. Bourdieu P (1993) The field of cultural production: Essays on art and literature. Polity Press, Cambridge
7. Bourdieu P (1990[1984]) Homo Academicus. Cambridge: Polity Press
8. Brenneis D (1987) Performing passions: aesthetics and politics in an occasionally egalitarian community. Am Ethnol 14(2):236–250

9. Collins JL (2003) Threads: gender, labor, and power in the global apparel industry. University of Chicago Press, Chicago
10. Collins JL (2016) Expanding the labor theory of value. Dialect Anthropol 40(2):103–123
11. Crane D (2000) Fashion and its social agendas: class, gender, and identity in clothing. University of Chicago Press, Chicago
12. De Neve G (2016) Power, inequality, and corporate social responsibility: the politics of ethical compliance in the south indian garment industry. In: Dolan C, Rajak D (eds) The anthropology of corporate social responsibility. Berghahn Books, New York
13. DeNicola AO (2003) Creating borders, maintaining boundaries: traditional work and global markets in Bagru's handblock textile industry. Syracuse University, Anthropology, PhD Diss.
14. DeNicola AO (2016) Asymmetrical indications: negotiating creativity through geographical indications in North India. Econ Anthropol 3:283–303
15. DeNicola AO, Wilkinson-Weber C (eds) (2016) Critical craft: technology, globalization and capitalism. Bloomsbury, London
16. Entwistle J (2000) The fashioned body: fashion, dress and modern social theory. Polity Press, Cambridge, Oxford
17. Evans C (2013) The mechanical smile: Modernism and the first fashion shows in France and America, 1900–1929. Yale University Press, New Haven
18. Fernandez L (2000) Nationalizing 'The Global': media images, cultural politics and the middle class in India. Media Cult Soc 22(5):611–628
19. Graeber D (2001) Toward an anthropological theory of value: the false coin of our own dreams. Palgrave MacMillan, New York
20. Grasseni C (2007) Introduction. In: Grasseni C (ed) Skilled visions: between apprenticeship and standards. Berghahn, New York, pp 1–19
21. Greenough P (1995) Nation, economy, and tradition displayed: the Indian crafts museum, New Delhi. In: Breckenridge CA (ed) Consuming modernity: public culture in a South Asian World. University of Minnesota Press, Minneapolis, pp 216–249
22. Guha-Thakurta T (1992) The making of a new "Indian" art: artists, aesthetics, and nationalism in Bengal, c. 1850–1920. Cambridge, Cambridge University Press
23. Hatcher J, Tu TLN (2017) "Make What You Love": homework, the handmade, and the precarity of the maker movement. Women's Stud Quart 3–4:271–286
24. Herzfeld M (2004) The body impolitic: artisans and artifice in the global hierarchy of value. University of Chicago Press, Chicago
25. Ingold T, Hallam E (2007) Creativity and cultural improvisation: an introduction. In: Hallam E, Ingold T (eds) Creativity and cultural improvisation. Berg, London, pp 1–25
26. Ingold T (2001) Beyond art and technology: the anthropology of skill. In: Schiffer MB (ed) Anthropological perspectives on technology. Amerind Foundation, Dragoon, AZ, pp 17–32
27. Jain C, Dutta D (2009) Explore beyond the obvious: India's position in the global textile and clothing trade. Online J Int Fash Apparel Indus 4:1–3
28. Jansen A (2020) Fashion and the phantasmagoria of modernity: an introduction to decolonial fashion discourse. Fash Theory 24:815–836
29. Kaur R (2020) Brand new nation: capitalist dreams and nationalist designs in twenty-first-century India. Stanford University Press, Stanford
30. Kawlra A (2016) Recipes for re-enchantment: natural dyes and dyeing. Marg 67(4):78–87
31. Kawlra A (2014) Duplicating the local: GI and the politics of 'Place' in Kanchipuram, Tamil Nadu. NMML Occasional Paper, Perspectives in Indian Development, 29. Nehru memorial Museum and Library, New Delhi
32. Kawlra A (2020) Sari-Kimono and the making of a transnational craftscape. In: Narsimhan S (ed) India-Japan narratives: lesser known historical and cultural interactions. Mombusho Scholars Association of India, New Delhi, pp 138–152
33. Khaire M (2011) The Indian fashion industry and traditional Indian crafts. Business History Review 85(2):345–366
34. Kuldova T (2016) Luxury Indian fashion: a social critique. Bloomsbury, London

35. Kuldova T (2017) Traversing trade fairs and fashion weeks: on dependence and disavowal in the Indian fashion industry. In: Leivestad HH, Nyqvist A (eds) Ethnographies of conferences and trade fairs: shaping industries, creating professionals. Palgrave Macmillan, Cham, pp 147–169
36. Luvaas B (2010) Designer vandalism: Indonesian Indie fashion and the cultural practice of cut 'n' paste. Vis Anthropol Rev 26(1):1–16
37. Lynch JE (2016) The good of cloth: bringing ethics to market in India's handloom textile industry. University of Michigan, Ann Arbor, Anthropology, PhD Diss.
38. Marchand T (2016) Introduction: craftwork as problem solving. In: Marchand T (ed) Craftwork as problem solving: ethnographic studies of design and making. Ashgate, London, pp 1–32
39. Marx K (1992[1867]) Capital: Volume 1. Penguin, New York
40. Mathur S (2007) India by design: colonial history and cultural display. University of California Press, Berkeley
41. Mazzarella W (2003) "Very Bombay": contending with the global in an Indian advertising agency. Cult Anthropol 18(1):33–71
42. Mazzarella W (2005) Middle Class. In: Dwyer R (ed) South Asia keywords. Online. SOAS South Asia Institute. https://www.soas.ac.uk/south-asia-institute/keywords/file24808.pdf
43. McGowan A (2021) Mothers and godmothers of crafts: female leadership and the imagination of India as a crafts nation, 1947–67. South Asia J South Asian Stud 44(2):282–297
44. McRobbie A (1998) British fashion design: rag trade or image industry? Routledge, London
45. Mezzadri A (2017) The sweatshop regime: labouring bodies, exploitation, and garments made in India. Cambridge University Press, Cambridge
46. Mohsini M (2010) Becoming an "Asli Karigar": the production of authenticity among Old Delhi's Muslim Artisans. PhD Diss., School of African and Oriental Studies, University of London, Anthropology and Sociology
47. Moon CH (2011) Material intimacies: The labor of creativity in the global fashion industry. Yale University, Anthropology, PhD Diss.
48. Murphy KM (2012) Transmodality and temporality in design interactions. J Pragmat 44:1966–1981
49. Murphy KM (2015) Swedish design: an ethnography. Cornell University Press, Ithaca
50. Myers F (2004) Ontologies of the image and economies of exchange. Am Ethnol 31(1):5–20
51. Myers F (2001) Introduction: the empire of things. In: Myers F (ed) Empire of things: regimes of value and material culture. SAR Press, Santa Fe, pp 3–64
52. Nicholas CB (2014) Master class: the technical and ethical re-education of moroccan artisans. Princeton University, Anthropology, PhD Diss.
53. Ong A (1999) Flexible citizenship: the cultural logics of transnationality. Duke University Press, Durham
54. Partridge DJ (2011) Activist capitalism and supply-chain citizenship producing ethical regimes and ready-to-wear clothes. Curr Anthropol 52(S3):S97–S111
55. Paton E (2021) India's fashion artisans face 'extreme distress' in pandemic. New York Times, April 25
56. Pouillard V (2021) Paris to New York: the transatlantic fashion industry in the twentieth century. Harvard University Press, Cambridge, MA
57. Raina A (2019) NY designer says artisan leaked designs to rival. Mumbai Mirror, November 25
58. Rofel L, Yanagisako S (2018) Fabricating transnational capitalism: a collaborative ethnography of Italian-Chinese global fashion. Duke University Press, Durham
59. Roy T (1998) Economic reforms and textile industry in India. Econ Pol Wkly 33(32):2173–2182
60. Sandhu A (2020) Fashioning wellbeing through craft: a case study of Aneeth Arora's strategies for sustainable fashion and decolonizing design. Fash Pract 12(2):172–192
61. Schneider J (1987) The anthropology of cloth. Annu Rev Anthropol 16:409–448
62. Sennett R (2008) The craftsman. Yale University Press, New Haven
63. Shah A (2021) Crafting a future: stories of indian textiles and sustainable practices. Niyogi Books, Delhi
64. Statista (2021) Apparel market in India. Online. Retrieved June 1

65. Tarlo E (1996) Clothing matters: dress and identity in India. University of Chicago Press, Chicago
66. Thomas N (1991) Entangled objects: exchange, material culture and colonialism in the pacific. Harvard University Press, Cambridge, MA
67. Thomas K (2019) Cultures of sustainability in the fashion industry. Fash Theory 24(5):715–742
68. Trampusch C, Busemeyer MR (eds) (2012) The political economy of collective skill formation. Oxford University Press, Oxford
69. Varma M (2015) Making designs on fashion: Producing contemporary Indian aesthetics. PhD Diss., UCLA, Anthropology
70. Venkatesan S (2009) Rethinking agency: persons and things in the heterotopia of 'traditional indian craft.' JRAI 15(1):78–95
71. Weiner AB (1994) Cultural difference and the density of objects. Am Ethnol 21(2):391–403
72. Wilkinson-Weber CM (2004) Women, work and the imagination of craft in South Asia. Contemp South Asia 13(3):287–306

The Fashion Crossroad Method: Political and Epistemological Practices

Carla Costa, Carolina Casarin, João Dalla Rosa Júnior, Heloísa Helena de Olivera Santos, and Mi Medrado

Abstract The article aims to shed light on theoretical engagements in fashion and decoloniality and brings concepts conceived by the Fashion and Decoloniality: Global South Crossroads working group - CoMoDE. We are proposing a graphical representation of methodological steps to undertake the decolonial turn in/at fashion research. Based on a pluridisciplinary perspective, the fashion crossroad method consists of two concepts: social backstitch in fashion and tears up in fashion, which is represented graphically through the sewing machine bobbin. The fashion crossroad method as a concept and image yields broad knowledge of fashion gears. Its political and epistemological practices show how to transgress western canon and the kind of fashion studies taught in countries that are still facing colonialism.

Fashion and Decoloniality: Global South Crossroads working group - CoMoDE composed of Carla Costa. Carolina Casarin, João Dalla Rosa Júnior, Heloísa Helena de Olivera Santos, and Mi Medrado

---

C. Costa (✉)
Mestre em Artes Cênicas, UNIRIO, Rua André Cavalcanti 111, Rio de Janeiro RJ CEP 20.231-050, Brazil
e-mail: cah.costa84@gmail.com

C. Casarin
Doutora em Artes Visuais, UFRJ, Avenida Francisco Matarazzo, 232, Apt. 85. Água Branca, São Paulo, SP, Brazil
e-mail: c.casarinfh@gmail.com

J. D. R. Júnior
Doutora em Design, PUC-Rio, Rua José Bonifácio, 820, Ap. 808. Todos Os Santos, Rio de Janeiro, RJ, Brazil
e-mail: joaodrjr@gmail.com

H. H. de Olivera Santos
Doutora em Design, PUC-Rio, Rua Marechal Bitencourt, 90/202, Riachuelo, Rio de Janeiro/RJ 20950-200, Brazil
e-mail: heloisahosantos@gmail.com

M. Medrado
Mestre em Artes, UCLA, Rua Sao Luiz, 42, Barra, Salvador, BA, Brazil
e-mail: mi@mimedrado.com

M. Á. Gardetti and R. P. Larios-Francia (eds.), *Sustainability Challenges in the Fashion Industry*, Sustainable Textiles: Production, Processing, Manufacturing & Chemistry,
https://doi.org/10.1007/978-981-99-0349-8_18

1 Introduction

The Coletivo Moda e Decolonialidade: Encruzilhadas no Sul Global–CoMoDe, is a research workgroup that systematically studies and analyzes fashion from post-colonial and decolonial perspectives. The CoMoDe researchers and professors are from the social science, humanities, and arts disciplines; we first met on June 9, 2020, aiming to challenge Brazilian fashion from established discourses focusing on ways to decolonize it. Our meetings stimulate theoretical debates based on scholarly readings, literature, movies, and newspapers. Our analytical dialogue intends to elaborate epistemological tools to enable researchers and students to take a decolonial turn when examining fashion systems.

The fashion crossroad method: political and epistemological practices is an essay aiming to demonstrate the decoloniality movement requested to pursue an interdisciplinary approach to raise awareness and explore steps to undertake the decolonial turn-*in/at* fashion research. Therefore, we use the crossroad (*encruzilhadas* in Portuguese) as our reference and decided to draw the *fashion crossroad method* graphically. We believe that this kind of representation is embedded with a significant educational value; as you will see, the graphics invite the reader to access the subjectivities faced when studying the fashion field.

The content is organized in two sessions. First, we will introduce the *fashion crossroad method,* then demonstrate how it unfolds into the concepts of *social back-stitch in fashion* and *tears up in fashion* through graphic representations of fabric inside out and sewing machine bobbin.

2 Advocating for a Fashion Crossroad Method

The post-colonial debate begins with the decolonization of Africa and Asia in the mid-twentieth century. Intellectuals from these continents, such as Frantz Fanon, Aimé Césaire, Alber Memmi, and Edward Said, carried out fruitful work on the need to draw criticism on scientific production, which included (a) analyzing the impacts of colonization and imperialism on the lives of the colonized, which encompassed knowledge production and intellectual work; (b) hear Africans and Asians' voices about themselves, as well as their analytical perspectives on the world; (c) question the traditional methods—related theoretical apparatus—enforced by European intellectuals toward colonized societies. In conjunction with that decolonial thoughts are related to Latin American roots and the debate initiated in the mid-1990s. After that, in the breakthrough of the modernity/coloniality group, specifically with Aníbal Quijano and Walter Mignolo, the intent of the rupture with the European knowledge production chose to use concepts involved with references to the Andes, *Abya Yala*, and Latin American knowledges. On these grounds, we articulate our viewpoint with the modernity/coloniality group perspective to argue fashion as a field of knowledge in the Brazilian context looking at its relationship with the coloniality of power to propose two concepts for confronting the colonial matrix.

For CoMoDe, reflecting upon fashion involves recognizing the impact of race, gender, and class in their colonial/modern framework. Such a reassessment process must be called a *social backstitch in fashion*, situated with the post-colonial and decolonial approach that unveils social oppressions when aligned with the fashion system. The word *backstitch* was not randomly chosen. Even though the backstitch shares its meaning with regression and lateness in Portuguese, we want to emphasize that the stitches are made backward to the general direction of the sewing machine. For those who deal with the mass production of clothing, *backstitch* points to the well-known technique of stitching in which, once the seam is sewn, the quilter sews new stitches in reverse to ensure the forward stitches are positioned in place. The procedure design—also known as backstitching—reinforces the seam where the thread is sewn so that it does not unravel quickly and the clothing threads stay strong.

Our goal with the *social backstitch in fashion* notion is to offer ways to reconsider the universal knowledge formulated on what fashion is and on which theory and practices parallel the production of coloniality and power. It is necessary to shift the established interpretation of scientific-based on evolutionary ideas. It is time to recall common sense concepts, burn them up, and then continue a reflection that keeps on overlapping the latter. In such a scheme, instead of denying it, we recognize the importance of what is produced and emphasize the urgency of reassessing the concepts; as we backstitch when sewing, the notion of *social backstitch in fashion* represents a way to reinforce and stabilize the stitches of meaning conveyed in fashion.

In methodological terms, we affirm that the *social backstitch in fashion* concept effort is akin to what post-colonial authors offered to the field of criticism on knowledge. Several researchers from Africa and Asia who graduated from north axis universities adopted notions provided by a scholarship from the north to criticize traditional ideas conceived about the South. Post-colonial authors created and established innovative concepts to reflect on themselves, their societies of origin, and the global economic-political logic that promoted colonialism and imperialism, going beyond the structure of values that regimented modernity.

Therefore, to ensure that we are cracking such approaches, we acknowledge the importance of a conceptual rupture to assess to which extent we are drowning and reproducing the colonial logic in our knowledge production, teaching, and research practices. For that, we introduce the *tears up in fashion* concept. The concept is linked with the work of Latin American decolonial researchers as we approach what these authors from the modernity/coloniality group called the *decolonial turn* [7, 4].

Tears up in fashion are understood by the CoMoDe as a process of radical rupture in the axiomatic practices, theoretical teaching methods, and other scholarly productions that still portray colonial, hegemonic, and imperialist logic in the field of fashion. Different from cutting a piece of fabric with scissors or any other instrument to separate it into two or more parts, when we tear up a fabric, we act abruptly with our bare hands, and the roughness shows the characteristic of an unstable yarn.

The movement consists of abandoning concepts and pointing out the underlying structures of domination so that, with and from it, we can rethink and reinvent local notions and texts. It is about founding and grounding epistemologies from the South, researching and discussing concepts based and created in *favelas/quebradas* [neighborhoods], ghettos, samba schools, *baile funk* parties, Umbanda and Candomblé *terreiros* [temples], prostitution zones, in encruzilhadas e esquinas (crossroads and street corners). In other words, the term comprehends everything the liberal, political and intellectual order based on Northern grammar has discarded as legitimate.

Furthermore, *tears up in fashion* is about carrying out what Antônio Simas and Luiz Rufino called in *Pedagogia das Encruzilhadas* [9] called *rolê epistemológico/epistemological strolls*, which means "refusing to accept the condition of immobility propagated by these effects, we must transgress their parameters." Analytically the following definitions will offer the reader's lucubrations of these metaphors. We will use the notion of copying for both analyses:

(a) The *social backstitch in fashion*: The notion of copy is generally used in fashion studies research to define the relationship between non-north Axis countries and their local ways of dressing in opposition to European dress forms. When *backstitch* is necessary, firstly, to indicate geographically in which intellectual context such reflections were being produced. It is worth considering that much of the debate on fashion is established in a logic of authorization, and authors who engage in the concept of copying, comprehend that only one social group is authorized to have a relationship with fashion. They are the European and North American elite. If we accept this group's restricted notion, perhaps the concept of copying makes sense. However, to bring our reflections forward, we must challenge and think that **copying is one of the multiple ways social groups, such as Brazilian society, relate to fashion**. In the course of that, looking historically at the circulation of knowledge on clothes in Brazil in the 17th to nineteenth century, we find several enslaved people whose professional expertise was working in the clothes sector. Yet, we must also consider what indigenous, enslaved, and poor colonized populations wore and produced. As such understanding, the *backstitch* works as a way not to disregard the concept of copying but to strive and locate it socially. The core idea is to broaden our view of other productions in the territory and welcome local clothing in the fashion system. **It is known that authors who studied local clothing are aware that they are analyzing the elites**: we sense that the *backstitch* is crucial to stop seeing copying and the elites as the only legitimate means to clothing in the fashion system. We would also like to consider these elites adept at re-signifying that. Notwithstanding, comprehend that copying is also an integral act and part of the creative component of northern producers. We acknowledge that what Europeans and Northern elites claim to be a pure and original creation, focusing on the value and concept of originality, is also always a copy. At last, as a mimetic gesture, the act of copying is part of all creative processes.

(b) The *Tears up in fashion:* We take a different procedure. The idea of copying is considered a perception in line with a perspective that only understands the West as a fashion producer. In this way, the colonial logic of knowledge production is inverted, and we reach the idea that local clothing production has always had ties with world fashion, **just as all fashions in the world relate to each other**. There is no copying because copying is impossible in societies characterized by crossing/transiting. What we have here are the relationship and exchange.

3 Crossing Knowledge–Designing Images

In this section, we aim to demonstrate using visual representations how the *fashion crossroad method* unfolds into the: *social backstitch in fashion* and *tears up in fashion* concepts.

Fashion scholars from several geographies, social science, and art disciplines have emphasized the power dynamics inherent to the field of fashion. Niessen [6] highlights the way that an area of dispute was formed, in which the West hijacked fashion for itself and, chiefly, self-laimed authority to judge what is fashion and what is not, as well as decides who is capable of turning objects and ideas into fashion and who is solely able of copying which has the supremacy. Such structure defines the West as modern and innovative. It relates to the colonial spectrum: immersed in an evolutionary logic permeated by the scientific racism that still persists among us, the production by the peoples of the North is continually recognized as superior. Indeed, it is a common practice that when European designers appropriately insert their aesthetic references from other societies, the contemporary colonizer in the neo-fashion colonialism has a presumption of innocence protection. The appropriation *quid pro quo* under this syntagma guarantees that no guilt can be presumed when the West's fashion industry copies. If there are charges, they should be taken positively as inspiration or reinterpretation.

Going along with a legal justice metaphor, we may think of this fashion system as a court trial and its hearings. On the one hand, there is a fashion law court where European fashion producers are responsible for creating a fashion measurement and sentencing on the other. While on the other hand, the jury, the prosecution, and the judge share the same ideology. The strategy is violent. The judge who should apply an isonomic perspective to evaluate and assure equality before the law decides whom and which foreign fashion players will be associated with their party reproduces a fashion world-system prejudices. We see that when local professionals working as it were in European branches in the Global South feel entitled to this micro-power and inspect who is fashionable and who is not. You might be asking why [?] what are the gains or exchanges? Prestige and economic capital. Even though we face geographical discrepancies–when compared with the North. We feel we are seen as exotics—the others. We became evaluated and fragmented by the subalternity grammar, knowledge hierarchies, and the discursivities that control the inequalities defined by the North.

The resemblance to the colonial discourse of racialization and ethnicization is not a mere coincidence. Another way the European fashion law court defines fashion is

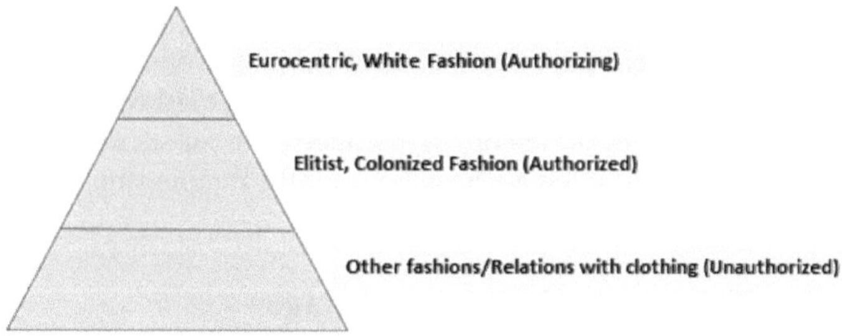

Fig. 1 Hierarchical structure of fashion pyramid representation. *Source* Fashion and Decoloniality: Global South Crossroads working group–CoMoDE (2020)

when they hunt for an endorsement from a local producer as the "fashion heiress." [1, 2, 8, 5] point out that the quest is usually sustained by the process of exoticization, along with a mark of inferiority. Consequently, Western fashion turned into "The Fashion," other societies have clothing systems with caveats. So, their fashion must be situated: Latin fashion, African fashion, Asian fashion, and distinguished–exotic, hot, wild, colorful, and sexy.

Therefore, to examine the gaze logic attached to his hierarchical structure of fashion, we present a pyramid chart (Fig. 1). You see the western fashionists on the top. Then, below you find those selected by the former and who is not "the West" chosen by the former. Finally, at the bottom are those without fashion. It is essential to say that these poles are not closed structures, i.e., there are several possible configurations between them, as society is much more complex than any ideal type [13] could try to grasp.

CoMoDE acknowledges such a hierarchical structure of fashion pyramid representation and proposes a methodological perspective on fashion to confront the long-established stratified arrangement. For us, it is about shifting our gaze to analyze European productions as equal and parallel to overseas creations.

(a) Grammarly, use *fashions* (in plural) instead of *fashion* (singular): All forms of relationship with clothing must be understood as *fashions*. Such perspective dialogues with the *social backstitch in fashion* concept.
(b) Then, *fashions* (in plural) must represent the idea that all kinds of relationships with clothes are legit, equal, and valid. In a given society, they can be only judged from within, evaluated from local development themselves, summing up with their relations with global production, connections of permanent contact, crossing, transit, exchange, and resignification, which inevitably include change.

The perspective led us to a circular perception of fashion and a vision that did not reproduce fashion in a vertical, uncritical way, where the West is seen as the universal and driving force of binarism. Fashion is understood as horizontal. Horizontal because it is also a geopolitical relationship in which colonial cartography placed Europe at the top of the map. We conceived it horizontally as an axis to look at productions and relationships with clothing of the same height. Circularity refers

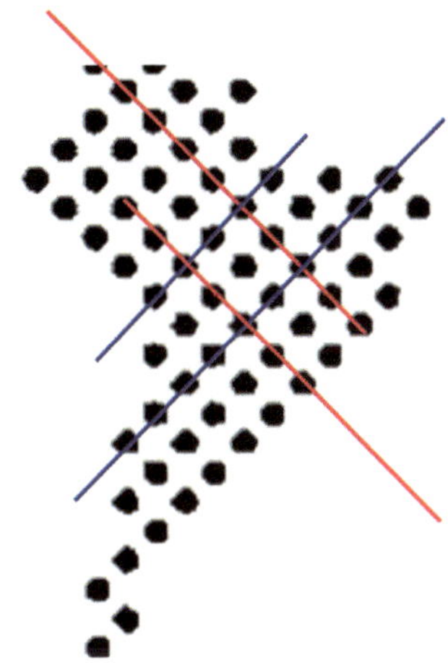

Fig. 2 *Fashions* and their crosses representation. *Source* Fashion and Decoloniality: Global South Crossroads working group–CoMoDE, (2020)

to perceiving all relationships with fashion as circular structures whose lines contour and are not close to the outside. They exchange, contact, and re-signify even though a circle is closed because they are individual units of varying sizes and complexities.

We redressed the analysis amid the counter-blows versed in the notion of a pedagogy of the crossroads of [9] "a political/epistemological/educational project whose main purpose is to unspell the curse of racism/colonialism by transgressing traditional western canon." For the group, the relationship we propose to establish between fashions, far beyond contact, exchange, and resignification, is a true crossroads of stories. In this sense, we understand that each circle or clothing structure is, using macro representation, crossed by lines (Fig. 2) that cut through the entire fashion world-system, as this is how established these relationships have been since the first contact between humans: in the encounters, in the crosses, in the delicate meeting between knowledge, which transforms what it touches, renewed and reborn.

The definition of crossroads proposed by Simas and Rufino [9], (p. 18) supports our methodological understanding and its graphical representation to research on fashion. The crossroads enable the notion of a journey; for that, we associate the path to line definition to allow a trace in a drawing. Thus, if the crossroad refers to the crossing of tracks, it can be graphically represented as a crossing of lines.

In drawing theory, a line is formed by numerous points. However, when we think about the practical action of using graphic material on a surface: as is the case with pencil on paper, we can say that the line is the trace, the path, the movement that a point makes on the surface plane. When applying the pencil to the paper, the slightest contact generates a unit defined as a point—the smallest and first visual element [3], p. 14. However, any displacement the point performs will form a line, which can be curved, straight, continuous, or dashed; its characteristics will depend on the movement and direction the point assumes.

Observing the shape of the crossroads, we can see that it is formed by crossing lines that correspond to different paths. The crossing itself marks a specific spot, which is the one where the two lines intersect. In the geometry parameters, an area alludes to the pair of X and Y coordinates, which we will take for our methodological metaphor, as the starting place of a point of view that we want to emphasize and highlight in the perspectives on fashion.

Acknowledging the notion that the entire graphic and epistemological understanding of crossroads takes place in a given space, we design a planar representation of the world map to visualize the world-system. Then, we can propose a methodology for research in fashion.

To represent the fashion crossroad method, we offer to see the world map from different perspectives. We will abstract their divisions and adopt the notion that spaces are formed by numerous points (Fig. 3). It will help us distance the colonial canon attributions and thoughts of stationary culture for non-European society. At the various crossings of intersections, the fashion paths can be drawn by countless lines that bring out the cultural hubs of clothing systems. This way, the dots of different spaces are connected by the threads that generate the crossings. Fashion crossroads are formed by the paths of other points of the clothing systems that meet through movement in space. In [10], (p. 18) words, "the cross is like becoming, an unfinished, salient, unordered, and inapprehensible movement. The cross relates to an erasure, fissure, contamination, catalysis, bricolage—a manifestation of Eshu [...]".

From a methodological point of view, thinking about fashion using crossroads means facing the tactical action that the cross represents.

> Crosses meet and demarcate border zones. These cross-border zones are places of emptiness that will be filled by bodies, sounds, and words. From these fillings, other possibilities for

Fig. 3 Embroidery of the world-system: fashion crossroad method. *Source* Fashion and Decoloniality: Global South Crossroads working group–CoMoDE

> the invention of life will emerge based on the tones of radical transformations, cognitive justice, and diversity of knowledge [9].

Working with the cross means "erasing the alleged universality of the western canon" [9], p. 19, i.e., transgressing it by crossing other perspectives. This transgression, for us, must be accomplished by a turn in understanding threads and the movements of points. For us, the best metaphor to relate is the sewing machine bobbin.

As we understand, the clothing production system is linked through factories outside the Euroamerican realm; their workforce struggles with low wages, unsafe conditions, and harassment, which seem to shape neo-slavery relationships due to subhuman conditions. The consequences are seen through men's and women's exploitation. Indeed, such a system ensures capitalist individualism values and logic, where some fashion items are available exclusively for some bodies. In contrast, the others are left behind, on the edge of the sacrifice zones. According to Niessen [6], "neighborhoods and cultures identified as non-producers [but workersforce] of fashion are so through colonial and capitalist constructs of this industry and should be recognized as 'fashion sacrifice zones.'" Metaphorically, we will expose our perceptions about this relationship we have just mentioned. For that, we will use examples excerpted from sewing machines, their accessories, and their purpose in garment-making.

Among fabrics, threads, and needles, there is the figure of a seamstress, a critical piece in the making of clothing. She must have at least one straight-stitch sewing machine to handle the daily production of her clothes. The straight-stitch sewing machine is one of the oldest machines ever created, dating from the industrial revolution when its first gears were developed, and significant transformations have taken place in clothing production. We aim to illustrate how these gears work and how this reflects in the final product. In addition, these gears, whose operators will be taken into account in our reflection through brief associations, are at the core of the debate.

Straightening machines greatly benefited from new technologies in the textile industry, with new devices added to meet the demands of productivity behind clothing confection. Modern engines work through various mechanisms and accessories, many of which are vital for making garments, needles, presser feet, etc.—accessories that accelerate and enhance the quality of the finished product.

We want to bring a fundamental part of this gear to our illustration, the bobbin case, and bobbin (Fig. 4).

The function of the sewing machine bobbin is to provide bottom stitches while stitching the top part, which is evident in the clothing and appreciated by lovers of good finishing, that is, a backstitch. Backstitches are an essential part of a well-made, valuable piece. Almost no one sees behind the beauty of a well-finished outfit are the "invisible" stitches fed by the bobbin. And how does this bobbin gear work, and what is the bobbin case of a sewing machine? We can say that the machine's heart, the rotation, maintains functionality on the vertical and horizontal axis. It is a cog in the gears, just like the crossroads of an operating system that ignores its existence. The bobbin regulates the stitches, which are adjusted and controlled by the upper mechanism of the machine. The bobbin does not operate by itself: as soon as it stops

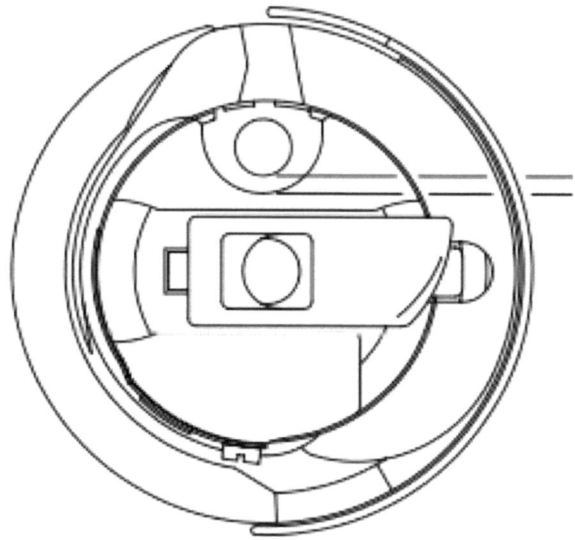

Fig. 4 The sewing machine bobbin and the bobbin case. *Source* Manual Siruba (p. 341)

spinning, the entire system is paralyzed. It is impossible to rotate the axis without what we understand as the heart of the machine. Even in older models, if the bobbin is not seated correctly, there are no clothes nor production. But what about the engine–is it not the heart of the machine? The answer is that engines came into existence at the end of the nineteenth century when sewing devices were still operated using cranks and pedals. Still, they have always needed what we understand today as a bobbin case and bobbin.

You may see from the description how the machine works and the role of the bobbin; the threads intersect during sewing, and each stitch causes tension as one thread is positioned at the top and another at the bottom. In Fig. 5, we recall the stitching process and associate it with an embroidery of the world map representing the crossroads. The map cut in half depicts the division caused by the hierarchy of threads, which is evoked by the colors. In our view, the threads can symbolize the asymmetry between North and South in fashion practices: the bobbin thread is hidden underneath the fabric, thus representing the entire part that sustains the fashion system itself. Although the northern part is delimited by a cut roughly equivalent to the equator, we believe that the division is even more asymmetric since what is considered North is not restricted to the geographic notion of the hemispheres. Different spots on the map above the Equator are considered south on the power scale of the world-system, and at the crossroads of fashion, they are made invisible by the upper seams.

In the face of this metaphor, it is worth mentioning that the role of sewists as agents who feed the entire production chain of fashion can also be understood through the function of the bobbin in the sewing machine. These professionals correspond to those who support the fashion system and are made invisible by the threads that sew the stitches. This image alludes to the one that Françoise [12] portrays about racialized

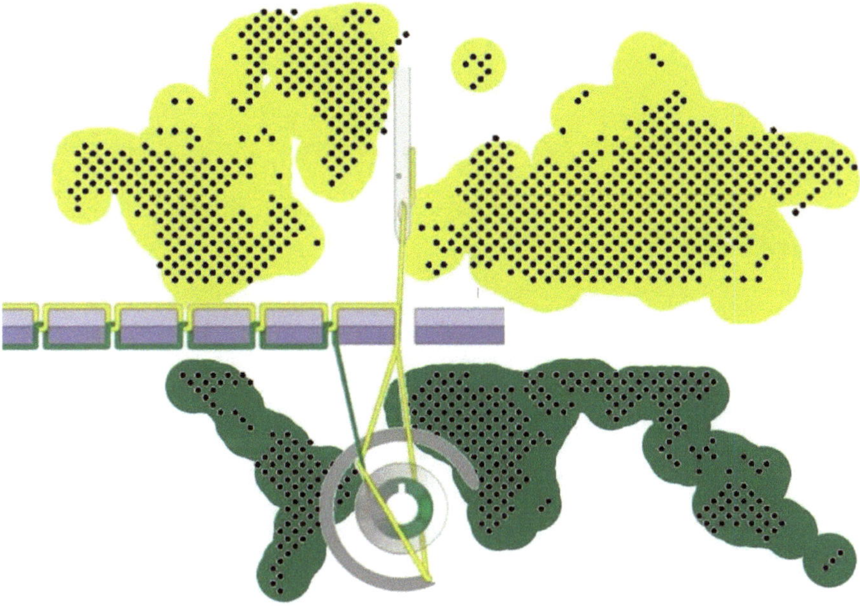

Fig. 5 "The world sewing" asymmetry: fashion South and North. *Source* Fashion and Decoloniality: Global South Crossroads working group–CoMoDE (2020)

women who do the cleaning jobs in companies in France. The proposal of a decolonial turn based on crossing implies turning these professionals into research agents. This means widening the range of fashion beyond what is visible to include all relations between agents of clothing production. Therefore, this widening corresponds to not looking at the stitches and crossroads from the right side of the fabric: instead, it is crucial to see the results of the upper threads that, in Fig. 6, correspond to the yellow lines.

For the group, the *turn* requires looking at the reverse side of the fabric: that side where the bobbin thread can be seen and identified as fundamental for the sewing to take place. The rotation causes the image to be seen in reverse, and the perception is similar to the provocation by Joaquín Torres García in 1943, as he drew an inverted map of South America, stating that "our north is the south." Seeing the reverse corresponds to changing the poles and highlighting the lines made invisible by the crossroads of fashion, as shown in Fig. 7. The method of crossroads presupposes contact with the place of tension: within the scope of the research, it indicates how the paths cross or, as a result, in sewing, how the threads intertwine. Analyzing the pressure on the machine thread, strength, type, and links on the mechanisms of fashion indicates widening the observation range in the paths of the crossroads.

The method of crossing has challenges as long as …

> Humanity has always faced crossed paths with awe and wonder. After all, crossroads are places of uncertainty, pathways, and amazement of realizing that living presupposes the risk

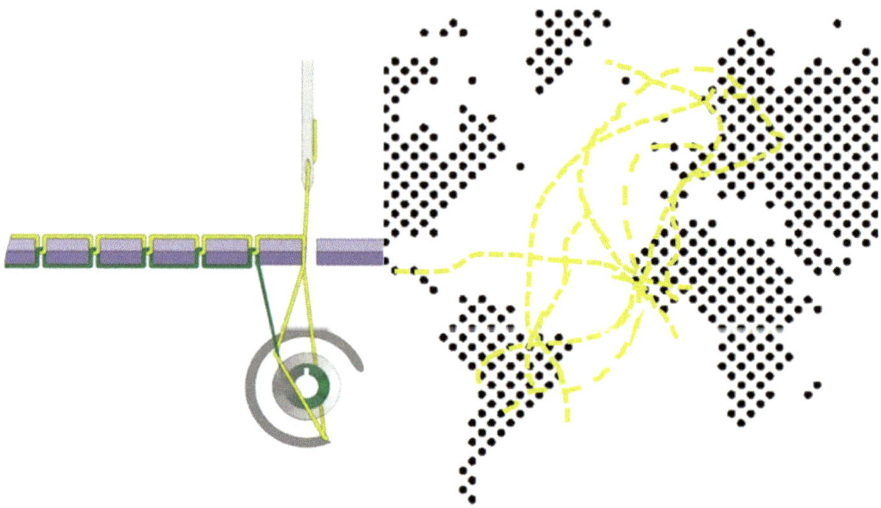

Fig. 6 Fabric right side and the upper threads. *Source* Fashion and Decoloniality: Global South Crossroads working group–CoMoDE (2020)

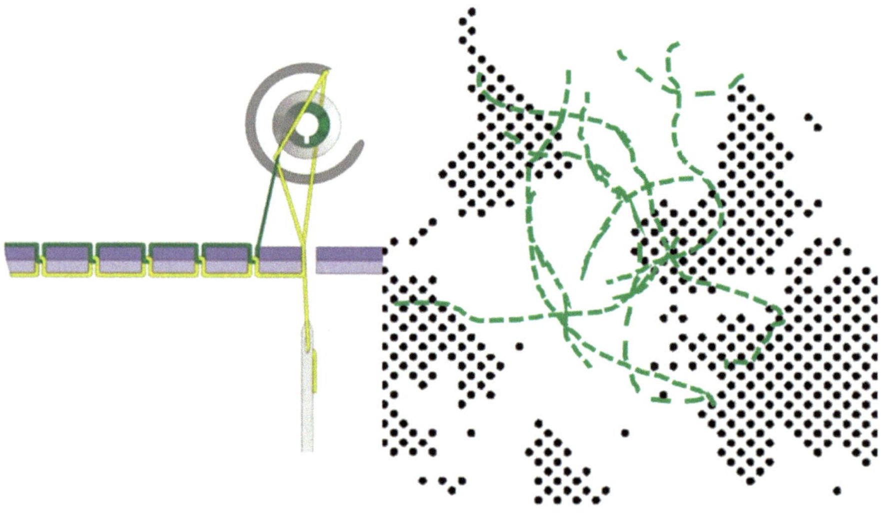

Fig. 7 Fabric reverse side and the invisible threads. *Source* Fashion and Decoloniality: Global South Crossroads working group–CoMoDE (2020)

of choices. Where to walk? A crossroad is a vexation–that is the thrill. We say about this whole story that we conduct our own lives. The rite must be practiced; we ask permission from the invisible, and we continue as minor heirs of the human spirit, with amazement being the guiding thread of luck. We who are from the crossroads are suspicious of those on a straight path [9].

For the Coletivo Moda e Decolonialidade: Encruzilhadas do Sul Global- CoMode, this theoretical-methodological proposition in fashion research denotes the urgency of connections and correlations that only by crossing and transiting paths will heal the colonial wounds that have regulated the ethical, aesthetic, imagery, and subjective aspects of producing, circulating and doing research in the field of Fashion.

We believe that in this way, we will be able to "get up off the epistemological sofa and divc into thc crossroads of altcrity" [9]. Bccausc it is through thc *social backstitch in fashion* and *tearing in fashion* that we will untie ourselves from the binarism between modernity and coloniality, thus being able to "weave genealogies that were disconnected" [11], to face what has been systematically silenced, although actively planned, behind being, and becoming invisible.

References

1. Allman J (2004) Fashioning Africa: power and the politics of dress. Indiana Press, Bloomington, IN
2. Craik J (2003) The face of fashion. In: Cultural studies in fashion. Routledge, Londres
3. Lupton E, Phillips JC (2008) Novos fundamentos do design. Cosac Naify, São Paulo
4. Maldonado-Torres N (2017) The decolonial turn. In: New approaches to Latin American studies: culture and power. Routledge, Londres, p.p 111–127
5. Niessen S, Leboshkowich AM et al (2003) Re-orienting fashion: the globalization of asian dress. Berg Publishers, Londres
6. Niessen S (2020) Fashion, its sacrifice zone, and sustainability. Fashion Theory
7. Quijano A (2005) Colonialidade do poder, eurocentrismo e América Latina. In: Lander E (org.). A colonialidade do saber: eurocentrismo e Ciências Sociais: perspectivas latino-americanas. Buenos Aires, CLACSO, pp. 105–127
8. Rabine LW (2002) The global circulation of African fashion. Nova Iorque, Berg
9. Simas LA, Rufino L (2018) Fogo no mato: a ciência encantada das macumbas, 1 edn. Mórula, Rio de Janeiro
10. Rufino L (2019) Pedagogia das encruzilhadas. Mórula, Rio de Janeiro
11. Vázquez R (2020) Vistas of modernity decolonial aesthesis and the end of contemporary. Mondrian Fund, Amsterdam
12. Vergès F (2020) Um feminismo decolonial. Ubu Editora, São Paulo
13. Weber M (1992) Metodologia das Ciências Sociais, vol 2. In: Trad: Augustin Wernet. Cortez e ' UNICAMP, São Paulo